D1216150

Ford Windstar Automotive Repair Manual

by Jay Storer, Jeff Kibler and John H Haynes

Member of the Guild of Motoring Writers

Models covered:

All Ford Windstar models
1995 through 2001

(8D5 - 36097)

ABCDE
FGHIJ
KLM

Haynes Publishing Group
Sparkford Nr Yeovil
Somerset BA22 7JJ England

Haynes North America, Inc
861 Lawrence Drive
Newbury Park
California 91320 USA

Acknowledgements

Wiring diagrams were provided exclusively for Haynes North America, Inc. by Valley Forge Technical Communications.

A book in the Haynes Automotive Repair Manual Series

Printed in the U.S.A.

ISBN 1 56392 388 2

Library of Congress Catalog Card Number 00-104200

While every attempt is made to ensure that the information in this manual is correct, no liability can be accepted by the authors or publishers for loss, damage or injury caused by any errors in, or omissions from, the information given.

01-288

Contents

Haynes mechanic, author and photographer with Ford Windstar

About this manual

Its purpose

The purpose of this manual is to help you get the best value from your vehicle. It can do so in several ways. It can help you decide what work must be done, even if you choose to have it done by a dealer service department or a repair shop; it provides information and procedures for routine maintenance and servicing; and it offers diagnostic and repair procedures to follow when trouble occurs.

We hope you use the manual to tackle the work yourself. For many simpler jobs, doing it yourself may be quicker than arranging an appointment to get the vehicle into a shop and making the trips to leave it and pick it up. More importantly, a lot of money can be saved by avoiding the expense the shop must pass on to you to cover its labor and overhead costs. An added benefit is the sense of satisfaction and accomplishment that you feel after doing the job yourself.

Using the manual

The manual is divided into Chapters. Each Chapter is divided into numbered Sections, which are headed in bold type between horizontal lines. Each Section consists of consecutively numbered paragraphs.

At the beginning of each numbered Section you will be referred to any illustrations which apply to the procedures in that Section. The reference numbers used in illustration captions pinpoint the pertinent Section and the Step within that Section. That is, illustration 3.2 means the illustration refers to Section 3 and Step (or paragraph) 2 within that Section.

Procedures, once described in the text, are not normally repeated. When it's necessary to refer to another Chapter, the reference will be given as Chapter and Section number. Cross references given without use of the word "Chapter" apply to Sections and/or paragraphs in the same Chapter. For example, "see Section 8" means in the same Chapter.

References to the left or right side of the vehicle assume you are sitting in the driver's seat, facing forward.

Even though we have prepared this manual with extreme care, neither the publisher nor the author can accept responsibility for any errors in, or omissions from, the information given.

NOTE

A **Note** provides information necessary to properly complete a procedure or information which will make the procedure easier to understand.

CAUTION

A **Caution** provides a special procedure or special steps which must be taken while completing the procedure where the Caution is found. Not heeding a Caution can result in damage to the assembly being worked on.

WARNING

A **Warning** provides a special procedure or special steps which must be taken while completing the procedure where the Warning is found. Not heeding a Warning can result in personal injury.

Introduction to the Ford Windstar

The Ford Windstar is available in a Minivan body style only.

All models are equipped with either a 3.0L or 3.8L V6 engine. The engines are equipped with a multi-port fuel injection system and a distributorless ignition system. The systems utilize the On Board Diagnostic Second-Generation (OBD-II) computerized engine management system that controls virtually every aspect of engine operation. OBD-II monitors the fuel and emissions system components for signs of degradation and engine operation for any malfunction that could affect engine operation and emissions, turning on the CHECK ENGINE light if any faults are detected.

All models are equipped with a transversely mounted four-speed automatic transaxle, driving the front wheels via independent driveaxles.

Independent suspension, featuring coil spring/strut damper units, is used on the front wheels, while a semi independent suspension using coil springs, shock absorbers and a trailing arm is used at the rear. An optional computer controlled rear air suspension is optional, utilizing air springs in place of the coil springs. The rack and pinion steering unit is mounted behind the engine with power-assist available as an option.

The brakes are disc on the front and either drum or disc on the rear wheels, with an Anti-Lock Brake System (ABS) standard on most models.

Vehicle identification numbers

Modifications are a continuing and unpublicized process in vehicle manufacturing. Since spare parts lists and manuals are compiled on a numerical basis, the individual vehicle numbers are necessary to correctly identify the component required.

Vehicle Identification Number (VIN)

This very important identification number is stamped on a plate attached to the dashboard inside the windshield on the driver's side of the vehicle (see illustration). The VIN also appears on the Vehicle Certificate of Title and Registration. It contains information such as where and when the vehicle was manufactured, the model year and the body style.

VIN engine and model year codes

Two particularly important pieces of information found in the VIN are the engine code and the model year code. Counting from the left, the engine code letter designation is the 8th digit and the model year code letter designation is the 10th digit.

On the models covered by this manual the engine codes are:

U .... 3.0L V6 (1995 through 1997)
V .... 3.0L V6 (1998 and later)
4 .... 3.8L V6

On the models covered by this manual the model year codes are:

S 1995
T 1996
V 1997
W 1998
X 1999
Y 2000

Vehicle Certification Label

The Vehicle Certification Label is attached to the driver's side door pillar (see illustration). Information on this label includes the name of the manufacturer, the month and year of production, as well as information on the options with which it is equipped. This label is especially useful for matching the color and type of paint for repair work.

Engine identification number

Labels containing the engine code, engine number and build date can be found on the valve cover (see illustration). The engine number is also stamped onto a machined pad on the external surface of the engine block.

Automatic transaxle identification number

The automatic transaxle ID number is affixed to a label on the top of the transaxle bellhousing (see illustration).

Vehicle Emissions Control Information label

This label is found in the engine compartment. See Chapter 6 for more information on this label.

The Vehicle Identification Number (VIN) is stamped into a metal plate fastened to the dashboard on the driver's side - it is visible through the windshield

The Vehicle Safety Certification label is affixed to the drivers door pillar

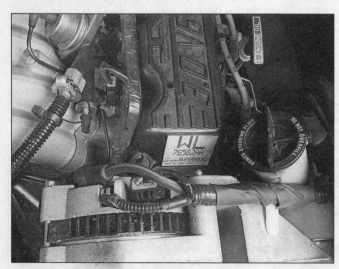

The engine identification label is located on the valve cover

The transaxle identification label is affixed to the top of the bellhousing

Buying parts

Replacement parts are available from many sources, which generally fall into one of two categories - authorized dealer parts departments and independent retail auto parts stores. Our advice concerning these parts is as follows:

Retail auto parts stores: Good auto parts stores will stock frequently needed components which wear out relatively fast, such as clutch components, exhaust systems, brake parts, tune-up parts, etc. These stores often supply new or reconditioned parts on an exchange basis, which can save a considerable amount of money. Discount auto parts stores are often very good places to buy materials and parts needed for general vehicle maintenance such as oil, grease, filters, spark plugs, belts, touch-up paint, bulbs, etc. They also usually sell tools and general accessories, have convenient hours, charge lower prices and can often be found not far from home.

Authorized dealer parts department: This is the best source for parts which are unique to the vehicle and not generally available elsewhere (such as major engine parts, transmission parts, trim pieces, etc.).

Warranty information: If the vehicle is still covered under warranty, be sure that any replacement parts purchased - regardless of the source - do not invalidate the warranty!

To be sure of obtaining the correct parts, have engine and chassis numbers available and, if possible, take the old parts along for positive identification.

Maintenance techniques, tools and working facilities

Maintenance techniques

There are a number of techniques involved in maintenance and repair that will be referred to throughout this manual. Application of these techniques will enable the home mechanic to be more efficient, better organized and capable of performing the various tasks properly, which will ensure that the repair job is thorough and complete.

Fasteners

Fasteners are nuts, bolts, studs and screws used to hold two or more parts together. There are a few things to keep in mind when working with fasteners. Almost all of them use a locking device of some type, either a lockwasher, locknut, locking tab or thread adhesive. All threaded fasteners should be clean and straight, with undamaged threads and undamaged corners on the hex head where the wrench fits. Develop the habit of replacing all damaged nuts and bolts with new ones. Special locknuts with nylon or fiber inserts can only be used once. If they are removed, they lose their locking ability and must be replaced with new ones.

Rusted nuts and bolts should be treated with a penetrating fluid to ease removal and prevent breakage. Some mechanics use turpentine in a spout-type oil can, which works quite well. After applying the rust penetrant, let it work for a few minutes before trying to loosen the nut or bolt. Badly rusted fasteners may have to be chiseled or sawed off or removed with a special nut breaker, available at tool stores.

If a bolt or stud breaks off in an assembly, it can be drilled and removed with a special tool commonly available for this purpose. Most automotive machine shops can perform this task, as well as other repair procedures, such as the repair of threaded holes that have been stripped out.

Flat washers and lockwashers, when removed from an assembly, should always be replaced exactly as removed. Replace any damaged washers with new ones. Never use a lockwasher on any soft metal surface (such as aluminum), thin sheet metal or plastic.

Fastener sizes

For a number of reasons, automobile manufacturers are making wider and wider use of metric fasteners. Therefore, it is important to be able to tell the difference between standard (sometimes called U.S. or SAE) and metric hardware, since they cannot be interchanged.

All bolts, whether standard or metric, are sized according to diameter, thread pitch and length. For example, a standard 1/2 - 13 x 1 bolt is 1/2 inch in diameter, has 13 threads per inch and is 1 inch long. An M12 - 1.75 x 25 metric bolt is 12 mm in diameter, has a thread pitch of 1.75 mm (the distance between threads) and is 25 mm long. The two bolts are nearly identical, and easily confused, but they are not interchangeable.

In addition to the differences in diameter, thread pitch and length, metric and standard bolts can also be distinguished by examining the bolt heads. To begin with, the distance across the flats on a standard bolt head is measured in inches, while the same dimension on a metric bolt is sized in millimeters (the same is true for nuts). As a result, a standard wrench should not be used on a metric bolt and a metric wrench should not be used on a standard bolt. Also, most standard bolts have slashes radiating out from the center of the head to denote the grade or strength of the bolt, which is an indication of the amount of torque that can be applied to it. The greater the number of slashes, the greater the strength of the bolt. Grades 0 through 5 are commonly used on automobiles. Metric bolts have a property class (grade) number, rather than a slash, molded into their heads to indicate bolt strength. In this case, the higher the number, the stronger the bolt. Property class numbers 8.8, 9.8 and 10.9 are commonly used on automobiles.

Strength markings can also be used to distinguish standard hex nuts from metric hex nuts. Many standard nuts have dots stamped into one side, while metric nuts are marked with a number. The greater the number of dots, or the higher the number, the greater the strength of the nut.

Metric studs are also marked on their ends according to property class (grade). Larger studs are numbered (the same as metric bolts), while smaller studs carry a geometric code to denote grade.

It should be noted that many fasteners, especially Grades 0 through 2, have no distinguishing marks on them. When such is the case, the only way to determine whether it is standard or metric is to measure the thread pitch or compare it to a known fastener of the same size.

Standard fasteners are often referred to as SAE, as opposed to metric. However, it should be noted that SAE technically refers to a non-metric fine thread fastener only. Coarse thread non-metric fasteners are referred to as USS sizes.

Grade 1 or 2   Grade 5   Grade 8

Bolt strength marking (standard/SAE/USS; bottom - metric)

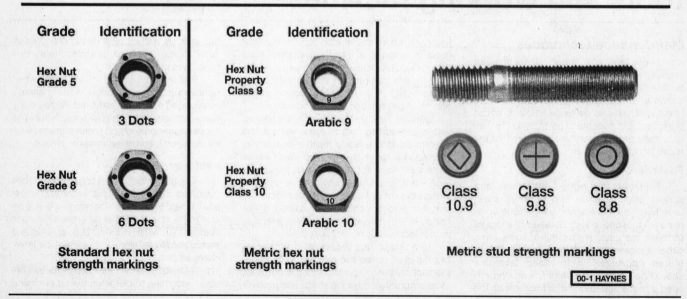

Grade	Identification	Grade	Identification
Hex Nut Grade 5	3 Dots	Hex Nut Property Class 9	Arabic 9
Hex Nut Grade 8	6 Dots	Hex Nut Property Class 10	Arabic 10

Class 10.9   Class 9.8   Class 8.8

Standard hex nut strength markings   **Metric hex nut strength markings**   **Metric stud strength markings**

Since fasteners of the same size (both standard and metric) may have different strength ratings, be sure to reinstall any bolts, studs or nuts removed from your vehicle in their original locations. Also, when replacing a fastener with a new one, make sure that the new one has a strength rating equal to or greater than the original.

Tightening sequences and procedures

Most threaded fasteners should be tightened to a specific torque value (torque is the twisting force applied to a threaded com-

ponent such as a nut or bolt). Overtightening the fastener can weaken it and cause it to break, while undertightening can cause it to eventually come loose. Bolts, screws and studs, depending on the material they are made of and their thread diameters, have specific torque values, many of which are noted in the Specifications at the beginning of each Chapter. Be sure to follow the torque recommendations closely. For fasteners not assigned a specific torque, a general torque value chart is presented here as a guide. These torque values are for dry (unlubricated) fasteners threaded into steel or cast iron (not

aluminum). As was previously mentioned, the size and grade of a fastener determine the amount of torque that can safely be applied to it. The figures listed here are approximate for Grade 2 and Grade 3 fasteners. Higher grades can tolerate higher torque values.

Fasteners laid out in a pattern, such as cylinder head bolts, oil pan bolts, differential cover bolts, etc., must be loosened or tightened in sequence to avoid warping the component. This sequence will normally be shown in the appropriate Chapter. If a specific pattern is not given, the following procedures can be used to prevent warping.

Metric thread sizes	Ft-lbs	Nm
M-6	6 to 9	9 to 12
M-8	14 to 21	19 to 28
M-10	28 to 40	38 to 54
M-12	50 to 71	68 to 96
M-14	80 to 140	109 to 154
Pipe thread sizes		
1/8	5 to 8	7 to 10
1/4	12 to 18	17 to 24
3/8	22 to 33	30 to 44
1/2	25 to 35	34 to 47
U.S. thread sizes		
1/4 - 20	6 to 9	9 to 12
5/16 - 18	12 to 18	17 to 24
5/16 - 24	14 to 20	19 to 27
3/8 - 16	22 to 32	30 to 43
3/8 - 24	27 to 38	37 to 51
7/16 - 14	40 to 55	55 to 74
7/16 - 20	40 to 60	55 to 81
1/2 - 13	55 to 80	75 to 108

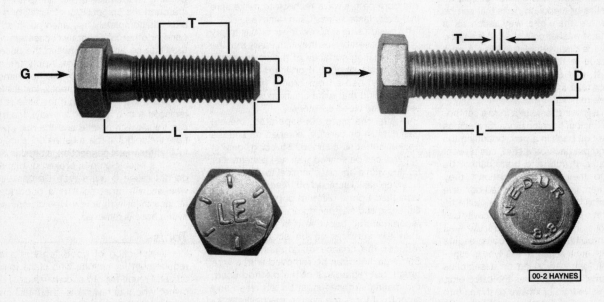

00-2 HAYNES

Standard (SAE and USS) bolt dimensions/grade marks

G   *Grade marks (bolt strength)*
L   *Length (in inches)*
T   *Thread pitch (number of threads per inch)*
D   *Nominal diameter (in inches)*

Metric bolt dimensions/grade marks

P   *Property class (bolt strength)*
L   *Length (in millimeters)*
T   *Thread pitch (distance between threads in millimeters)*
D   *Diameter*

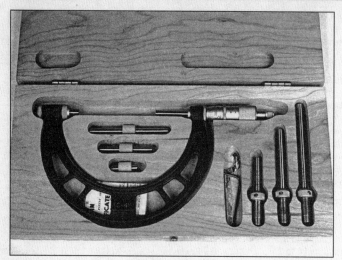

Micrometer set

Dial indicator set

Initially, the bolts or nuts should be assembled finger-tight only. Next, they should be tightened one full turn each, in a criss-cross or diagonal pattern. After each one has been tightened one full turn, return to the first one and tighten them all one-half turn, following the same pattern. Finally, tighten each of them one-quarter turn at a time until each fastener has been tightened to the proper torque. To loosen and remove the fasteners, the procedure would be reversed.

Component disassembly

Component disassembly should be done with care and purpose to help ensure that the parts go back together properly. Always keep track of the sequence in which parts are removed. Make note of special characteristics or marks on parts that can be installed more than one way, such as a grooved thrust washer on a shaft. It is a good idea to lay the disassembled parts out on a clean surface in the order that they were removed. It may also be helpful to make sketches or take instant photos of components before removal.

When removing fasteners from a component, keep track of their locations. Sometimes threading a bolt back in a part, or putting the washers and nut back on a stud, can prevent mix-ups later. If nuts and bolts cannot be returned to their original locations, they should be kept in a compartmented box or a series of small boxes. A cupcake or muffin tin is ideal for this purpose, since each cavity can hold the bolts and nuts from a particular area (i.e. oil pan bolts, valve cover bolts, engine mount bolts, etc.). A pan of this type is especially helpful when working on assemblies with very small parts, such as the carburetor, alternator, valve train or interior dash and trim pieces. The cavities can be marked with paint or tape to identify the contents.

Whenever wiring looms, harnesses or connectors are separated, it is a good idea to identify the two halves with numbered pieces of masking tape so they can be easily reconnected.

Gasket sealing surfaces

Throughout any vehicle, gaskets are used to seal the mating surfaces between two parts and keep lubricants, fluids, vacuum or pressure contained in an assembly.

Many times these gaskets are coated with a liquid or paste-type gasket sealing compound before assembly. Age, heat and pressure can sometimes cause the two parts to stick together so tightly that they are very difficult to separate. Often, the assembly can be loosened by striking it with a soft-face hammer near the mating surfaces. A regular hammer can be used if a block of wood is placed between the hammer and the part. Do not hammer on cast parts or parts that could be easily damaged. With any particularly stubborn part, always recheck to make sure that every fastener has been removed.

Avoid using a screwdriver or bar to pry apart an assembly, as they can easily mar the gasket sealing surfaces of the parts, which must remain smooth. If prying is absolutely necessary, use an old broom handle, but keep in mind that extra clean up will be necessary if the wood splinters.

After the parts are separated, the old gasket must be carefully scraped off and the gasket surfaces cleaned. Stubborn gasket material can be soaked with rust penetrant or treated with a special chemical to soften it so it can be easily scraped off. A scraper can be fashioned from a piece of copper tubing by flattening and sharpening one end. Copper is recommended because it is usually softer than the surfaces to be scraped, which reduces the chance of gouging the part. Some gaskets can be removed with a wire brush, but regardless of the method used, the mating surfaces must be left clean and smooth. If for some reason the gasket surface is gouged, then a gasket sealer thick enough to fill scratches will have to be used during reassembly of the components. For most applications, a non-drying (or semi-drying) gasket sealer should be used.

Hose removal tips

Warning: *If the vehicle is equipped with air conditioning, do not disconnect any of the A/C hoses without first having the system depressurized by a dealer service department or a service station.*

Hose removal precautions closely parallel gasket removal precautions. Avoid scratching or gouging the surface that the hose mates against or the connection may leak. This is especially true for radiator hoses. Because of various chemical reactions, the rubber in hoses can bond itself to the metal spigot that the hose fits over. To remove a hose, first loosen the hose clamps that secure it to the spigot. Then, with slip-joint pliers, grab the hose at the clamp and rotate it around the spigot. Work it back and forth until it is completely free, then pull it off. Silicone or other lubricants will ease removal if they can be applied between the hose and the outside of the spigot. Apply the same lubricant to the inside of the hose and the outside of the spigot to simplify installation.

As a last resort (and if the hose is to be replaced with a new one anyway), the rubber can be slit with a knife and the hose peeled from the spigot. If this must be done, be careful that the metal connection is not damaged.

If a hose clamp is broken or damaged, do not reuse it. Wire-type clamps usually weaken with age, so it is a good idea to replace them with screw-type clamps whenever a hose is removed.

Tools

A selection of good tools is a basic requirement for anyone who plans to maintain and repair his or her own vehicle. For the owner who has few tools, the initial investment might seem high, but when compared to the spiraling costs of professional auto maintenance and repair, it is a wise one.

To help the owner decide which tools are needed to perform the tasks detailed in this manual, the following tool lists are offered: *Maintenance and minor repair,*

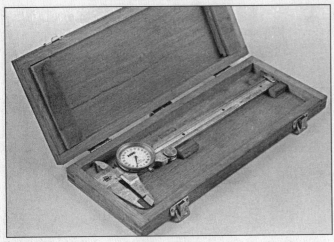

Dial caliper

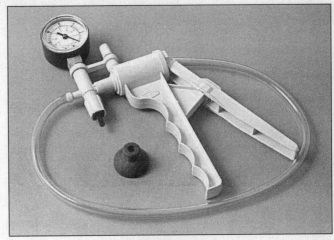

Hand-operated vacuum pump

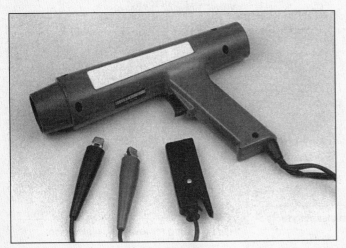

Timing light

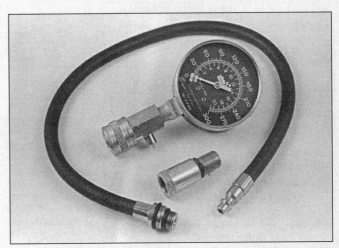

Compression gauge with spark plug hole adapter

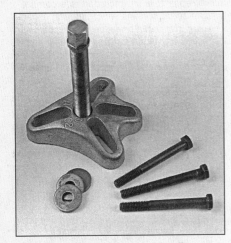

Damper/steering wheel puller

General purpose puller

Hydraulic lifter removal tool

Repair/overhaul and *Special.*

The newcomer to practical mechanics should start off with the *maintenance and minor repair* tool kit, which is adequate for the simpler jobs performed on a vehicle. Then, as confidence and experience grow, the owner can tackle more difficult tasks, buying additional tools as they are needed.

Eventually the basic kit will be expanded into the *repair and overhaul* tool set. Over a period of time, the experienced do-it-yourselfer will assemble a tool set complete enough for most repair and overhaul procedures and will add tools from the special category when it is felt that the expense is justified by the frequency of use.

Maintenance and minor repair tool kit

The tools in this list should be considered the minimum required for performance of routine maintenance, servicing and minor repair work. We recommend the purchase of combination wrenches (box-end and open-

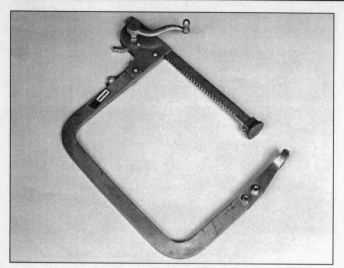

Valve spring compressor

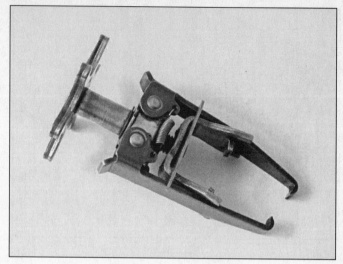

Valve spring compressor

Ridge reamer

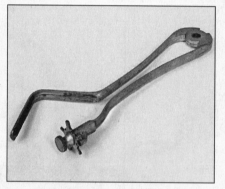

Piston ring groove cleaning tool

Ring removal/installation tool

end combined in one wrench). While more expensive than open end wrenches, they offer the advantages of both types of wrench.

> Combination wrench set (1/4-inch to 1 inch or 6 mm to 19 mm)
> Adjustable wrench, 8 inch
> Spark plug wrench with rubber insert
> Spark plug gap adjusting tool
> Feeler gauge set
> Brake bleeder wrench

Ring compressor

> Standard screwdriver (5/16-inch x 6 inch)
> Phillips screwdriver (No. 2 x 6 inch)
> Combination pliers - 6 inch
> Hacksaw and assortment of blades
> Tire pressure gauge
> Grease gun
> Oil can
> Fine emery cloth
> Wire brush
> Battery post and cable cleaning tool
> Oil filter wrench
> Funnel (medium size)
> Safety goggles
> Jackstands (2)
> Drain pan

Note: If basic tune-ups are going to be part of routine maintenance, it will be necessary to purchase a good quality stroboscopic timing light and combination tachometer/dwell meter. Although they are included in the list of special tools, it is mentioned here because they are absolutely necessary for tuning most vehicles properly.

Repair and overhaul tool set

These tools are essential for anyone who plans to perform major repairs and are in addition to those in the maintenance and

minor repair tool kit. Included is a comprehensive set of sockets which, though expensive, are invaluable because of their versatility, especially when various extensions and drives are available. We recommend the 1/2-inch drive over the 3/8-inch drive. Although the larger drive is bulky and more expensive, it has the capacity of accepting a very wide range of large sockets. Ideally, however, the mechanic should have a 3/8-inch drive set and a 1/2-inch drive set.

> Socket set(s)
> Reversible ratchet
> Extension - 10 inch
> Universal joint
> Torque wrench (same size drive as sockets)
> Ball peen hammer - 8 ounce
> Soft-face hammer (plastic/rubber)
> Standard screwdriver (1/4-inch x 6 inch)
> Standard screwdriver (stubby - 5/16-inch)
> Phillips screwdriver (No. 3 x 8 inch)
> Phillips screwdriver (stubby - No. 2)
> Pliers - vise grip
> Pliers - lineman's
> Pliers - needle nose
> Pliers - snap-ring (internal and external)
> Cold chisel - 1/2-inch

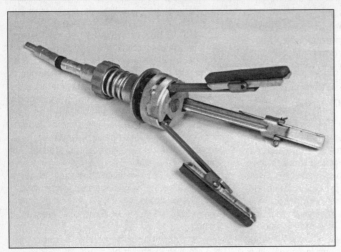

Cylinder hone

Brake hold-down spring tool

Scribe
Scraper (made from flattened copper
    tubing)
Centerpunch
Pin punches (1/16, 1/8, 3/16-inch)
Steel rule/straightedge - 12 inch
Allen wrench set (1/8 to 3/8-inch or
    4 mm to 10 mm)
A selection of files
Wire brush (large)
Jackstands (second set)
Jack (scissor or hydraulic type)

Note: Another tool which is often useful is an electric drill with a chuck capacity of 3/8-inch and a set of good quality drill bits.

Special tools

The tools in this list include those which are not used regularly, are expensive to buy, or which need to be used in accordance with their manufacturer's instructions. Unless these tools will be used frequently, it is not very economical to purchase many of them. A consideration would be to split the cost and use between yourself and a friend or friends. In addition, most of these tools can be obtained from a tool rental shop on a temporary basis.

This list primarily contains only those tools and instruments widely available to the public, and not those special tools produced by the vehicle manufacturer for distribution to dealer service departments. Occasionally, references to the manufacturer's special tools are included in the text of this manual. Generally, an alternative method of doing the job without the special tool is offered. However, sometimes there is no alternative to their use. Where this is the case, and the tool cannot be purchased or borrowed, the work should be turned over to the dealer service department or an automotive repair shop.

Valve spring compressor
Piston ring groove cleaning tool
Piston ring compressor
Piston ring installation tool
Cylinder compression gauge
Cylinder ridge reamer
Cylinder surfacing hone
Cylinder bore gauge
Micrometers and/or dial calipers
Hydraulic lifter removal tool
Balljoint separator
Universal-type puller
Impact screwdriver
Dial indicator set

Stroboscopic timing light (inductive
    pick-up)
Hand operated vacuum/pressure pump
Tachometer/dwell meter
Universal electrical multimeter
Cable hoist
Brake spring removal and installation
    tools
Floor jack

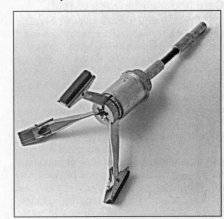

Brake cylinder hone

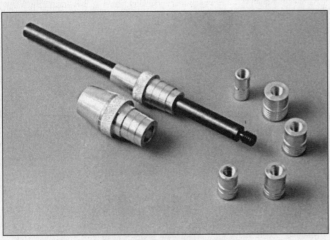

Clutch plate alignment tool

Tap and die set

Buying tools

For the do-it-yourselfer who is just starting to get involved in vehicle maintenance and repair, there are a number of options available when purchasing tools. If maintenance and minor repair is the extent of the work to be done, the purchase of individual tools is satisfactory. If, on the other hand, extensive work is planned, it would be a good idea to purchase a modest tool set from one of the large retail chain stores. A set can usually be bought at a substantial savings over the individual tool prices, and they often come with a tool box. As additional tools are needed, add-on sets, individual tools and a larger tool box can be purchased to expand the tool selection. Building a tool set gradually allows the cost of the tools to be spread over a longer period of time and gives the mechanic the freedom to choose only those tools that will actually be used.

Tool stores will often be the only source of some of the special tools that are needed, but regardless of where tools are bought, try to avoid cheap ones, especially when buying screwdrivers and sockets, because they won't last very long. The expense involved in replacing cheap tools will eventually be greater than the initial cost of quality tools.

Care and maintenance of tools

Good tools are expensive, so it makes sense to treat them with respect. Keep them clean and in usable condition and store them properly when not in use. Always wipe off any dirt, grease or metal chips before putting them away. Never leave tools lying around in the work area. Upon completion of a job, always check closely under the hood for tools that may have been left there so they won't get lost during a test drive.

Some tools, such as screwdrivers, pliers, wrenches and sockets, can be hung on a panel mounted on the garage or workshop wall, while others should be kept in a tool box or tray. Measuring instruments, gauges, meters, etc. must be carefully stored where they cannot be damaged by weather or impact from other tools.

When tools are used with care and stored properly, they will last a very long time. Even with the best of care, though, tools will wear out if used frequently. When a tool is damaged or worn out, replace it. Subsequent jobs will be safer and more enjoyable if you do.

How to repair damaged threads

Sometimes, the internal threads of a nut or bolt hole can become stripped, usually from overtightening. Stripping threads is an all-too-common occurrence, especially when working with aluminum parts, because aluminum is so soft that it easily strips out.

Usually, external or internal threads are only partially stripped. After they've been cleaned up with a tap or die, they'll still work. Sometimes, however, threads are badly damaged. When this happens, you've got three choices:

1) Drill and tap the hole to the next suitable oversize and install a larger diameter bolt, screw or stud.

2) Drill and tap the hole to accept a threaded plug, then drill and tap the plug to the original screw size. You can also buy a plug already threaded to the original size. Then you simply drill a hole to the specified size, then run the threaded plug into the hole with a bolt and jam nut. Once the plug is fully seated, remove the jam nut and bolt.

3) The third method uses a patented thread repair kit like Heli-Coil or Slimsert. These easy-to-use kits are designed to repair damaged threads in straight-through holes and blind holes. Both are available as kits which can handle a variety of sizes and thread patterns. Drill the hole, then tap it with the special included tap. Install the Heli-Coil and the hole is back to its original diameter and thread pitch.

Regardless of which method you use, be sure to proceed calmly and carefully. A little impatience or carelessness during one of these relatively simple procedures can ruin your whole day's work and cost you a bundle if you wreck an expensive part.

Working facilities

Not to be overlooked when discussing tools is the workshop. If anything more than routine maintenance is to be carried out, some sort of suitable work area is essential.

It is understood, and appreciated, that many home mechanics do not have a good workshop or garage available, and end up removing an engine or doing major repairs outside. It is recommended, however, that the overhaul or repair be completed under the cover of a roof.

A clean, flat workbench or table of comfortable working height is an absolute necessity. The workbench should be equipped with a vise that has a jaw opening of at least four inches.

As mentioned previously, some clean, dry storage space is also required for tools, as well as the lubricants, fluids, cleaning solvents, etc. which soon become necessary.

Sometimes waste oil and fluids, drained from the engine or cooling system during normal maintenance or repairs, present a disposal problem. To avoid pouring them on the ground or into a sewage system, pour the used fluids into large containers, seal them with caps and take them to an authorized disposal site or recycling center. Plastic jugs, such as old antifreeze containers, are ideal for this purpose.

Always keep a supply of old newspapers and clean rags available. Old towels are excellent for mopping up spills. Many mechanics use rolls of paper towels for most work because they are readily available and disposable. To help keep the area under the vehicle clean, a large cardboard box can be cut open and flattened to protect the garage or shop floor.

Whenever working over a painted surface, such as when leaning over a fender to service something under the hood, always cover it with an old blanket or bedspread to protect the finish. Vinyl covered pads, made especially for this purpose, are available at auto parts stores.

Booster battery (jump) starting

Observe the following precautions when using a booster battery to start a vehicle:

a) *Before connecting the booster battery, make sure the ignition switch is in the Off position.*

b) *Ensure that all electrical equipment (lights, heater, wipers etc.) are switched off.*

c) *Make sure that the booster battery is the same voltage as the discharged battery in the vehicle.*

d) *If the battery is being jump started from the battery in another vehicle, the two vehicles MUST NOT TOUCH each other.*

e) *Make sure the transaxle is in Neutral (manual transaxle) or Park (automatic transaxle).*

f) *Wear eye protection when jump starting a vehicle.*

Connect one jumper lead between the positive (+) terminals of the two batteries. Connect the other jumper lead first to the negative (-) terminal of the booster battery, then to a good engine ground on the vehicle to be started **(see illustration)**. Attach the lead at least 18 inches from the battery, if possible. Make sure that the jumper leads will not contact the fan, drivebelt of other moving parts of the engine.

Start the engine using the booster battery and allow the engine idle speed to stabilize. Disconnect the jumper leads in the reverse order of connection.

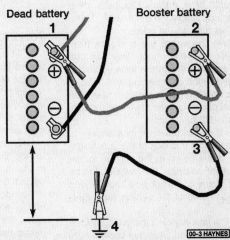

Make the booster battery cable connections in the numerical order shown (note that the negative cable of the booster battery is NOT attached to the negative terminal of the dead battery)

Jacking and towing

Jacking

Warning: *The jack supplied with the vehicle should only be used for changing a tire or placing jackstands under the frame. Never work under the vehicle or start the engine while this jack is being used as the only means of support.*

The vehicle should be on level ground. Place the shift lever in Park. Block the wheel diagonally opposite the wheel being changed. Set the parking brake.

Remove the spare tire and jack from stowage. Remove the wheel cover and trim ring (if so equipped) with the tapered end of the lug nut wrench by inserting and twisting the handle and then prying against the back of the wheel cover. Loosen the wheel lug nuts about 1/4-to-1/2 turn each.

Place the scissors-type jack under the side of the vehicle and adjust the jack height until it fits in the notch in the vertical rocker panel flange nearest the wheel to be changed. There is a front and rear jacking point on each side of the vehicle **(see illustration)**.

Turn the jack handle clockwise until the tire clears the ground. Remove the lug nuts and pull the wheel off. Replace it with the spare.

Install the lug nuts with the beveled edges facing in. Tighten them snugly. Don't

attempt to tighten them completely until the vehicle is lowered or it could slip off the jack. Turn the jack handle counterclockwise to lower the vehicle. Remove the jack and tighten the lug nuts in a diagonal pattern.

Install the cover (and trim ring, if used) and be sure it's snapped into place all the way around.

Stow the tire, jack and wrench. Unblock the wheels.

Towing

As a general rule, the vehicle should be towed from the front with the front (drive) wheels off the ground. If the vehicle must be towed from the rear, place the front wheels on a towing dolly. **Caution:** *Never tow a front wheel drive vehicle from the rear with the front wheels on the ground.*

Equipment specifically designed for towing should be used. It should be attached to the main structural members of the vehicle, not the bumpers or brackets. Do not use the tie-down hook loops at the front or the rear of the vehicle for towing. These hooks loops are designed for securing the vehicle during transport, if used for towing, damage to the front or rear bumper may occur.

The ignition key must be in the ACC position, since the steering lock mechanism isn't strong enough to hold the front wheels

The jack fits under the rocker panel (there are two jacking points on each side of the vehicle, indicated by a notch in the rocker panel flange)

straight while towing. Pace the shift lever in neutral and release the parking brake.

Safety is a major consideration when towing and all applicable state and local laws must be obeyed. A safety chain system must be used at all times.

Automotive chemicals and lubricants

A number of automotive chemicals and lubricants are available for use during vehicle maintenance and repair. They include a wide variety of products ranging from cleaning solvents and degreasers to lubricants and protective sprays for rubber, plastic and vinyl.

Cleaners

Carburetor cleaner and choke cleaner is a strong solvent for gum, varnish and carbon. Most carburetor cleaners leave a dry-type lubricant film which will not harden or gum up. Because of this film it is not recommended for use on electrical components.

Brake system cleaner is used to remove grease and brake fluid from the brake system, where clean surfaces are absolutely necessary. It leaves no residue and often eliminates brake squeal caused by contaminants.

Electrical cleaner removes oxidation, corrosion and carbon deposits from electrical contacts, restoring full current flow. It can also be used to clean spark plugs, carburetor jets, voltage regulators and other parts where an oil-free surface is desired.

Demoisturants remove water and moisture from electrical components such as alternators, voltage regulators, electrical connectors and fuse blocks. They are non-conductive, non-corrosive and non-flammable.

Degreasers are heavy-duty solvents used to remove grease from the outside of the engine and from chassis components. They can be sprayed or brushed on and, depending on the type, are rinsed off either with water or solvent.

Lubricants

Motor oil is the lubricant formulated for use in engines. It normally contains a wide variety of additives to prevent corrosion and reduce foaming and wear. Motor oil comes in various weights (viscosity ratings) from 0 to 50. The recommended weight of the oil depends on the season, temperature and the demands on the engine. Light oil is used in cold climates and under light load conditions. Heavy oil is used in hot climates and where high loads are encountered. Multi-viscosity oils are designed to have characteristics of both light and heavy oils and are available in a number of weights from 5W-20 to 20W-50.

Gear oil is designed to be used in differentials, manual transmissions and other areas where high-temperature lubrication is required.

Chassis and wheel bearing grease is a heavy grease used where increased loads and friction are encountered, such as for wheel bearings, balljoints, tie-rod ends and universal joints.

High-temperature wheel bearing grease is designed to withstand the extreme temperatures encountered by wheel bearings in disc brake equipped vehicles. It usually contains molybdenum disulfide (moly), which is a dry-type lubricant.

White grease is a heavy grease for metal-to-metal applications where water is a problem. White grease stays soft under both low and high temperatures (usually from -100 to +190-degrees F), and will not wash off or dilute in the presence of water.

Assembly lube is a special extreme pressure lubricant, usually containing moly, used to lubricate high-load parts (such as main and rod bearings and cam lobes) for initial start-up of a new engine. The assembly lube lubricates the parts without being squeezed out or washed away until the engine oiling system begins to function.

Silicone lubricants are used to protect rubber, plastic, vinyl and nylon parts.

Graphite lubricants are used where oils cannot be used due to contamination problems, such as in locks. The dry graphite will lubricate metal parts while remaining uncontaminated by dirt, water, oil or acids. It is electrically conductive and will not foul electrical contacts in locks such as the ignition switch.

Moly penetrants loosen and lubricate frozen, rusted and corroded fasteners and prevent future rusting or freezing.

Heat-sink grease is a special electrically non-conductive grease that is used for mounting electronic ignition modules where it is essential that heat is transferred away from the module.

Sealants

RTV sealant is one of the most widely used gasket compounds. Made from silicone, RTV is air curing, it seals, bonds, waterproofs, fills surface irregularities, remains flexible, doesn't shrink, is relatively easy to remove, and is used as a supplementary sealer with almost all low and medium temperature gaskets.

Anaerobic sealant is much like RTV in that it can be used either to seal gaskets or to form gaskets by itself. It remains flexible, is solvent resistant and fills surface imperfections. The difference between an anaerobic sealant and an RTV-type sealant is in the curing. RTV cures when exposed to air, while an anaerobic sealant cures only in the absence of air. This means that an anaerobic sealant cures only after the assembly of parts, sealing them together.

Thread and pipe sealant is used for sealing hydraulic and pneumatic fittings and vacuum lines. It is usually made from a Teflon compound, and comes in a spray, a paint-on liquid and as a wrap-around tape.

Chemicals

Anti-seize compound prevents seizing, galling, cold welding, rust and corrosion in fasteners. High-temperature ant-seize, usually made with copper and graphite lubricants, is used for exhaust system and exhaust manifold bolts.

Anaerobic locking compounds are used to keep fasteners from vibrating or working loose and cure only after installation, in the absence of air. Medium strength locking compound is used for small nuts, bolts and screws that may be removed later. High-strength locking compound is for large nuts, bolts and studs which aren't removed on a regular basis.

Oil additives range from viscosity index improvers to chemical treatments that claim to reduce internal engine friction. It should be noted that most oil manufacturers caution against using additives with their oils.

Gas additives perform several functions, depending on their chemical makeup. They usually contain solvents that help dissolve gum and varnish that build up on carburetor, fuel injection and intake parts. They also serve to break down carbon deposits that form on the inside surfaces of the combustion chambers. Some additives contain upper cylinder lubricants for valves and piston rings, and others contain chemicals to remove condensation from the gas tank.

Miscellaneous

Brake fluid is specially formulated hydraulic fluid that can withstand the heat and pressure encountered in brake systems. Care must be taken so this fluid does not come in contact with painted surfaces or plastics. An opened container should always be resealed to prevent contamination by water or dirt.

Weatherstrip adhesive is used to bond weatherstripping around doors, windows and trunk lids. It is sometimes used to attach trim pieces.

Undercoating is a petroleum-based, tar-like substance that is designed to protect metal surfaces on the underside of the vehicle from corrosion. It also acts as a sound-deadening agent by insulating the bottom of the vehicle.

Waxes and polishes are used to help protect painted and plated surfaces from the weather. Different types of paint may require the use of different types of wax and polish. Some polishes utilize a chemical or abrasive cleaner to help remove the top layer of oxidized (dull) paint on older vehicles. In recent years many non-wax polishes that contain a wide variety of chemicals such as polymers and silicones have been introduced. These non-wax polishes are usually easier to apply and last longer than conventional waxes and polishes.

Conversion factors

Length (distance)

Inches (in)	X	25.4	= Millimetres (mm)	X 0.0394	= Inches (in)
Feet (ft)	X	0.305	= Metres (m)	X 3.281	= Feet (ft)
Miles	X	1.609	= Kilometres (km)	X 0.621	= Miles

Volume (capacity)

Cubic inches (cu in; in³)	X	16.387	= Cubic centimetres (cc; cm³)	X 0.061	= Cubic inches (cu in; in³)
Imperial pints (Imp pt)	X	0.568	= Litres (l)	X 1.76	= Imperial pints (Imp pt)
Imperial quarts (Imp qt)	X	1.137	= Litres (l)	X 0.88	= Imperial quarts (Imp qt)
Imperial quarts (Imp qt)	X	1.201	= US quarts (US qt)	X 0.833	= Imperial quarts (Imp qt)
US quarts (US qt)	X	0.946	= Litres (l)	X 1.057	= US quarts (US qt)
Imperial gallons (Imp gal)	X	4.546	= Litres (l)	X 0.22	= Imperial gallons (Imp gal)
Imperial gallons (Imp gal)	X	1.201	= US gallons (US gal)	X 0.833	= Imperial gallons (Imp gal)
US gallons (US gal)	X	3.785	= Litres (l)	X 0.264	= US gallons (US gal)

Mass (weight)

Ounces (oz)	X	28.35	= Grams (g)	X 0.035	= Ounces (oz)
Pounds (lb)	X	0.454	= Kilograms (kg)	X 2.205	= Pounds (lb)

Force

Ounces-force (ozf; oz)	X	0.278	= Newtons (N)	X 3.6	= Ounces-force (ozf; oz)
Pounds-force (lbf; lb)	X	4.448	= Newtons (N)	X 0.225	= Pounds-force (lbf; lb)
Newtons (N)	X	0.1	= Kilograms-force (kgf; kg)	X 9.81	= Newtons (N)

Pressure

Pounds-force per square inch (psi; lbf/in²; lb/in²)	X	0.070	= Kilograms-force per square centimetre (kgf/cm²; kg/cm²)	X 14.223	= Pounds-force per square inch (psi; lbf/in²; lb/in²)
Pounds-force per square inch (psi; lbf/in²; lb/in²)	X	0.068	= Atmospheres (atm)	X 14.696	= Pounds-force per square inch (psi; lbf/in²; lb/in²)
Pounds-force per square inch (psi; lbf/in²; lb/in²)	X	0.069	= Bars	X 14.5	= Pounds-force per square inch (psi; lbf/in²; lb/in²)
Pounds-force per square inch (psi; lbf/in²; lb/in²)	X	6.895	= Kilopascals (kPa)	X 0.145	= Pounds-force per square inch (psi; lbf/in²; lb/in²)
Kilopascals (kPa)	X	0.01	= Kilograms-force per square centimetre (kgf/cm²; kg/cm²)	X 98.1	= Kilopascals (kPa)

Torque (moment of force)

Pounds-force inches (lbf in; lb in)	X	1.152	= Kilograms-force centimetre (kgf cm; kg cm)	X 0.868	= Pounds-force inches (lbf in; lb in)
Pounds-force inches (lbf in; lb in)	X	0.113	= Newton metres (Nm)	X 8.85	= Pounds-force inches (lbf in; lb in)
Pounds-force inches (lbf in; lb in)	X	0.083	= Pounds-force feet (lbf ft; lb ft)	X 12	= Pounds-force inches (lbf in; lb in)
Pounds-force feet (lbf ft; lb ft)	X	0.138	= Kilograms-force metres (kgf m; kg m)	X 7.233	= Pounds-force feet (lbf ft; lb ft)
Pounds-force feet (lbf ft; lb ft)	X	1.356	= Newton metres (Nm)	X 0.738	= Pounds-force feet (lbf ft; lb ft)
Newton metres (Nm)	X	0.102	= Kilograms-force metres (kgf m; kg m)	X 9.804	= Newton metres (Nm)

Vacuum

Inches mercury (in. Hg)	X	3.377	= Kilopascals (kPa)	X 0.2961	= Inches mercury
Inches mercury (in. Hg)	X	25.4	= Millimeters mercury (mm Hg)	X 0.0394	= Inches mercury

Power

Horsepower (hp)	X	745.7	= Watts (W)	X 0.0013	= Horsepower (hp)

Velocity (speed)

Miles per hour (miles/hr; mph)	X	1.609	= Kilometres per hour (km/hr; kph)	X 0.621	= Miles per hour (miles/hr; mph)

Fuel consumption*

Miles per gallon, Imperial (mpg)	X	0.354	= Kilometres per litre (km/l)	X 2.825	= Miles per gallon, Imperial (mpg)
Miles per gallon, US (mpg)	X	0.425	= Kilometres per litre (km/l)	X 2.352	= Miles per gallon, US (mpg)

Temperature

Degrees Fahrenheit = (°C x 1.8) + 32

Degrees Celsius (Degrees Centigrade; °C) = (°F - 32) x 0.56

*It is common practice to convert from miles per gallon (mpg) to litres/100 kilometres (l/100km),
where mpg (Imperial) x l/100 km = 282 and mpg (US) x l/100 km = 235

Safety first!

Regardless of how enthusiastic you may be about getting on with the job at hand, take the time to ensure that your safety is not jeopardized. A moment's lack of attention can result in an accident, as can failure to observe certain simple safety precautions. The possibility of an accident will always exist, and the following points should not be considered a comprehensive list of all dangers. Rather, they are intended to make you aware of the risks and to encourage a safety conscious approach to all work you carry out on your vehicle.

Essential DOs and DON'Ts

DON'T rely on a jack when working under the vehicle. Always use approved jackstands to support the weight of the vehicle and place them under the recommended lift or support points.

DON'T attempt to loosen extremely tight fasteners (i.e. wheel lug nuts) while the vehicle is on a jack - it may fall.

DON'T start the engine without first making sure that the transmission is in Neutral (or Park where applicable) and the parking brake is set.

DON'T remove the radiator cap from a hot cooling system - let it cool or cover it with a cloth and release the pressure gradually.

DON'T attempt to drain the engine oil until you are sure it has cooled to the point that it will not burn you.

DON'T touch any part of the engine or exhaust system until it has cooled sufficiently to avoid burns.

DON'T siphon toxic liquids such as gasoline, antifreeze and brake fluid by mouth, or allow them to remain on your skin.

DON'T inhale brake lining dust - it is potentially hazardous (see *Asbestos* below).

DON'T allow spilled oil or grease to remain on the floor - wipe it up before someone slips on it.

DON'T use loose fitting wrenches or other tools which may slip and cause injury.

DON'T push on wrenches when loosening or tightening nuts or bolts. Always try to pull the wrench toward you. If the situation calls for pushing the wrench away, push with an open hand to avoid scraped knuckles if the wrench should slip.

DON'T attempt to lift a heavy component alone - get someone to help you.

DON'T rush or take unsafe shortcuts to finish a job.

DON'T allow children or animals in or around the vehicle while you are working on it.

DO wear eye protection when using power tools such as a drill, sander, bench grinder, etc. and when working under a vehicle.

DO keep loose clothing and long hair well out of the way of moving parts.

DO make sure that any hoist used has a safe working load rating adequate for the job.

DO get someone to check on you periodically when working alone on a vehicle.

DO carry out work in a logical sequence and make sure that everything is correctly assembled and tightened.

DO keep chemicals and fluids tightly capped and out of the reach of children and pets.

DO remember that your vehicle's safety affects that of yourself and others. If in doubt on any point, get professional advice.

Asbestos

Certain friction, insulating, sealing, and other products - such as brake linings, brake bands, clutch linings, torque converters, gaskets, etc. - may contain asbestos. Extreme care must be taken to avoid inhalation of dust from such products, since it is hazardous to health. If in doubt, assume that they do contain asbestos.

Fire

Remember at all times that gasoline is highly flammable. Never smoke or have any kind of open flame around when working on a vehicle. But the risk does not end there. A spark caused by an electrical short circuit, by two metal surfaces contacting each other, or even by static electricity built up in your body under certain conditions, can ignite gasoline vapors, which in a confined space are highly explosive. Do not, under any circumstances, use gasoline for cleaning parts. Use an approved safety solvent.

Always disconnect the battery ground (-) cable at the battery before working on any part of the fuel system or electrical system. Never risk spilling fuel on a hot engine or exhaust component. It is strongly recommended that a fire extinguisher suitable for use on fuel and electrical fires be kept handy in the garage or workshop at all times. Never try to extinguish a fuel or electrical fire with water.

Fumes

Certain fumes are highly toxic and can quickly cause unconsciousness and even death if inhaled to any extent. Gasoline vapor falls into this category, as do the vapors from some cleaning solvents. Any draining or pouring of such volatile fluids should be done in a well ventilated area.

When using cleaning fluids and solvents, read the instructions on the container carefully. Never use materials from unmarked containers.

Never run the engine in an enclosed space, such as a garage. Exhaust fumes contain carbon monoxide, which is extremely poisonous. If you need to run the engine, always do so in the open air, or at least have the rear of the vehicle outside the work area.

If you are fortunate enough to have the use of an inspection pit, never drain or pour gasoline and never run the engine while the vehicle is over the pit. The fumes, being heavier than air, will concentrate in the pit with possibly lethal results.

The battery

Never create a spark or allow a bare light bulb near a battery. They normally give off a certain amount of hydrogen gas, which is highly explosive.

Always disconnect the battery ground (-) cable at the battery before working on the fuel or electrical systems.

If possible, loosen the filler caps or cover when charging the battery from an external source (this does not apply to sealed or maintenance-free batteries). Do not charge at an excessive rate or the battery may burst.

Take care when adding water to a non maintenance-free battery and when carrying a battery. The electrolyte, even when diluted, is very corrosive and should not be allowed to contact clothing or skin.

Always wear eye protection when cleaning the battery to prevent the caustic deposits from entering your eyes.

Household current

When using an electric power tool, inspection light, etc., which operates on household current, always make sure that the tool is correctly connected to its plug and that, where necessary, it is properly grounded. Do not use such items in damp conditions and, again, do not create a spark or apply excessive heat in the vicinity of fuel or fuel vapor.

Secondary ignition system voltage

A severe electric shock can result from touching certain parts of the ignition system (such as the spark plug wires) when the engine is running or being cranked, particularly if components are damp or the insulation is defective. In the case of an electronic ignition system, the secondary system voltage is much higher and could prove fatal.

Troubleshooting

Contents

Engine

1 Engine will not rotate when attempting to start

1 Battery terminal connections loose or corroded. Check the cable terminals at the battery; tighten cable clamp and/or clean off corrosion as necessary (see Chapter 1).
2 Battery discharged or faulty. If the cable ends are clean and tight on the battery posts, turn the key to the On position and switch on the headlights or windshield wipers. If they won't run, the battery is discharged.
3 Automatic transmission not engaged in park (P) or Neutral (N).
4 Broken, loose or disconnected wires in the starting circuit. Inspect all wires and connectors at the battery, starter solenoid and ignition switch (on steering column).
5 Starter motor pinion jammed in driveplate ring gear. Remove starter (Chapter 5) and inspect pinion and driveplate (Chapter 2).
6 Starter solenoid faulty (Chapter 5).
7 Starter motor faulty (Chapter 5).
8 Ignition switch faulty (Chapter 12).
9 Engine seized. Try to turn the crankshaft with a large socket and breaker bar on the pulley bolt.
10 Starter relay faulty (Chapter 5).
11 Transmission Range (TR) sensor out of adjustment or defective (Chapter 6).

2 Engine rotates but will not start

1 Fuel tank empty.
2 Battery discharged (engine rotates slowly).
3 Battery terminal connections loose or corroded.
4 Fuel not reaching fuel injectors. Check for clogged fuel filter or lines and defective fuel pump. Also make sure the tank vent lines aren't clogged (Chapter 4).
5 Low cylinder compression. Check as described in Chapter 2.
6 Water in fuel. Drain tank and fill with new fuel.
7 Defective ignition coil(s) (Chapter 5).
8 Dirty or clogged fuel injector(s) (Chapter 4).
9 Wet or damaged ignition components (Chapters 1 and 5).
10 Worn, faulty or incorrectly gapped spark plugs (Chapter 1).
11 Broken, loose or disconnected wires in the starting circuit (see previous Section).
12 Broken, loose or disconnected wires at the ignition coil or faulty coil (Chapter 5).
13 Timing chain failure or wear affecting valve timing (Chapter 2).
14 Fuel injection or engine control systems failure (Chapters 4 and 6).
15 Defective MAF sensor (Chapter 6).

3 Starter motor operates without turning engine

1 Starter pinion sticking. Remove the starter (Chapter 5) and inspect.
2 Starter pinion or driveplate teeth worn or broken. Remove the inspection cover and inspect.

4 Engine hard to start when cold

1 Battery discharged or low. Check as described in Chapter 1.
2 Fuel not reaching the fuel injectors. Check the fuel filter, lines and fuel pump (Chapters 1 and 4).
3 Defective spark plugs (Chapter 1).
4 Defective engine coolant temperature sensor (Chapter 6).
5 Fuel injection or engine control systems malfunction (Chapters 4 and 6).

5 Engine hard to start when hot

1 Air filter dirty (Chapter 1).
2 Bad engine ground connection.
3 Fuel injection or engine control systems malfunction (Chapters 4 and 6).

6 Starter motor noisy or engages roughly

1 Pinion or driveplate teeth worn or broken. Remove the inspection cover on the left side of the engine and inspect.
2 Starter motor mounting bolts loose or missing.

7 Engine starts but stops immediately

1 Loose or damaged wire harness connections at distributor, coil or alternator.
2 Intake manifold vacuum leaks. Make sure all mounting bolts/nuts are tight and all vacuum hoses connected to the manifold are attached properly and in good condition.
3 Insufficient fuel pressure (see Chapter 4).
4 Fuel injection or engine control systems malfunction (Chapters 4 and 6).

8 Engine 'lopes' while idling or idles erratically

1 Vacuum leaks. Check mounting bolts at the intake manifold for tightness. Make sure that all vacuum hoses are connected and in good condition. Use a stethoscope or a length of fuel hose held against your ear to listen for vacuum leaks while the engine is running. A hissing sound will be heard. A soapy water solution will also detect leaks. Check the intake manifold gasket surfaces.
2 Leaking EGR valve or plugged PCV valve (see Chapters 1 and 6).
3 Air filter clogged (Chapter 1).
4 Fuel pump not delivering sufficient fuel (Chapter 4).
5 Leaking head gasket. Perform a cylinder compression check (Chapter 2).
6 Timing chain(s) worn (Chapter 2).
7 Camshaft lobes worn (Chapter 2).
8 Valves burned or otherwise leaking (Chapter 2).
9 Ignition timing out of adjustment (Chapter 5).
10 Ignition system not operating properly (Chapters 1 and 5).
11 Fuel injection or engine control systems malfunction (Chapters 4 and 6).

9 Engine misses at idle speed

1 Spark plugs faulty or not gapped properly (Chapter 1).
2 Faulty spark plug wires (Chapter 1).
3 Wet or damaged ignition components (Chapter 5).
4 Short circuits in ignition, coil or spark plug wires.
5 Sticking or faulty emissions systems (see Chapter 6).
6 Clogged fuel filter and/or foreign matter in fuel. Remove the fuel filter (Chapter 1) and inspect.
7 Vacuum leaks at intake manifold or hose connections. Check as described in Section 9.
8 Low or uneven cylinder compression. Check as described in Chapter 2.
9 Fuel injection or engine control systems malfunction (Chapters 4 and 6).

10 Excessively high idle speed

1 Sticking throttle linkage (Chapter 4).
2 Vacuum leaks at intake manifold or hose connections. Check as described in Section 8.
3 Fuel injection or engine control systems malfunction (Chapters 4 and 6).

11 Battery will not hold a charge

1 Alternator drivebelt defective or not adjusted properly (Chapter 1).
2 Battery cables loose or corroded (Chapter 1).
3 Alternator not charging properly (Chapter 5).
4 Loose, broken or faulty wires in the charging circuit (Chapter 5).
5 Short circuit causing a continuous drain on the battery.
6 Battery defective internally.

12 Alternator light stays on

1   Fault in alternator or charging circuit (Chapter 5).
2   Alternator drivebelt defective or not properly adjusted (Chapter 1).

13 Alternator light fails to come on when key is turned on

1   Faulty bulb (Chapter 12).
2   Defective alternator (Chapter 5).
3   Fault in the printed circuit, dash wiring or bulb holder (Chapter 12).

14 Engine misses throughout driving speed range

1   Fuel filter clogged and/or impurities in the fuel system. Check fuel filter (Chapter 1) or clean system (Chapter 4).
2   Faulty or incorrectly gapped spark plugs (Chapter 1).
3   Incorrect ignition timing (Chapter 5).
4   Defective spark plug wires (Chapter 1).
5   Emissions system components faulty (Chapter 6).
6   Low or uneven cylinder compression pressures. Check as described in Chapter 2.
7   Weak or faulty ignition coil(s) (Chapter 5).
8   Weak or faulty ignition system (Chapter 5).
9   Vacuum leaks at intake manifold or vacuum hoses (see Section 8).
10   Dirty or clogged fuel injector(s) (Chapter 4).
11   Leaky EGR valve (Chapter 6).
12   Fuel injection or engine control systems malfunction (Chapters 4 and 6).

15 Hesitation or stumble during acceleration

1   Ignition system not operating properly (Chapter 5).
2   Dirty or clogged fuel injector(s) (Chapter 4).
3   Low fuel pressure. Check for proper operation of the fuel pump and for restrictions in the fuel filter and lines (Chapter 4).
4   Fuel injection or engine control systems malfunction (Chapters 4 and 6).

16 Engine stalls

1   Idle speed incorrect (Chapter 4).
2   Fuel filter clogged and/or water and impurities in the fuel system (Chapter 1).
3   Damaged or wet distributor cap and wires.
4   Emissions system components faulty

(Chapter 6).
5   Faulty or incorrectly gapped spark plugs (Chapter 1). Also check the spark plug wires (Chapter 1).
6   Vacuum leak at the intake manifold or vacuum hoses. Check as described in Section 8.
7   Fuel injection or engine control systems malfunction (Chapters 4 and 6).

17 Engine lacks power

1   Incorrect ignition timing (Chapter 5).
2   Faulty or incorrectly gapped spark plugs (Chapter 1).
3   Air filter dirty (Chapter 1).
4   Faulty ignition coil(s) (Chapter 5).
5   Brakes binding (Chapters 1 and 10).
6   Automatic transmission fluid level incorrect, causing slippage (Chapter 1).
7   Fuel filter clogged and/or impurities in the fuel system (Chapters 1 and 4).
8   EGR system not functioning properly (Chapter 6).
9   Use of sub-standard fuel. Fill tank with proper octane fuel.
10   Low or uneven cylinder compression pressures. Check as described in Chapter 2.
11   Vacuum leak at intake manifold or vacuum hoses (check as described in Section 8).
12   Dirty or clogged fuel injector(s) (Chapters 1 and 4).
13   Fuel injection or engine control systems malfunction (Chapters 4 and 6).
14   Restricted exhaust system (Chapter 4).

18 Engine backfires

1   EGR system not functioning properly (Chapter 6).
2   Ignition timing incorrect (Chapter 5).
3   Damaged valve springs or sticking valves (Chapter 2).
4   Vacuum leak at the intake manifold or vacuum hoses (see Section 8).

19 Engine surges while holding accelerator steady

1   Vacuum leak at the intake manifold or vacuum hoses (see Section 8).
2   Restricted air filter (Chapter 1).
3   Fuel pump or pressure regulator defective (Chapter 4).
4   Fuel injection or engine control systems malfunction (Chapters 4 and 6).

20 Pinging or knocking engine sounds when engine is under load

1   Incorrect grade of fuel. Fill tank with fuel of the proper octane rating.

2   Ignition timing incorrect (Chapter 5).
3   Carbon build-up in combustion chambers. Remove cylinder head(s) and clean combustion chambers (Chapter 2).
4   Incorrect spark plugs (Chapter 1).
5   Fuel injection or engine control systems malfunction (Chapters 4 and 6).
6   Restricted exhaust system (Chapter 4).

21 Engine diesels (continues to run) after being turned off

1   Idle speed too high (Chapter 4).
2   Ignition timing incorrect (Chapter 5).
3   Incorrect spark plug heat range (Chapter 1).
4   Vacuum leak at the intake manifold or vacuum hoses (see Section 8).
5   Carbon build-up in combustion chambers. Remove the cylinder head(s) and clean the combustion chambers (Chapter 2).
6   Valves sticking (Chapter 2).
7   EGR system not operating properly (Chapter 6).
8   Fuel injection or engine control systems malfunction (Chapters 4 and 6).
9   Check for causes of overheating (Section 27).

22 Low oil pressure

1   Improper grade of oil.
2   Oil pump worn or damaged (Chapter 2).
3   Engine overheating (refer to Section 27).
4   Clogged oil filter (Chapter 1).
5   Clogged oil strainer (Chapter 2).
6   Oil pressure gauge not working properly (Chapter 2).

23 Excessive oil consumption

1   Loose oil drain plug.
2   Loose bolts or damaged oil pan gasket (Chapter 2).
3   Loose bolts or damaged front cover gasket (Chapter 2).
4   Front or rear crankshaft oil seal leaking (Chapter 2).
5   Loose bolts or damaged valve cover gasket (Chapter 2).
6   Loose oil filter (Chapter 1).
7   Loose or damaged oil pressure switch (Chapter 2).
8   Pistons and cylinders excessively worn (Chapter 2).
9   Piston rings not installed correctly on pistons (Chapter 2).
10   Worn or damaged piston rings (Chapter 2).
11   Intake and/or exhaust valve oil seals worn or damaged (Chapter 2).
12   Worn or damaged valves/guides (Chapter 2).
13   Faulty or incorrect PCV valve allowing

too much crankcase airflow.

14   Leak at remote oil filter hose (Expedition/Navigator only).

24   Excessive fuel consumption

1   Dirty or clogged air filter element (Chapter 1).
2   Incorrect ignition timing (Chapter 5).
3   Incorrect idle speed (Chapter 4).
4   Low tire pressure or incorrect tire size (Chapter 10).
5   Inspect for binding brakes.
6   Fuel leakage. Check all connections, lines and components in the fuel system (Chapter 4).
7   Dirty or clogged fuel injectors (Chapter 4).
8   Fuel injection or engine control systems malfunction (Chapters 4 and 6).
9   Thermostat stuck open or not installed.
10   Improperly operating transmission.

25   Fuel odor

1   Fuel leakage. Check all connections, lines and components in the fuel system (Chapter 4).
2   Fuel tank overfilled. Fill only to automatic shut-off.
3   Charcoal canister filter in Evaporative Emissions Control system clogged (Chapter 1).
4   Vapor leaks from Evaporative Emissions Control system lines (Chapter 6).

26   Miscellaneous engine noises

1   A strong dull noise that becomes more rapid as the engine accelerates indicates worn or damaged crankshaft bearings or an unevenly worn crankshaft. To pinpoint the trouble spot, remove the spark plug wire from one plug at a time and crank the engine over. If the noise stops, the cylinder with the removed plug wire indicates the problem area. Replace the bearing and/or service or replace the crankshaft (Chapter 2).
2   A similar (yet slightly higher pitched) noise to the crankshaft knocking described in the previous paragraph, that becomes more rapid as the engine accelerates, indicates worn or damaged connecting rod bearings (Chapter 2). The procedure for locating the problem cylinder is the same as described in Paragraph 1.
3   An overlapping metallic noise that increases in intensity as the engine speed increases, yet diminishes as the engine warms up indicates abnormal piston and cylinder wear (Chapter 2). To locate the problem cylinder, use the procedure described in Paragraph 1.
4   A rapid clicking noise that becomes faster as the engine accelerates indicates a worn piston pin or piston pin hole. This sound will happen each time the piston hits the highest and lowest points in the stroke (Chapter 2). The procedure for locating the problem piston is described in Paragraph 1.
5   A metallic clicking noise coming from the water pump indicates worn or damaged water pump bearings or pump. Replace the water pump with a new one (Chapter 3).
6   A rapid tapping sound or clicking sound that becomes faster as the engine speed increases indicates "valve tapping." This can be identified by holding one end of a section of hose to your ear and placing the other end at different spots along the valve cover. The point where the sound is loudest indicates the problem valve. If the pushrod and rocker arm components are in good shape, you likely have a collapsed valve lifter. Changing the engine oil and adding a high viscosity oil treatment will sometimes cure a stuck lifter problem. If the problem persists, the lifters, pushrods and rocker arms must be removed for inspection (see Chapter 2).
7   A steady metallic rattling or rapping sound coming from the area of the timing chain cover indicates a worn, damaged or out-of-adjustment timing chain. Service or replace the chain and related components (Chapter 2).

Cooling system

27   Overheating

1   Insufficient coolant in system (Chapter 1).
2   Water pump drivebelt defective or out of adjustment (Chapter 1).
3   Radiator core blocked or grille restricted (Chapter 3).
4   Thermostat faulty (Chapter 3).
5   Electric cooling fan blades broken or cracked (Chapter 3).
6   Cooling fan electrical problem (Chapter 3).
7   Radiator cap not maintaining proper pressure (Chapter 3).

28   Overcooling

Faulty thermostat (Chapter 3).

29   External coolant leakage

1   Deteriorated/damaged hoses or loose clamps (Chapters 1 and 3).
2   Water pump seal defective (Chapters 1 and 3).
3   Leakage from radiator core or header tank (Chapter 3).
4   Engine drain or water jacket core plugs leaking (Chapter 2).
5   Leak at engine oil cooler (Chapter 3).

30   Internal coolant leakage

1   Leaking cylinder head gasket (Chapter 2).
2   Cracked cylinder bore or cylinder head (Chapter 2).

31   Coolant loss

1   Too much coolant in system (Chapter 1).
2   Coolant boiling away because of overheating (Chapter 3).
3   Internal or external leakage (Chapter 3).
4   Faulty radiator cap (Chapter 3).

32   Poor coolant circulation

1   Inoperative water pump (Chapter 3).
2   Restriction in cooling system (Chapters 1 and 3).
3   Water pump drivebelt defective or out of adjustment (Chapter 1).
4   Thermostat sticking (Chapter 3).

Automatic transaxle

Note: *Due to the complexity of the automatic transaxle, it's difficult for the home mechanic to properly diagnose and service this component. For problems other than the following, the vehicle should be taken to a dealer service department or a transmission shop.*

33   Fluid leakage

1   Automatic transmission fluid is a deep red color. Fluid leaks should not be confused with engine oil, which can easily be blown by air flow to the transaxle.
2   To pinpoint a leak, first remove all built-up dirt and grime from the transaxle housing with degreasing agents and/or steam cleaning. Drive the vehicle at low speeds so air flow will not blow the leak far from its source. Raise the vehicle and determine where the leak is coming from. Common areas of leakage are:

a) Pan (Chapters 1 and 7).
b) Filler pipe (Chapter 7).
c) Transaxle oil lines (Chapter 7).
d) Speedometer gear or sensor (Chapter 7).

34   Transaxle fluid brown or has a burned smell

Transaxle overheated. Change the fluid (Chapter 1).

35   General shift mechanism problems

1   Chapter 7 deals with checking and

adjusting the shift linkage on automatic transaxles. Common problems which may be attributed to poorly adjusted linkage are:

a) *Engine starting in gears other than Park or Neutral.*
b) *Indicator on shifter pointing to a gear other than the one actually being used.*
c) *Vehicle moves when in Park.*

2   Refer to Chapter 7 for the shift linkage adjustment procedure.

36   Transaxle will not downshift with accelerator pedal pressed to the floor

Transmission Range Sensor (Chapter 7).

37   Engine will start in gears other than Park or Neutral

Park/Neutral switch malfunctioning (Chapter 7).

38   Transaxle slips, shifts roughly, is noisy or has no drive in forward or reverse gears

There are many probable causes for the above problems, but the home mechanic should be concerned with only one possibility - fluid level. Before taking the vehicle to a repair shop, check the level and condition of the fluid as described in Chapter 1.

Correct the fluid level as necessary or change the fluid and filter if needed. If the problem persists, have a professional diagnose the probable cause with a factory scan tool.

Driveaxles

39   Clicking noise in turns

Worn or damaged outer CV joint. Check for cut or damaged boots (Chapter 1). Repair as necessary (Chapter 8).

40   Knock or clunk when accelerating after coasting

Worn or damaged CV joint. Check for cut or damaged boots (Chapter 1). Repair as necessary (Chapter 8).

41   Shudder or vibration during acceleration

1   Excessive inner CV joint angle. Check and correct as necessary (Chapter 8).

2   Worn or damaged CV joints. Repair or replace as necessary (Chapter 8).
3   Sticking inboard joint assembly. Correct or replace as necessary (Chapter 8).

Brakes

Note: *Before assuming that a brake problem exists, make sure the tires are in good condition and properly inflated (Chapter 1), the front end alignment is correct (Chapter 10), and the vehicle isn't loaded with weight in an unequal manner. All service procedures for the brakes are included in Chapter 9, unless otherwise noted.*

42   Vehicle pulls to one side during braking

1   Incorrect tire pressures (Chapter 1).
2   Front end out of alignment (have the front end aligned).
3   Unmatched tires on same axle.
4   Restricted brake lines or hoses (Chapter 9).
5   Malfunctioning brake assembly (Chapter 9).
6   Loose suspension parts (Chapter 10).
7   Loose brake calipers (Chapter 9).
8   Contaminated brake linings (Chapters 1 and 9).

43   Noise (high-pitched squeal when the brakes are applied)

Front disc brake pads worn out. The noise comes from the wear sensor rubbing against the disc. Replace pads with new ones immediately (Chapter 9).

44   Brake roughness or chatter (pedal pulsates)

Note: *Some brake pedal pulsation during operation of the Anti-Lock Brake System (ABS) is normal.*
1   Excessive front brake disc lateral runout (Chapter 9).
2   Parallelism not within specifications (Chapter 9).
3   Uneven pad wear caused by caliper not sliding due to improper clearance or dirt (Chapter 9).
4   Defective brake disc (Chapter 9).

45   Excessive pedal effort required to stop vehicle

1   Malfunctioning power brake booster (Chapter 9).
2   Partial system failure (Chapter 9).

3   Excessively worn pads (Chapter 9).
4   One or more caliper pistons or wheel cylinders seized or sticking (Chapter 9).
5   Brake pads contaminated with oil or grease (Chapter 9).
6   New pads installed and not yet seated. It will take a while for the new material to seat.

46   Excessive brake pedal travel

1   Partial brake system failure (Chapter 9).
2   Insufficient fluid in master cylinder (Chapters 1 and 9).
3   Air trapped in system (Chapters 1 and 9).
4   Excessively worn rear shoes (Chapter 9).

47   Dragging brakes

1   Master cylinder pistons not returning correctly (Chapter 9).
2   Restricted brakes lines or hoses (Chapters 1 and 9).
3   Incorrect parking brake adjustment (Chapter 9).
4   Sticking pistons in calipers (Chapter 9).

48   Grabbing or uneven braking action

1   Malfunction of proportioner valves (Chapter 9).
2   Malfunction of power brake booster unit (Chapter 9).
3   Binding brake pedal mechanism (Chapter 9).
4   Sticking pistons in calipers (Chapter 9).

49   Brake pedal feels spongy when depressed

1   Air in hydraulic lines (Chapter 9).
2   Master cylinder mounting bolts loose (Chapter 9).
3   Master cylinder defective (Chapter 9).

50   Brake pedal travels to the floor with little resistance

Little or no fluid in the master cylinder reservoir caused by leaking caliper or wheel cylinder pistons, loose, damaged or disconnected brake lines (Chapter 9).

51   Parking brake does not hold

Check the parking brake (Chapter 9).

Suspension and steering systems

Note: *Before attempting to diagnose the suspension and steering systems, perform the following preliminary checks:*
- *a) Check the tire pressures and look for uneven wear.*
- *b) Check the steering universal joints or coupling from the column to the steering gear for loose fasteners and wear.*
- *c) Check the front and rear suspension and the steering gear assembly for loose and damaged parts.*
- *d) Look for out-of-round or out-of-balance tires, bent rims and loose and/or rough wheel bearings.*

52 Vehicle pulls to one side

1   Mismatched or uneven tires (Chapter 10).
2   Broken or sagging springs (Chapter 10).
3   Front wheel alignment incorrect (Chapter 10).
4   Front brakes dragging (Chapter 9).

53 Abnormal or excessive tire wear

1   Front wheel alignment incorrect (Chapter 10).
2   Sagging or broken springs (Chapter 10).
3   Tire out-of-balance (Chapter 10).
4   Worn shock absorber (Chapter 10).
5   Overloaded vehicle.
6   Tires not rotated regularly.

54 Wheel makes a "thumping" noise

1   Blister or bump on tire (Chapter 1).
2   Improper shock absorber action (Chapter 10).

55 Shimmy, shake or vibration

1   Tire or wheel out-of-balance or out-of-round (Chapter 10).
2   Loose or worn wheel bearings (Chapter 10).
3   Worn tie-rod ends (Chapter 10).
4   Worn balljoints (Chapter 10).
5   Excessive wheel runout (Chapter 10).
6   Blister or bump on tire (Chapter 1).

56 Hard steering

1   Lack of lubrication at balljoints, tie-rod ends and steering gear assembly (Chapter 10).
2   Front wheel alignment incorrect (Chapter 10).
3   Low tire pressure (Chapter 1).

57 Steering wheel does not return to center position correctly

1   Lack of lubrication at balljoints and tie-rod ends (Chapter 10).
2   Binding in steering column (Chapter 10).
3   Defective rack-and-pinion assembly (Chapter 10).
4   Front wheel alignment problem (Chapter 10).

58 Abnormal noise at the front end

1   Lack of lubrication at balljoints and tie-rod ends (Chapter 1).
2   Loose upper strut mount (Chapter 10).
3   Worn tie-rod ends (Chapter 10).
4   Loose stabilizer bar (Chapter 10).
5   Loose wheel lug nuts (Chapter 1).
6   Loose suspension bolts (Chapter 10).

59 Wander or poor steering stability

1   Mismatched or uneven tires (Chapter 10).
2   Lack of lubrication at balljoints or tie-rod ends (Chapters 1 and 10).
3   Worn shock absorbers (Chapter 10).
4   Loose stabilizer bar (Chapter 10).
5   Broken or sagging springs (Chapter 10).
6   Front wheel alignment incorrect (Chapter 10).
7   Worn steering gear clamp bushings (Chapter 10).

60 Erratic steering when braking

1   Wheel bearings worn (Chapters 8 and 10).
2   Broken or sagging springs (Chapter 10).
3   Leaking wheel cylinder or caliper (Chapter 9).
4   Warped brake discs (Chapter 9).
5   Worn steering gear clamp bushings (Chapter 10).

61 Excessive pitching and/or rolling around corners or during braking

1   Loose stabilizer bar (Chapter 10).
2   Worn shock absorbers or mounts (Chapter 10).
3   Broken or sagging springs (Chapter 10).
4   Overloaded vehicle.
5   Malfunction in the air-suspension system (if equipped) (Chapter 10).

62 Suspension bottoms

1   Overloaded vehicle.

2   Worn shock absorbers (Chapter 10).
3   Incorrect, broken or sagging springs (Chapter 10).
4   Malfunction in the air-suspension system (if equipped) (Chapter 10).

63 Cupped tires

1   Front wheel alignment incorrect (Chapter 10).
2   Worn shock absorbers (Chapter 10).
3   Wheel bearings worn (Chapters 8 and 10).
4   Excessive tire or wheel runout (Chapter 10).
5   Worn balljoints (Chapter 10).

64 Excessive tire wear on outside edge

1   Inflation pressures incorrect (Chapter 1).
2   Excessive speed in turns.
3   Front end alignment incorrect (excessive toe-in or positive camber). Have professionally aligned.
4   Suspension arm bent or twisted (Chapter 10).

65 Excessive tire wear on inside edge

1   Inflation pressures incorrect (Chapter 1).
2   Front end alignment incorrect (toe-out or excessive negative camber). Have professionally aligned.
3   Loose or damaged steering components (Chapter 10).

66 Tire tread worn in one place

1   Tires out-of-balance.
2   Damaged or buckled wheel. Inspect and replace if necessary.
3   Defective tire (Chapter 1).

67 Excessive play or looseness in steering system

1   Wheel bearings worn (Chapter 10).
2   Tie-rod end loose or worn (Chapter 10).
3   Steering gear loose (Chapter 10).

68 Rattling or clicking noise in rack and pinion

Steering gear clamps loose (Chapter 10).

Notes

Chapter 1
Tune-up and routine maintenance

Contents

Specifications

Recommended lubricants and fluids

Note: *Listed here are manufacturer recommendations at the time this manual was written. Manufacturers occasionally upgrade their fluid and lubricant specifications, so check with your local auto parts store for current recommendations.*

Engine oil
    Type .... API grade SH or SH/CC multigrade and fuel efficient oil
    Viscosity .... See accompanying chart
Fuel .... Unleaded gasoline, 87 octane or higher
Engine coolant .... 50/50 mixture of ethylene glycol based antifreeze and water
Brake fluid .... DOT 3 heavy duty brake fluid
Power steering fluid .... Ford premium power steering fluid or equivalent
Automatic transaxle fluid .... MERCON automatic transmission fluid

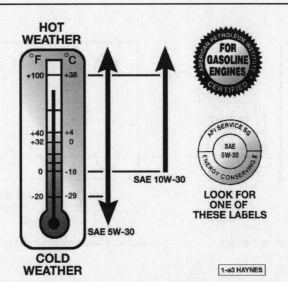

HOT WEATHER

°F °C

+100   +38

+40   +4
+32   0

0   -18

-20   -29

SAE 10W-30

SAE 5W-30

COLD WEATHER

FOR GASOLINE ENGINES

API SERVICE SG
SAE 5W-30
ENERGY CONSERVING II

LOOK FOR ONE OF THESE LABELS

Recommended engine oil viscosity

1-a3 HAYNES

Capacities*

Engine oil (with filter change)
   1995 through 1997 .... 4.5 qts
   1998 on
      3.0L engine .... 4.5 qts
      3.8L engine .... 4.2 qts
Fuel tank
   GL models
      Standard .... 20.0 gallons
      Optional .... 25.0 gallons
   LX/LTD models
      Standard .... 25.0 gallons
Cooling system
   Standard heater.. 12.0 qts
   Auxiliary rear heater.. 14.0 qts
Automatic transaxle
   Drain and refill .... 4.5 qts
   Dry .... 12.25 qts

* All capacities approximate. Add as necessary to bring to appropriate level.

General

Radiator cap pressure rating .... 16 psi
Disc brake pad thickness (minimum)..................................... 1/8 inch
Drum brake shoe thickness (minimum)
   Bonded.. 1/8 inch
   Riveted .... 1/16 inch

Ignition system

Spark plug type and gap
   1995
      3.0L engine.. Motorcraft AWSF-32PP
         or equivalent @ 0.044 inch
      3.8L engine.. Motorcraft AWSF-44PP
         or equivalent @ 0.054 inch
   1996 through 2000
      3.0L engine.. Motorcraft AWSF-32PP
         or equivalent @ 0.044 inch
      3.8L engine.. Motorcraft AWSF-42EE
         or equivalent @ 0.054 inch
Firing order (all models) 1-4-2-5-3-6

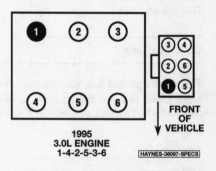

**1995
3.0L ENGINE
1-4-2-5-3-6**

HAYNES-36097-SPECS

**Cylinder location and coil terminal
identification diagram**

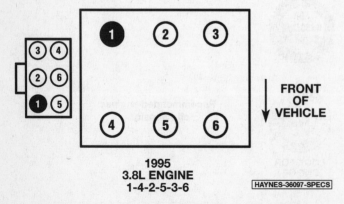

**1995
3.8L ENGINE
1-4-2-5-3-6**

HAYNES-36097-SPECS

Cylinder location and coil terminal identification diagram

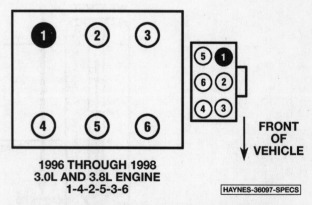

**1996 THROUGH 1998
3.0L AND 3.8L ENGINE
1-4-2-5-3-6**

HAYNES-36097-SPECS

Cylinder location and coil terminal identification diagram

Torque specifications

	Ft-lbs (unless otherwise noted)
Wheel lug nuts	83 to 113
Spark plugs..	7 to 15
Oil pan drain plug	97 to 141 in-lbs
Automatic transaxle pan bolts	8 to 11
Drivebelt tensioner retaining bolt	
3.0L engine	30 to 40
3.8L engine	52 to 70

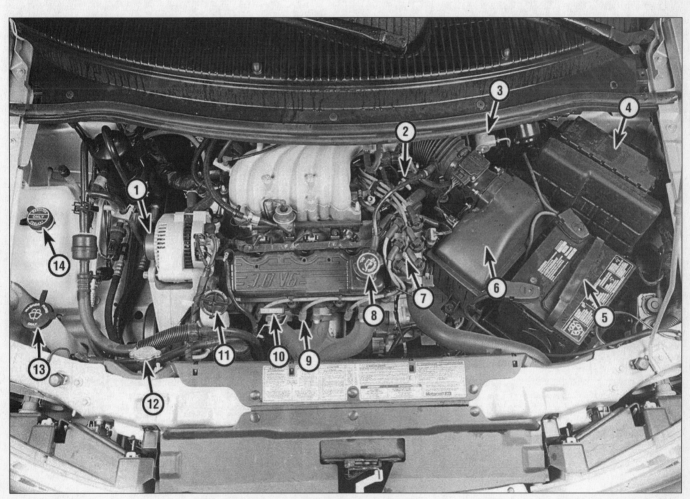

Typical engine compartment components (3.0L engine, typical)

1	Drivebelt	6	Air filter housing	11	Power steering fluid reservoir
2	Automatic transaxle dipstick	7	Ignition coil pack	12	Radiator cap
3	Brake master cylinder reservoir	8	Engine oil filler cap	13	Windshield washer fluid reservoir
4	Engine compartment fuse box	9	Front spark plugs	14	Engine coolant reservoir
5	Battery	10	Engine oil dipstick		

Typical engine compartment components (1996 through 1998 3.8L engine shown, 1995 similar)

1	Drivebelt	6	Automatic transaxle dipstick	11	Power steering fluid reservoir
2	Brake master cylinder reservoir	7	Ignition coil pack	12	Radiator cap
3	Air filter housing	8	Engine oil filler cap	13	Windshield washer fluid reservoir
4	Battery	9	Engine oil dipstick	14	Engine coolant reservoir
5	Engine compartment fuse box	10	Spark plugs (left bank)		

**Typical engine
compartment
underside
components**

1   Exhaust
    system
2   Oil filter
3   Automatic
    transaxle fluid
    pan
4   Lower ball joint
5   Strut rod
    bushings
6   Engine oil drain
    plug
7   Driveaxle
8   Brake disc
9   Brake caliper
10 Stabilizer bar

1

**Typical rear
underside
components**

1   Exhaust pipe
2   Brake drum
3   Rear shock
    absorber
4   Axle beam
5   Fuel tank
6   Parking brake
    cable
7   Brake shoes

1 Ford Windstar Maintenance schedule

The following maintenance intervals are based on the assumption that the vehicle owner will be doing the maintenance or service work, as opposed to having a dealer service department or other repair shop do the work. Although the time/mileage intervals are loosely based on factory recommendations, most have been shortened to ensure, for example, that such items as lubricants and fluids are checked/changed at intervals that promote maximum engine/driveline service life. Also, subject to the preference of the individual owner interested in keeping his or her vehicle in peak condition at all times, and with the vehicle's ultimate resale in mind, many of the maintenance procedures may be performed more often than recommended in the following schedule. We encourage such owner initiative.

When the vehicle is new it should be serviced initially by a factory authorized dealer service department to protect the factory warranty. In many cases the initial maintenance check is done at no cost to the owner (check with your dealer service department for more information).

Every 250 miles or weekly, whichever comes first

Check the engine oil level (Section 4)
Check the engine coolant level (Section 4)
Check the windshield washer fluid level (Section 4)
Check the brake level (Section 4)
Check the tires and tire pressures (Section 5)

Every 3000 miles or 3 months, whichever comes first

All items listed above, plus:
Check the power steering fluid level (Section 6)
Check the automatic transaxle fluid level (Section 7)
Change the engine oil and oil filter (Section 8)

Every 6000 miles or 6 months, whichever comes first

All items listed above, plus:
Check and service the battery (Section 9)

Inspect and replace, if necessary, the windshield wiper blades (Section 10)
Rotate the tires (Section 11)
Inspect the exhaust system (Section 12)
Check the seat belt operation (Section 13)

Every 15,000 miles or 12 months, whichever comes first

All items listed above, plus:
Inspect and replace, if necessary, all underhood hoses (Section 14)
Inspect the cooling system (Section 15)
Check the fuel system (Section 16)
Replace the fuel filter (Section 17)
Inspect the steering and suspension components (Section 18)
Inspect the brakes (Section 19)

Every 30,000 miles or 24 months, whichever comes first

Replace the air filter (Section 20)*
Service the cooling system (drain, flush and refill) (Section 21)
Change the automatic transaxle fluid and filter (Section 22)**
Change the brake fluid (Section 23)

Every 60,000 miles or 48 months, whichever comes first

Check the engine drivebelt(s) (Section 24)
Check the PCV valve (Section 25)
Replace the spark plugs (Section 26)
Inspect and replace, if necessary, the ignition system components (Section 27)

*** If the vehicle is operated in continuous stop-and-go driving or in mountainous areas, change at 15,000 miles*
* Replace more often if is the vehicle is driven in dusty areas

2 Introduction

This Chapter is designed to help the home mechanic maintain the Ford Windstar with the goals of maximum performance, economy, safety and reliability in mind.

Included is a master maintenance schedule (page 1-6), followed by procedures dealing specifically with each item on the schedule. Visual checks, adjustments, component replacement and other helpful items are included. Refer to the **accompanying illustrations** of the engine compartment and the underside of the vehicle for the locations of various components.

Servicing the vehicle, in accordance with the mileage/time maintenance schedule and the step-by-step procedures will result in a planned maintenance program that should produce a long and reliable service life. Keep in mind that it is a comprehensive plan, so maintaining some items but not others at the specified intervals will not produce the same results.

As you service the vehicle, you will discover that many of the procedures can - and should - be grouped together because of the nature of the particular procedure you're performing or because of the close proximity of two otherwise unrelated components to one another.

For example, if the vehicle is raised for chassis lubrication, you should inspect the exhaust, suspension, steering and fuel systems while you're under the vehicle. When you're rotating the tires, it makes good sense to check the brakes since the wheels are already removed. Finally, let's suppose you have to borrow or rent a torque wrench. Even if you only need it to tighten the spark plugs, you might as well check the torque of as many critical fasteners as time allows.

The first step in this maintenance program is to prepare yourself before the actual work begins. Read through all the procedures you're planning to do, then gather up all the parts and tools needed. If it looks like you might run into problems during a particular job, seek advice from a mechanic or an experienced do-it-yourselfer.

3 Tune-up general information

The term tune-up is used in this manual to represent a combination of individual operations rather than one specific procedure.

If, from the time the vehicle is new, the routine maintenance schedule is followed closely and frequent checks are made of fluid levels and high wear items, as suggested throughout this manual, the engine will be kept in relatively good running condition and the need for additional work will be minimized.

More likely than not, however, there will be times when the engine is running poorly due to lack of regular maintenance. This is even more likely if a used vehicle, which has not received regular and frequent maintenance checks, is purchased. In such cases, an engine tune-up will be needed outside of the regular routine maintenance intervals.

The first step in any tune-up or diagnostic procedure to help correct a poor running engine is a cylinder compression check. A compression check (see Chapter 2) will help determine the condition of internal engine components and should be used as a guide for tune-up and repair procedures. If, for instance, a compression check indicates serious internal engine wear, a conventional tune-up will not improve the performance of the engine and would be a waste of time and money. Because of its importance, the compression check should be done by someone with the right equipment and the knowledge to use it properly.

The following procedures are those most often needed to bring a generally poor running engine back into a proper state of tune.

Minor tune-up

Check all engine related fluids (Section 4)
Clean, inspect and test the battery
  (Section 9)
Check all underhood hoses (Section 14)
Check the cooling system (Section 15)
Check the fuel system (Section 16)
Check the air filter (Section 20)

Major tune-up

All items listed under Minor tune-up, plus . . .
Replace the fuel filter (Section 17)
Replace the air filter (Section 20)
Check the drivebelt (Section 24)
Replace the PCV valve (Section 25)
Replace the spark plugs (Section 26)
Replace the spark plug wires,
  (Section 27)
Check the charging system (Chapter 5)

4 Fluid level checks (every 250 miles or weekly)

1   Fluids are an essential part of the lubrication, cooling, brake and windshield washer systems. Because the fluids gradually become depleted and/or contaminated during normal operation of the vehicle, they must be periodically replenished. See *Recommended lubricants and fluids* at the beginning of this Chapter before adding fluid to any of the following components. **Note:** *The vehicle must be on level ground when fluid levels are checked.*

Engine oil

Refer to illustrations 4.2, 4.4 and 4.6

2   The engine oil level is checked with a dipstick, which is located on the front side of the engine **(see illustration)**. The dipstick extends through a metal tube down into the oil pan.

3   The oil level should be checked before the vehicle has been driven, or about 15 minutes after the engine has been shut off. If the oil is checked immediately after driving the vehicle, some of the oil will remain in the upper part of the engine, resulting in an inaccurate reading on the dipstick.

4   Pull the dipstick out of the tube and wipe all the oil from the end with a clean rag or paper towel. Insert the clean dipstick all the way back into the tube and pull it out again. Note the oil at the end of the dipstick. At its highest point, the level should be above the ADD mark and within the crosshatched section of the dipstick **(see illustration)**.

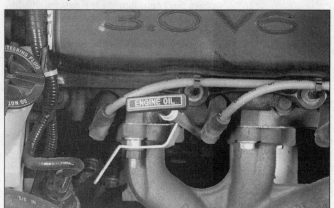

4.2 The engine oil dipstick is located on the front side of the engine on all models

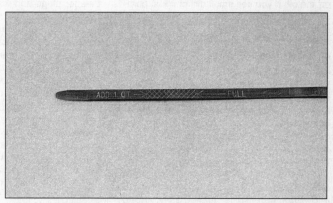

4.4 The oil level must be maintained between the marks at all times - it takes one quart of oil to raise the level from the ADD mark to the FULL mark

4.6 The engine oil filler cap is clearly marked and is located on the valve cover which faces the front of the engine compartment on all models - turn the oil filler cap counterclockwise to remove it

4.8 The coolant reservoir is located on the right side (passenger's side) of the engine compartment - the coolant level can be checked by observing it through the translucent reservoir

4.15 The brake fluid reservoir is located on the left side (driver's side) of the engine compartment - the fluid level should be kept at or near the MAX line on the side of the translucent plastic reservoir

5   It takes one quart of oil to raise the level from the ADD mark to the FULL mark on the dipstick. Do not allow the level to drop below the ADD mark or oil starvation may cause engine damage. Conversely, overfilling the engine (adding oil above the FULL mark) may cause oil fouled spark plugs, oil leaks or oil seal failures.

6   To add oil, remove the filler cap from the valve cover **(see illustration)**. After adding oil, wait a few minutes to allow the level to stabilize, then pull out the dipstick and check the level again. Add more oil if required. Install the filler cap and tighten it by hand only.

7   Checking the oil level is an important preventive maintenance step. A consistently low oil level indicates oil leakage through damaged seals, defective gaskets or past worn rings or valve guides. If the oil looks milky in color or has water droplets in it, the cylinder head gasket(s) may be blown or the head(s) or block may be cracked. The engine should be checked immediately. The condition of the oil should also be checked. Whenever you check the oil level, slide your thumb and index finger up the dipstick before wiping off the oil. If you see small dirt or metal particles clinging to the dipstick, the oil should be changed (see Section 8).

Engine coolant

Refer to illustration 4.8

Warning: *Do not allow antifreeze to come in contact with your skin or painted surfaces of the vehicle. Flush contaminated areas immediately with plenty of water. Don't store new coolant or leave old coolant lying around where it's accessible to children or pets - they're attracted by its sweet smell and may drink it. Ingestion of even a small amount of coolant can be fatal! Wipe up garage floor and drip pan spills immediately. Keep antifreeze containers covered and repair cooling system leaks as soon as they're noticed.*

8   All vehicles covered by this manual are equipped with a pressurized coolant recovery system. A white plastic coolant reservoir located in the engine compartment is connected by a hose to the radiator filler neck **(see illustration)**. If the engine overheats, coolant escapes through a valve in the radiator cap and travels through the hose into the reservoir. As the engine cools, the coolant is automatically drawn back into the cooling system to maintain the correct level.

9   The coolant level in the reservoir should be checked regularly. **Warning:** *Do not remove the radiator cap to check the coolant level when the engine is warm. The level in the reservoir varies with the temperature of the engine. When the engine is cold, the coolant level should be at or slightly above the FULL COLD mark on the reservoir. If it isn't, allow the engine to cool, then remove the cap from the reservoir and add a 50/50 mixture of ethylene glycol-based antifreeze and water.*

10   If the coolant level drops within a short time after replenishment, there may be a leak in the system. Inspect the radiator, hoses, engine coolant filler cap, drain plugs, air bleeder plugs and water pump. If no leak is evident, have the radiator cap pressure tested by your dealer. **Warning:** *Never remove the radiator cap or the coolant recovery reservoir cap when the engine is running or has just been shut down, because the cooling system is hot. Escaping steam and scalding liquid could cause serious injury.*

11   If it is necessary to open the radiator cap, wait until the system has cooled completely, then wrap a thick cloth around the cap and turn it to the first stop. If any steam escapes, wait until the system has cooled further, then remove the cap.

12   When checking the coolant level, always note its condition. It should be relatively clear. If it is brown or rust colored, the system should be drained, flushed and refilled. Even if the coolant appears to be normal, the cor-

rosion inhibitors wear out with use, so it must be replaced at the specified intervals.

13   Do not allow antifreeze to come in contact with your skin or painted surfaces of the vehicle. Flush contacted areas immediately with plenty of water.

Brake fluid

Refer to illustration 4.15

14   The brake fluid level is checked by looking through the plastic reservoir mounted on the master cylinder. The master cylinder is mounted on the front of the power booster unit in the left (driver's side) rear corner of the engine compartment.

15   The fluid level should be at or near the MAX line on the side of the reservoir **(see illustration)**. Add fluid if the level is 1/4 inch or more below the MAX line.

16   If the fluid level is low, wipe the top of the reservoir and the cap with a clean rag to prevent contamination of the system as the cap is unscrewed.

17   Add only the specified brake fluid to the reservoir (refer to *Recommended lubricants and fluids* at the front of this Chapter or your owner's manual). Mixing different types of brake fluid can damage the system. Fill the reservoir to the MAX line. **Warning:** *Brake fluid can harm your eyes and damage painted surfaces, so use extreme caution when handling or pouring it. Do not use brake fluid that has been standing open or is more than one year old. Brake fluid absorbs moisture from the air, which can cause a dangerous loss of braking effectiveness.*

18   While the reservoir cap is off, check the master cylinder reservoir for contamination. If rust deposits, dirt particles or water droplets are present, the system should be drained and refilled by a dealer service department or repair shop.

19   After filling the reservoir to the proper level, make sure the cap is seated to prevent fluid leakage and/or contamination.

20   The fluid level in the master cylinder will

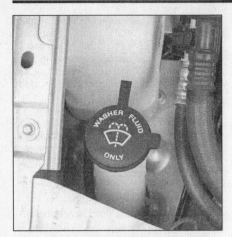

4.22 The windshield washer reservoir is located at the right front corner of the engine compartment - flip the windshield washer fluid cap up to add fluid

empty, the brake system should be bled (see Chapter 9).

Windshield washer fluid

Refer to illustration 4.22

22   Fluid for the windshield washer system is stored in a plastic reservoir located at the right (passenger) side of the engine compartment **(see illustration)**.

23   In milder climates, plain water can be used in the reservoir, but it should be kept no more than 2/3 full to allow for expansion if the water freezes. In colder climates, use windshield washer system antifreeze, available at any auto parts store, to lower the freezing point of the fluid. Mix the antifreeze with water in accordance with the manufacturer's directions on the container. **Caution:** *Do not use cooling system antifreeze - it will damage the vehicle's paint.*

---

5   Tire and tire pressure checks (every 250 miles or weekly)

Refer to illustrations 5.2, 5.3, 5.4a, 5.4b and 5.8

1   Periodic inspection of the tires may spare you the inconvenience of being stranded with a flat tire. It can also provide you with vital information regarding possible problems in the steering and suspension sys-

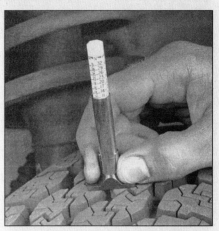

5.2 Use a tire tread depth indicator to monitor tire wear - they are available at auto parts stores and service stations and cost very little

tems before major damage occurs.

2   The original tires on this vehicle are equipped with 1/2-inch wide bands that will appear when tread depth reaches 1/16-inch, at which point they can be considered worn out. Tread wear can be monitored with a simple, inexpensive device known as a tread depth indicator **(see illustration)**.

3   Note any abnormal tread wear **(see illustration)**. Tread pattern irregularities such

drop slightly as the brake shoes or pads at each wheel wear down during normal operation. If the brake fluid level drops consistently, check the entire system for leaks immediately. Examine all brake lines, hoses and connections, along with the calipers, wheel cylinders and master cylinder (see Section 19).

21   When checking the fluid level, if you discover one or both reservoirs empty or nearly

1

UNDERINFLATION

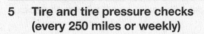

CUPPING

Cupping may be caused by:

- Underinflation and/or mechanical irregularities such as out-of-balance condition of wheel and/or tire, and bent or damaged wheel.
- Loose or worn steering tie-rod or steering idler arm.
- Loose, damaged or worn front suspension parts.

OVERINFLATION

INCORRECT TOE-IN OR EXTREME CAMBER

FEATHERING DUE TO MISALIGNMENT

5.3 This chart will help you determine the condition of the tires, the probable cause(s) of abnormal wear and the corrective action necessary

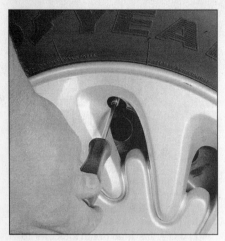

5.4a If a tire loses air on a steady basis, check the valve stem core first to make sure it's snug (special inexpensive wrenches are commonly available at auto parts stores)

5.4b If the valve stem core is tight, raise the corner of the vehicle with the low tire and spray a soapy water solution onto the tread as the tire is turned slowly - leaks will cause small bubbles to appear

as cupping, flat spots and more wear on one side than the other are indications of front end alignment and/or balance problems. If any of these conditions are noted, take the vehicle to a tire shop or service station to correct the problem.

4   Look closely for cuts, punctures and embedded nails or tacks. Sometimes a tire will hold air pressure for a short time or leak down very slowly after a nail has embedded itself in the tread. If a slow leak persists, check the valve stem core to make sure it is tight **(see illustration)**. Examine the tread for an object that may have embedded itself in the tire or for a "plug" that may have begun to leak (radial tire punctures are repaired with a plug that is installed in a puncture). If a puncture is suspected, it can be easily verified by spraying a solution of soapy water onto the puncture area **(see illustration)**. The soapy solution will bubble if there is a leak. Unless the puncture is unusually large, a tire shop or service station can usually repair the tire.

5   Carefully inspect the inner sidewall of each tire for evidence of brake fluid leakage. If you see any, inspect the brakes immediately.

6   Correct air pressure adds miles to the life span of the tires, improves mileage and enhances overall ride quality. Tire pressure cannot be accurately estimated by looking at a tire, especially if it's a radial. A tire pressure gauge is essential. Keep an accurate gauge in the glove compartment. The pressure gauges attached to the nozzles of air hoses at gas stations are often inaccurate.

7   Always check tire pressure when the tires are cold. Cold, in this case, means the vehicle has not been driven over a mile in the three hours preceding a tire pressure check. A pressure rise of four to eight pounds is not uncommon once the tires are warm.

8   Unscrew the valve cap protruding from the wheel or hubcap and push the gauge firmly onto the valve stem **(see illustration)**. Note the reading on the gauge and compare

the figure to the recommended tire pressure shown on the tire placard on the driver's side door. Be sure to reinstall the valve cap to keep dirt and moisture out of the valve stem mechanism. Check all four tires and, if necessary, add enough air to bring them up to the recommended pressure.

9   Don't forget to keep the spare tire inflated to the specified pressure (refer to your owner's manual or the decal attached to the right door pillar). Note that the pressure recommended for the temporary (mini) spare is higher than for the tires on the vehicle.

6   Power steering fluid level check (every 3000 miles or 3 months)

Refer to illustrations 6.2, 6.5a and 6.5b

1   Check the power steering fluid level periodically to avoid steering system problems, such as damage to the pump. **Caution:** *DO NOT hold the steering wheel against either stop (extreme left or right turn) for more than five seconds. If you do, the power steering pump could be damaged.*

2   The power steering pump is located at the front corner of the engine **(see illustration)** and is equipped with a twist-off cap with an integral fluid level dipstick.

3   Park the vehicle on level ground and apply the parking brake.

4   Run the engine until it has reached normal operating temperature. With the engine at idle, turn the steering wheel back-and-forth several times to get any air out of the steering system. Shut the engine off, remove the cap by turning it counterclockwise, wipe the dipstick clean and reinstall the cap. (Make sure it is seated).

5   Remove the cap again and note the fluid level. It must be between the two lines designating the FULL HOT or FULL COLD range. Be sure to use the proper temperature range on the dipstick when checking the fluid level - the FULL COLD lines on the reverse side of the dipstick are only usable when the engine

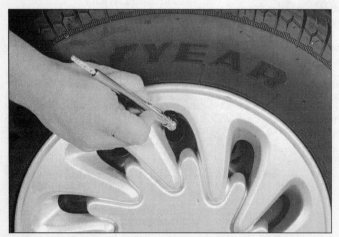

5.8 To extend the life of the tires, check the air pressure at least once a week with an accurate gauge (don't forget the spare!)

6.2 The power steering fluid dipstick (arrow) is located in the power steering pump reservoir - turn the cap counterclockwise to remove it

6.5a The dipstick is marked on both sides so the fluid can be checked hot . . .

6.5b . . . or cold

is cold **(see illustrations)**.

6   Add small amounts of fluid until the level is correct. **Caution:** *Do not overfill the pump. If too much fluid is added, remove the excess with a clean syringe or suction pump.*

7   If additional fluid is required, pour the specified type directly into the reservoir, using a funnel to prevent spills.

7   Automatic transaxle fluid level check (every 3000 miles or 3 months)

Refer to illustrations 7.4a, 7.4b and 7.6

1   The automatic transaxle fluid level should be carefully maintained. Low fluid level can lead to slipping or loss of drive, while overfilling can cause foaming and loss of fluid. Either condition can cause transaxle damage.

2   Since transaxle fluid expands as it heats up, the fluid level should only be checked when the transaxle is warm (at normal operating temperature). If the vehicle has just

been driven over 20 miles (32 km), the transaxle can be considered warm. **Caution:** *If the vehicle has just been driven for a long time at high speed or in city traffic in hot weather, or if it has been pulling a trailer, an accurate fluid level reading cannot be obtained.* Allow the transaxle to cool down for about 30 minutes. You can also check the transaxle fluid level when the transaxle is cold. If the vehicle has not been driven for over five hours and the fluid is about room temperature (70 to 95-degrees F), the transaxle is cold. However, the fluid level is normally checked with the transaxle warm to ensure accurate results.

3   Immediately after driving the vehicle, park it on a level surface, set the parking brake and start the engine. While the engine is idling, depress the brake pedal and move the selector lever through all the gear ranges, beginning and ending in Park.

4   Locate the automatic transaxle dipstick tube in the engine compartment **(see illustrations)**.

5   With the engine still idling, pull the dipstick from the tube, wipe it off with a clean rag, push it all the way back into the tube and

withdraw it again, then note the fluid level.

6   If the transaxle is cold, the level should be in the room temperature range on the dipstick (between the two holes); if it's warm, the fluid level should be in the operating temperature range (in the cross-hatched area) **(see illustration)**. If the level is low, add the specified automatic transmission fluid through the dipstick tube - use a funnel to prevent spills.

7   Add just enough of the recommended fluid to fill the transaxle to the proper level. It takes about one pint to raise the level from the low mark to the high mark when the fluid is hot, so add the fluid a little at a time and keep checking the level until it's correct.

8   The condition of the fluid should also be checked along with the level. If the fluid is black or a dark reddish-brown color, or if it smells burned, it should be changed (see Section 22). If you are in doubt about its condition, purchase some new fluid and compare the two for color and smell.

7.4a The automatic transaxle dipstick (arrow) on 3.0L and 1995 3.8L engines is located on the driver's side of the vehicle just behind the ignition coil pack

7.4b The automatic transaxle dipstick on 1996 and later 3.8L engines is located on the driver's side of the vehicle at the front of the engine compartment

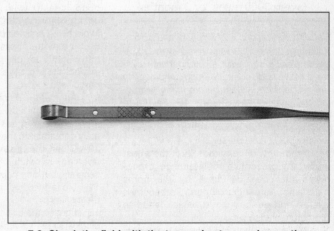

7.6 Check the fluid with the transaxle at normal operating temperature - the level should be kept in the HOT range in the cross-hatched area (don't add fluid if the level is anywhere in the cross-hatched area)

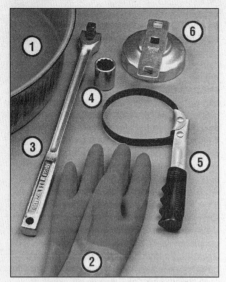

8.2 These tools are required when changing the engine oil and filter

1   *Drain pan* - It should be fairly shallow in depth, but wide to prevent spills
2   *Rubber gloves* - When removing the drain plug and filter, you will get oil on your hands (the gloves will prevent burns)
3   *Breaker bar* - Sometimes the oil drain plug is tight, and a long breaker bar is needed to loosen it
4   *Socket* – To be used with the breaker bar or a ratchet (must be the correct size to fit the drain plug - six-point preferred)
5   *Filter wrench* - This is a metal band-type wrench, which requires clearance around the filter to be effective
6   *Filter wrench* - This type fits on the bottom of the filter and can be turned with a ratchet or breaker bar (different-size wrenches are available for different types of filters)

8   Engine oil and filter change (every 3000 miles or 3 months)

Refer to illustrations 8.2, 8.7, 8.12 and 8.15
Warning: *Some models covered by this manual are equipped with air suspension systems. Always disconnect electrical power to the suspension system before lifting or towing the vehicle (see Chapter 10). Failure to perform this procedure may result in unexpected shifting or movement of the vehicle which could cause personal injury.*

1   Frequent oil changes are the most important preventive maintenance procedures that can be done by the home mechanic. As engine oil ages, in becomes diluted and contaminated, which leads to premature engine wear.
2   Make sure that you have all the necessary tools before you begin this procedure **(see illustration)**. You should also have plenty of rags or newspapers handy for mopping up oil spills.

8.7 Use a proper size wrench or socket to remove the oil drain plug and avoid rounding it off

3   Access to the oil drain plug and filter will be improved if the vehicle can be lifted on a hoist, driven onto ramps or supported by jackstands. **Warning:** *Do not work under a vehicle supported only by a jack - always use jackstands!*
4   If you haven't changed the oil on this vehicle before, get under it and locate the oil drain plug and the oil filter. The exhaust components will be warm as you work, so note how they are routed to avoid touching them when you are under the vehicle.
5   Start the engine and allow it to reach normal operating temperature - oil and sludge will flow out more easily when warm. If new oil, a filter or tools are needed, use the vehicle to go get them and warm up the engine/oil at the same time. Park on a level surface and shut off the engine when it's warmed up. Remove the oil filler cap from the valve cover.
6   Raise the vehicle and support it on jackstands. Make sure it is safely supported!
7   Being careful not to touch the hot exhaust components, position a drain pan under the plug in the bottom of the engine, then remove the plug **(see illustration)**. It's a good idea to wear a rubber glove while unscrewing the plug the final few turns to avoid being scalded by hot oil.
8   It may be necessary to move the drain pan slightly as oil flow slows to a trickle. Inspect the old oil for the presence of metal particles.
9   After all the oil has drained, wipe off the drain plug with a clean rag. Any small metal particles clinging to the plug would immediately contaminate the new oil.
10   Clean the area around the drain plug opening, reinstall the plug and tighten it securely, but don't strip the threads.
11   Move the drain pan into position under the oil filter.
12   Loosen the oil filter by turning it counterclockwise with a filter wrench **(see illustration)**. Any standard filter wrench will work.
13   Once the filter is loose, use your hands to unscrew it from the block. Just as the filter is detached from the block, immediately tilt

8.12 The oil filter is usually on very tight and will require a special oil filter wrench to remove it - DO NOT use the wrench to tighten the new filter

the open end up to prevent the oil inside the filter from spilling out.
14   Using a clean rag, wipe off the mounting surface on the block. Also, make sure that none of the old gasket remains stuck to the mounting surface. It can be removed with a scraper if necessary.
15   Compare the old filter with the new one to make sure they are the same type. Smear some engine oil on the rubber gasket of the new filter and screw it into place **(see illustration)**. Overtightening the filter will damage the gasket, so don't use a filter wrench. Most filter manufacturers recommend tightening the filter by hand only. Normally they should be tightened 3/4-turn after the gasket contacts the block, but be sure to follow the directions on the filter or container.
16   Remove all tools and materials from under the vehicle, being careful not to spill the oil in the drain pan, then lower the vehicle.
17   Add new oil to the engine through the oil filler cap. Use a funnel to prevent oil from spilling onto the top of the engine. Pour four quarts of fresh oil into the engine. Wait a few minutes to allow the oil to drain into the pan, then check the level on the dipstick (see Section 4 if necessary). If the oil level is in the OK

8.15 Lubricate the oil filter gasket with clean engine oil before installing the filter on the engine

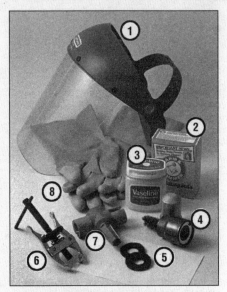

9.1 Tools and materials required for battery maintenance

1 **Face shield/safety goggles** - *When removing corrosion with a brush, the acidic particles can easily fly up into your eyes*

2 **Baking soda** - *A solution of baking soda and water can be used to neutralize corrosion*

3 **Petroleum jelly** - *A layer of this on the battery posts will help prevent corrosion*

4 **Battery post/cable cleaner** - *This wire brush cleaning tool will remove all traces of corrosion from the battery posts and cable clamps*

5 **Treated felt washers** - *Placing one of these on each post, directly under the cable clamps, will help prevent corrosion*

6 **Puller** - *Sometimes the cable clamps are very difficult to pull off the posts, even after the nut/bolt has been completely loosened. This tool pulls the clamp straight up and off the post without damage*

7 **Battery post/cable cleaner** - *Here is another cleaning tool which is a slightly different version of Number 4 above, but it does the same thing*

8 **Rubber gloves** - *Another safety item to consider when servicing the battery; remember that's acid inside the battery!*

range (hatched area), install the filler cap.

18 Start the engine and run it for about a minute. While the engine is running, look under the vehicle and check for leaks at the oil pan drain plug and around the oil filter. If either one is leaking, stop the engine and tighten the plug or filter slightly.

19 Wait a few minutes, then recheck the level on the dipstick. Add oil as necessary to bring the level into the OK range.

20 During the first few trips after an oil change, make it a point to check frequently for leaks and proper oil level.

21 The old oil drained from the engine can-

9.6a Battery terminal corrosion usually appears as light, fluffy powder

not be reused in its present state and should be discarded. Oil reclamation centers, auto repair shops and gas stations will normally accept the oil, which can be recycled. After the oil has cooled, it can be drained into a container (plastic jugs, bottles, milk cartons, etc.) for transport to a disposal site.

9 Battery check, maintenance and charging (every 6000 miles or 6 months)

Refer to illustrations 9.1, 9.6a, 9.6b, 9.7a and 9.7b

Warning: *Certain precautions must be followed when checking and servicing the battery. Hydrogen gas, which is highly flammable, is always present in the battery cells, so keep lighted tobacco and all other open flames and sparks away from the battery. The electrolyte inside the battery is actually dilute sulfuric acid, which will cause injury if splashed on your skin or in your eyes. It will also ruin clothes and painted surfaces. When removing the battery cables, always detach the negative cable first and hook it up last!*

1 A routine preventive maintenance program for the battery in your vehicle is the only way to ensure quick and reliable starts. But before performing any battery maintenance, make sure that you have the proper equipment necessary to work safely around the battery **(see illustration)**.

2 There are also several precautions that should be taken whenever battery maintenance is performed. Before servicing the battery, always turn the engine and all accessories off and disconnect the cable from the negative terminal of the battery.

3 The battery produces hydrogen gas, which is both flammable and explosive. Never create a spark, smoke or light a match around the battery. Always charge the battery in a ventilated area.

4 Electrolyte contains poisonous and corrosive sulfuric acid. Do not allow it to get in your eyes, on your skin on your clothes. Never ingest it. Wear protective safety

9.6b Removing the cable from a battery post with a wrench - sometimes special battery pliers are required for this procedure if corrosion has caused deterioration of the nut hex (always remove the ground cable first and hook it up last!)

glasses when working near the battery. Keep children away from the battery.

5 Note the external condition of the battery. If the positive terminal and cable clamp on your vehicle's battery is equipped with a rubber protector, make sure that it's not torn or damaged. It should completely cover the terminal. Look for any corroded or loose connections, cracks in the case or cover or loose hold-down clamps. Also check the entire length of each cable for cracks and frayed conductors.

6 If corrosion, which looks like white, fluffy deposits **(see illustration)** is evident, particularly around the terminals, the battery should be removed for cleaning. Loosen the cable clamp bolts with a wrench, being careful to remove the ground cable first, and slide them off the terminals **(see illustration)**. Then disconnect the hold-down clamp bolt and nut, remove the clamp and lift the battery from the engine compartment.

7 Clean the cable clamps thoroughly with

9.7a When cleaning the cable clamps, all corrosion must be removed (the inside of the clamp is tapered to match the taper on the post, so don't remove too much material)

9.7b Regardless of the type of tool used on the battery posts, a clean, shiny surface should be the result

10.3 Tilt the trim cap back and check the tightness of the wiper arm retaining nuts

a battery brush or a terminal cleaner and a solution of warm water and baking soda **(see illustration)**. Wash the terminals and the top of the battery case with the same solution but make sure that the solution doesn't get into the battery When cleaning the cables, terminals and battery top, wear safety goggles and rubber gloves to prevent any solution from coming in contact with your eyes or hands. Wear old clothes too - even diluted, sulfuric acid splashed onto clothes will burn holes in them. If the terminals have been extensively corroded, clean them up with a terminal cleaner **(see illustration)**. Thoroughly wash all cleaned areas with plain water.

8   Make sure that the battery tray is in good condition and the hold-down clamp bolts are tight. If the battery is removed from the tray, make sure no parts remain in the bottom of the tray when the battery is reinstalled. When reinstalling the hold-down clamp bolts, do not overtighten them.

9   Information on removing and installing the battery can be found in Chapter 5. Information on jump starting can be found at the front of this manual. For more detailed battery checking procedures, refer to the *Haynes Automotive Electrical Manual*.

Cleaning

10   Corrosion on the hold-down components, battery case and surrounding areas can be removed with a solution of water and baking soda. Thoroughly rinse all cleaned areas with plain water.

11   Any metal parts of the vehicle damaged by corrosion should be covered with a zinc-based primer, then painted.

Charging

Warning: *When batteries are being charged, hydrogen gas, which is very explosive and flammable, is produced. Do not smoke or allow open flames near a charging or a recently charged battery. Wear eye protection when near the battery during charging. Also, make sure the charger is unplugged before connecting or disconnecting the battery from the charger.*

12   Slow-rate charging is the best way to restore a battery that's discharged to the point where it will not start the engine. It's also a good way to maintain the battery charge in a vehicle that's only driven a few miles between starts. Maintaining the battery charge is particularly important in the winter when the battery must work harder to start

the engine and electrical accessories that drain the battery are in greater use.

13   It's best to use a one or two-amp battery charger (sometimes called a "trickle" charger). They are the safest and put the least strain on the battery. They are also the least expensive. For a faster charge, you can use a higher amperage charger, but don't use one rated more than 1/10th the amp/hour rating of the battery. Rapid boost charges that claim to restore the power of the battery in one to two hours are hardest on the battery and can damage batteries not in good condition. This type of charging should only be used in emergency situations.

14   The average time necessary to charge a battery should be listed in the instructions that come with the charger. As a general rule, a trickle charger will charge a battery in 12 to 16 hours.

10   Windshield wiper blade inspection and replacement (every 6000 miles or 6 months)

Refer to illustrations 10.3, 10.5 and 10.6

1   The windshield wiper and blade assembly should be inspected periodically for damage, loose components and cracked or worn blade elements.

2   Road film can build up on the wiper blades and affect their efficiency, so they should be washed regularly with a mild detergent solution.

3   The action of the wiping mechanism can loosen bolts, nuts and fasteners, so they should be checked and tightened, as necessary **(see illustration)**, at the same time the wiper blades are checked.

4   If the wiper blade elements are cracked, worn or warped, or no longer clean adequately, they should be replaced with new ones.

5   Lift the arm assembly away from the glass for clearance, press on the release lever, then slide the wiper blade assembly out of the hook in the end of the arm **(see illustration)**.

6   Use needle-nose pliers to compress the

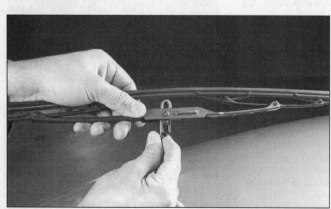

10.5 Press on the release tab and push the blade assembly down and away from the hook in the arm

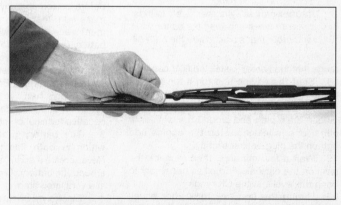

10.6 Use needle-nose pliers to compress the rubber element, then slide the element out - slide the new element in and lock the blade assembly fingers into the notches of the wiper element

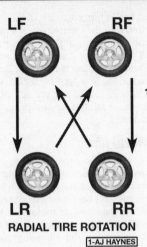

11.2 The recommended tire rotation pattern for these vehicles

RADIAL TIRE ROTATION

12.2a Check the exhaust flex tube flange (arrow) connections . . .

blade element, then slide the element out of the frame and discard it **(see illustration)**.

7   Installation is the reverse of removal.

11   Tire rotation (every 6000 miles or 6 months)

Refer to illustration 11.2
Warning: *Some models covered by this manual are equipped with air suspension systems. Always disconnect electrical power to the suspension system before lifting or towing the vehicle (see Chapter 10). Failure to perform this procedure may result in unexpected shifting or movement of the vehicle which could cause personal injury.*

1   The tires should be rotated at the specified intervals and whenever uneven wear is noticed. Since the vehicle will be raised and the tires removed anyway, check the brakes also (see Section 19).

2   Radial tires must be rotated in a specific pattern **(see illustration)**. If your vehicle has a compact spare tire, don't include it in the rotation pattern.

3   Refer to the information in *Jacking and towing* at the front of this manual for the proper procedure to follow when raising the vehicle and changing a tire. If the brakes must be checked, don't apply the parking brake as stated.

4   The vehicle must be raised on a hoist or supported on jackstands to get all four wheels off the ground. Make sure the vehicle is safely supported!

5   After the rotation procedure is finished, check and adjust the tire pressures as necessary and be sure to check the lug nut tightness.

12   Exhaust system check (every 6000 miles or 6 months)

Refer to illustrations 12.2a, 12.2b and 12.2c
Warning: *Some models covered by this manual are equipped with air suspension systems. Always disconnect electrical power to the suspension system before lifting or towing the vehicle (see Chapter 10). Failure to perform this procedure may result in unexpected shifting or movement of the vehicle which could cause personal injury.*

1   With the engine cold (at least three hours after the vehicle has been driven), check the complete exhaust system from the engine to the end of the tailpipe. Ideally, the inspection should be done with the vehicle on a hoist to permit unrestricted access. If a hoist isn't available, raise the vehicle and support it securely on jackstands.

2   Check the exhaust pipes and connections for evidence of leaks, severe corrosion and damage. Make sure that all brackets and hangers are in good condition and tight **(see illustrations)**.

3   At the same time, inspect the underside of the body for holes, corrosion, open seams, etc. which may allow exhaust gases to enter the passenger compartment. Seal all body openings with silicone or body putty.

4   Rattles and other noises can often be traced to the exhaust system, especially the mounts and hangers. Try to move the pipes, muffler and catalytic converter. If the components can come in contact with the body or suspension parts, secure the exhaust system with new mounts.

5   Check the running condition of the engine by inspecting inside the end of the tailpipe. The exhaust deposits here are an indication of engine state-of-tune. If the pipe is black and sooty or coated with white deposits, the engine may need a tune-up, including a thorough fuel system inspection and adjustment.

1

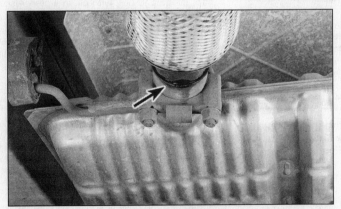

12.2b . . . and the exhaust pipe connections (arrow) for exhaust leaks - also check that the retaining nuts or bolts are securely tightened

12.2c Check the exhaust system hangers (arrows) for damage and cracks

13   Seat belt check (every 6000 miles or 6 months)

1   Check seat belts, buckles, latch plates and guide loops for obvious damage and signs of wear.
2   See if the seat belt reminder light comes on when the key is turned to the Run or Start position. A chime should also sound. On passive restraint systems, the shoulder belt should move into position in the A-pillar.
3   The seat belts are designed to lock up during a sudden stop or impact, yet allow free movement during normal driving. Make sure the retractors return the belt against your chest while driving and rewind the belt fully when the buckle is unlatched.
4   If any of the above checks reveal problems with the seat belt system, replace parts as necessary.

14   Underhood hose check and replacement (every 15,000 miles or 12 months)

Warning: *Replacement of air conditioning hoses must be left to a dealer service department or air conditioning shop that has the equipment to depressurize the system safely. Never remove air conditioning components or hoses until the system has been depressurized.*

General

1   High temperatures under the hood can cause deterioration of the rubber and plastic hoses used for engine, accessory and emission systems operation. Periodic inspection should be made for cracks, loose clamps, material hardening and leaks.
2   Information specific to the cooling system hoses can be found in Section 15.
3   Most (but not all) hoses are secured to the fittings with clamps. Where clamps are used, check to be sure they haven't lost their tension, allowing the hose to leak. If clamps aren't used, make sure the hose has not expanded and/or hardened where it slips over the fitting, allowing it to leak.

PCV system hose

4   To reduce hydrocarbon emissions, crankcase blow-by gas is vented through the PCV valve in the rocker arm cover to the intake manifold via a rubber hose on most models. The blow-by gases mix with incoming air in the intake manifold before being burned in the combustion chambers.
5   Check the PCV hose for cracks, leaks and other damage. Disconnect it from the valve cover and the intake manifold and check the inside for obstructions. If it's clogged, clean it out with solvent.

Vacuum hoses

6   It's quite common for vacuum hoses,
especially those in the emissions system, to be color coded or identified by colored stripes molded into them. Various systems require hoses with different wall thickness, collapse resistance and temperature resistance. When replacing hoses, be sure the new ones are made of the same material.
7   Often the only effective way to check a hose is to remove it completely from the vehicle. If more than one hose is removed, be sure to label the hoses and fittings to ensure correct installation.
8   When checking vacuum hoses, be sure to include any plastic T-fittings in the check. Inspect the fittings for cracks and the hose where it fits over each fitting for distortion, which could cause leakage.
9   A small piece of vacuum hose (1/4-inch inside diameter) can be used as a stethoscope to detect vacuum leaks. Hold one end of the hose to your ear and probe around vacuum hoses and fittings, listening for the "hissing" sound characteristic of a vacuum leak. **Warning:** *When probing with the vacuum hose stethoscope, be careful not to come into contact with moving engine components such as drivebelts, the cooling fan, etc.*

Fuel hose

Warning: *Gasoline is extremely flammable, so take extra precautions when you work on any part of the fuel system. Don't smoke or allow open flames or bare light bulbs near the work area, and don't work in a garage where a natural gas-type appliance (such as a water heater or clothes dryer) with a pilot light is present. Since gasoline is carcinogenic, wear latex gloves when there's a possibility of being exposed to fuel, and, if you spill any fuel on your skin, rinse it off immediately with soap and water. Mop up any spills immediately and do not store fuel-soaked rags where they could ignite. The fuel system is under constant pressure, so, if any fuel lines are to be disconnected, the fuel pressure in the system must be relieved first (see Chapter 4 for more information). When you perform any kind of work on the fuel system, wear safety glasses and have a Class B type fire extinguisher on hand.*
10   The fuel lines are usually under pressure, so if any fuel lines are to be disconnected be prepared to catch spilled fuel.
Warning: *Your vehicle is equipped with fuel injection and you must relieve the fuel system pressure before servicing the fuel lines. Refer to Chapter 4 for the fuel system pressure relief procedure.*
11   Check all flexible fuel lines for deterioration and chafing. Check especially for cracks in areas where the hose bends and just before fittings, such as where a hose attaches to the fuel pump, fuel filter and fuel injection unit.
12   When replacing a hose, use only hose that is specifically designed for your fuel injection system.
13   Spring-type clamps are sometimes used
on fuel return or vapor lines. These clamps often lose their tension over a period of time, and can be "sprung" during removal. Replace all spring-type clamps with screw clamps whenever a hose is replaced. Some fuel lines use spring-lock type couplings, which require a special tool to disconnect. See Chapter 4 for more information on these type of couplings.

Metal lines

14   Sections of metal line are often used for fuel line between the fuel pump and the fuel injection unit. Check carefully to make sure the line isn't bent, crimped or cracked.
15   If a section of metal fuel line must be replaced, use seamless steel tubing only, since copper and aluminum tubing do not have the strength necessary to withstand vibration caused by the engine.
16   Check the metal brake lines where they enter the master cylinder and brake proportioning unit (if used) for cracks in the lines and loose fittings. Any sign of brake fluid leakage calls for an immediate thorough inspection of the brake system.

15   Cooling system check (every 15,000 miles or 12 months)

Refer to illustration 15.4
1   Many major engine failures can be attributed to a faulty cooling system. If the vehicle is equipped with an automatic transaxle, the cooling system also cools the transaxle fluid and thus plays an important role in prolonging transaxle life.
2   The cooling system should be checked with the engine cold. Do this before the vehicle has been driven for the day or after the engine has been shut off for at least three hours.
3   Remove the radiator cap by turning it to the left until it reaches a stop. If you hear a hissing sound (indicating there is still pressure in the system), wait until it stops. Now press down on the cap with the palm of your hand and continue turning to the left until the cap can be removed. Thoroughly clean the cap, inside and out, with clean water. Also clean the filler neck on the radiator. All traces of corrosion should be removed. The coolant inside the radiator should be relatively transparent. If it's rust colored, the system should be drained and refilled (see Section 21). If the coolant level isn't up to the top, add additional antifreeze/coolant mixture (see Section 4).
4   Carefully check the large upper and lower radiator hoses along with the smaller diameter heater hoses which run from the engine to the firewall. Inspect each hose along its entire length, replacing any hose which is cracked, swollen or shows signs of deterioration. Cracks may become more apparent if the hose is squeezed **(see illustration)**. Regardless of condition, it's a good idea to replace hoses with new ones every two years.

Check for a chafed area that could fail prematurely.

Check for a soft area indicating the hose has deteriorated inside.

Overtightening the clamp on a hardened hose will damage the hose and cause a leak.

Check each hose for swelling and oil-soaked ends. Cracks and breaks can be located by squeezing the hose.

15.4 Hoses, like drivebelts, have a habit of failing at the worst possible time - to prevent the inconvenience of a blown radiator or heater hose, inspect them carefully as shown here

5   Make sure that all hose connections are tight. A leak in the cooling system will usually show up as white or rust colored deposits on the areas adjoining the leak. If wire-type clamps are used at the ends of the hoses, it may be a good idea to replace them with more secure screw-type clamps.

6   Use compressed air or a soft brush to remove bugs, leaves, etc. from the front of the radiator or air conditioning condenser. Be careful not to damage the delicate cooling fins or cut yourself on them.

7   Every other inspection, or at the first indication of cooling system problems, have the cap and system pressure tested. If you don't have a pressure tester, most gas stations and repair shops will do this for a minimal charge.

16   Fuel system check (every 15,000 miles or 12 months)

Refer to illustrations 16.5 and 16.6
Warning 1: *Gasoline is extremely flammable, so take extra precautions when you work on any part of the fuel system. Don't smoke or allow open flames or bare light bulbs near the work area, and don't work in a garage where a natural gas-type appliance (such as a water heater or clothes dryer) with a pilot light is present. Since gasoline is carcinogenic, wear latex gloves when there's a possibility of being exposed to fuel, and, if you spill any fuel on your skin, rinse it off immediately with soap and water. Mop up any spills immediately and do not store fuel-soaked rags where they could ignite. When you perform any kind of work on the fuel system, wear safety glasses and have a Class B type fire extinguisher on hand. The fuel system is under constant pressure, so, before any lines are disconnected, the fuel system pressure must be relieved. See Chapter 4.*
Warning 2: *Some models covered by this manual are equipped with air suspension systems. Always disconnect electrical power to the suspension system before lifting or tow-*

ing the vehicle (see Chapter 10). Failure to perform this procedure may result in unexpected shifting or movement of the vehicle which could cause personal injury.

1   If you smell gasoline while driving or after the vehicle has been sitting in the sun, inspect the fuel system immediately.

2   Remove the gas filler cap and inspect if for damage and corrosion. The gasket should have an unbroken sealing imprint. If the gasket is damaged or corroded, install a new cap.

3   Inspect the fuel feed and return lines for cracks. Make sure that the connections between the fuel lines and the fuel injection system and between the fuel lines and the in-line fuel filter are tight. **Warning:** *Your vehicle is fuel injected, so you must relieve the fuel system pressure before servicing fuel system components. The fuel system pressure relief procedure is outlined in Chapter 4.*

4   Since some components of the fuel system - the fuel tank and part of the fuel feed and return lines, for example - are underneath the vehicle, they can be inspected more easily with the vehicle raised on a hoist. If that's not possible, raise the vehicle and support it on jackstands.

5   With the vehicle raised and safely supported, inspect the gas tank and filler neck for punctures, cracks and other damage. The connection between the filler neck and the tank is particularly critical. Sometimes a rubber filler neck will leak because of loose clamps or deteriorated rubber **(see illustration)**. Inspect all fuel tank mounting brackets and straps to be sure that the tank is securely attached to the vehicle. **Warning:** *Do not, under any circumstances, try to repair a fuel tank (except rubber components). A welding torch or any open flame can easily cause fuel vapors inside the tank to explode.*

6   Carefully check all rubber hoses and metal lines leading away from the fuel tank **(see illustration).** Check for loose connections, deteriorated hoses, crimped lines and other damage. Repair or replace damaged sections as necessary (see Chapter 4).

1

16.5 Inspect fuel filler hoses for cracks and make sure the clamps (arrows) are tight

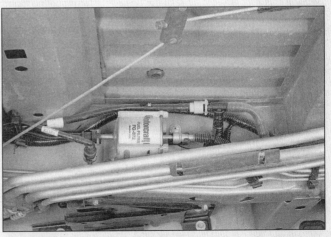

16.6 Carefully inspect fuel line couplings for damage

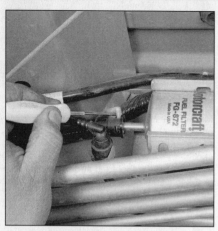

17.4 Use a small screwdriver to pry the fuel line fitting retaining clip off

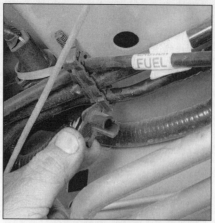

17.5a Later models are equipped with a fuel line disconnect tool that is located approximately ten inches behind the fuel filter - use a pair of diagonal cutters to detach it from the fuel line retaining clip - store the tool in the glove box after use

17.5b Detach the safety clip from the outlet port of the fuel filter

17 Fuel filter replacement (every 15,000 miles or 12 months)

Refer to illustrations 17.4, 17.5a, 17.5b and 17.5c

Warning 1: *Gasoline is extremely flammable, so take extra precautions when you work on any part of the fuel system. Don't smoke or allow open flames or bare light bulbs near the work area, and don't work in a garage where a natural gas-type appliance (such as a water heater or clothes dryer) with a pilot light is present. Since gasoline is carcinogenic, wear latex gloves when there's a possibility of being exposed to fuel, and, if you spill any fuel on your skin, rinse it off immediately with soap and water. Mop up any spills immediately and do not store fuel-soaked rags where they could ignite. When you perform any kind of work on the fuel system, wear safety glasses and have a Class B type fire extinguisher on hand.*

Warning 2: *Some models covered by this manual are equipped with air suspension systems. Always disconnect electrical power to the suspension system before lifting or towing the vehicle (see Chapter 10). Failure to perform this procedure may result in unexpected shifting or movement of the vehicle which could cause personal injury.*

Note: *This procedure requires a special fuel line disconnect tool which is available at most auto part stores. Some models may be equipped with a fuel line disconnect tool which is installed on the vehicle about ten inches behind the fuel filter.*

1   The fuel filter is mounted under the vehicle on the inside of the left (driver's side) frame rail in front of the fuel tank.
2   Inspect the hose fittings at both ends of the filter to see if they're clean. If more than a light coating of dust is present, clean the fittings before proceeding.
3   Relieve the fuel system pressure (see Chapter 4).
4   Disconnect the plastic push connect fitting from the inlet port by spreading the two clip legs apart about 1/8-inch to disengage

them, then push in on them. Pull on the other end of the clip to detach it from the fitting **(see illustration). Caution:** *If the new filter doesn't include new clips, don't use any tools or you may damage the plastic clips, which will have to be reused. Use your fingers only.*
5   Detach the safety clip from the outlet port, then install the special fuel line disconnect tool onto the fuel filter outlet port. Push the disconnect tool into the fuel line coupling until it releases itself from the fuel filter **(see illustrations).**
6   After both couplings have been released, grasp the fuel hoses, one at a time, and pull them straight off the filter. Be prepared for fuel spillage.
7   After the hoses have been detached, check the clips for damage and distortion. If they were damaged in any way during removal, new ones must be used when the hoses are reattached to the new filter (if new clips are packaged with the filter, be sure to use them in place of the originals).
8   Use a screwdriver to loosen the fuel filter mounting clamp, while noting the direction the fuel filter is installed.
9   Remove the filter from the bracket and install the new fuel filter in the same direction.
10   Carefully push each hose onto the filter until it's seated against the collar on the fitting, then install the clips. Make sure the clips are securely attached to the hose fittings - if they come off, the hoses could back off the filter and a fire could result!
11   Start the engine and check for fuel leaks.

18 Suspension, steering and driveaxle boot check (every 15,000 miles or 12 months)

Warning: *Some models covered by this manual are equipped with air suspension sys-*

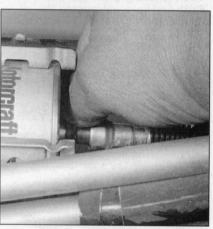

17.5c Install the fuel line disconnect tool onto the fuel filter port, then push the disconnect tool into the coupling until the fuel line releases itself

tems. Always disconnect electrical power to the suspension system before lifting or towing the vehicle (see Chapter 10). Failure to perform this procedure may result in unexpected shifting or movement of the vehicle which could cause personal injury.

Note: *The steering linkage and suspension components should be checked periodically. Worn or damaged suspension and steering linkage components can result in excessive and abnormal tire wear, poor ride quality and vehicle handling and reduced fuel economy. For detailed illustrations of the steering and suspension components, refer to Chapter 10.*

Shock absorber check

Refer to illustration 18.6

1   Park the vehicle on level ground, turn the engine off and set the parking brake. Check the tire pressures.
2   Push down at one corner of the vehicle, then release it while noting the movement of the body. It should stop moving and come to rest in a level position within one or two bounces.
3   If the vehicle continues to move up-and-

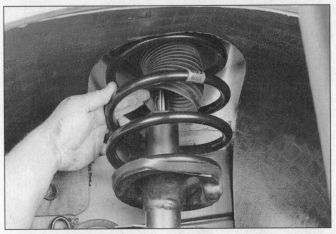

18.6 Check the front struts and rear shocks for leakage at the indicated area

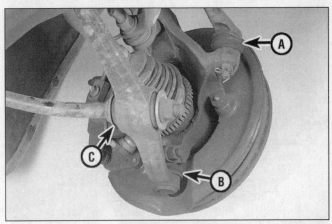

18.9a Inspect the tie rod ends (A) and the lower balljoints (B) for torn grease seals - Inspect the lower control arm bushings (C) for deteriorated bushings

18.9b Check the steering gear boots for cracks and leaking steering fluid

18.9c Check the stabilizer bar bushings (arrows) for deterioration at the front and the rear of the vehicle

down or if it fails to return to its original position, a worn or weak shock absorber is probably the reason.

4   Repeat the above check at each of the three remaining corners of the vehicle.

5   Raise the vehicle and support it securely on jackstands.

6   Check the shock absorbers for evidence of fluid leakage **(see illustration)**. A light film of fluid is no cause for concern. Make sure that any fluid noted is from the shocks and not from some other source. If leakage is noted, replace the shocks as a set.

7   Check the shocks to be sure that they are securely mounted and undamaged. Check the upper mounts for damage and wear. If damage or wear is noted, replace the shocks as a set (front or rear).

8   If the shocks must be replaced, refer to Chapter 10 for the procedure.

Steering and suspension check

Refer to illustrations 18.9a, 18.9b, 18.9c and 18.11

9   Visually inspect the steering and suspension components for damage and distor-

tion. Look for damaged seals, boots and bushings and leaks of any kind **(see illustrations)**.

10   Clean the lower end of the steering knuckle. Have an assistant grasp the lower edge of the tire and move the wheel in-and-out while you look for movement at the steering knuckle-to-control arm balljoint. If there is any movement the suspension balljoint(s) must be replaced.

11   Grasp each front tire at the front and rear edges, push in at the front, pull out at the rear and feel for play in the steering system components If any freeplay is noted, check the idler arm and the tie-rod ends for looseness **(see illustration)**.

12   Additional steering and suspension system information and illustrations can be found in Chapter 10.

Driveaxle boot check)

Refer to illustration 18.14

13   The driveaxle boots are very important because they prevent dirt, water and foreign material from entering and damaging the constant velocity (CV) joints. Oil and grease can cause the boot material to deteriorate prematurely, so it's a good idea to wash the

18.11 With the steering wheel in the lock position and the vehicle raised, grasp the front tire as shown and try to move it back-and-forth - if any play is noted, check the steering gear mounts and tie rod ends for looseness

boots with soap and water. Because it constantly pivots back and forth following the steering action of the front hub, the outer CV

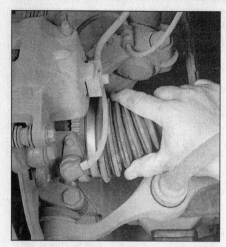

18.14 Flex the inner and outer driveaxle boots by hand to check for cracks and/or leaking grease

19.6a You will find an inspection hole like this in each caliper - placing a ruler across the hole should enable you to determine the thickness of remaining pad material on the inner pad

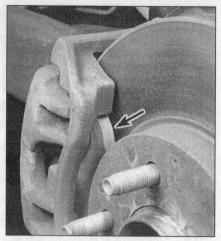

19.6b The outer pad (arrow) can be viewed and inspected from the outer side of the caliper

boot wears out sooner and should be inspected regularly.

14   Inspect the boots for tears and cracks as well as loose clamps **(see illustration)**. If there is any evidence of cracks or leaking lubricant, they must be replaced as described in Chapter 8.

19 Brake check (every 15,000 miles or 12 months)

Warning 1: *The dust created by the brake system may contain asbestos, which is harmful to your health. Never blow it out with compressed air and don't inhale any of it. An approved filtering mask should be worn when working on the brakes. Do not, under any circumstances, use petroleum-based solvents to clean brake parts. Use brake system cleaner only! Try to use non-asbestos replacement parts whenever possible.*

Warning 2: *Some models covered by this manual are equipped with air suspension systems. Always disconnect electrical power to the suspension system before lifting or towing the vehicle (see Chapter 10). Failure to perform this procedure may result in unexpected shifting or movement of the vehicle which could cause personal injury.*

Note: *For detailed photographs of the brake system, refer to Chapter 9.*

1   In addition to the specified intervals, the brakes should be inspected every time the wheels are removed or whenever a defect is suspected.

2   Any of the following symptoms could indicate a potential brake system defect: The vehicle pulls to one side when the brake pedal is depressed; the brakes make squealing or dragging noises when applied; brake pedal travel is excessive; the pedal pulsates; brake fluid leaks, usually onto the inside of the tire or wheel.

3   Loosen the wheel lug nuts.

4   Raise the vehicle and place it securely

on jackstands.

5   Remove the wheels (see *Jacking and towing* at the front of this book, or your owner's manual, if necessary).

Disc brakes

Refer to illustrations 19.6a, 19.6b and 19.11

6   There are two pads (an outer and an inner) in each caliper. The pads are visible through inspection holes in each caliper **(see illustrations)**.

7   Check the pad thickness by looking at each end of the caliper and through the inspection hole in the caliper body. If the lining material is less than the thickness listed in this Chapter's Specifications, replace the pads. **Note:** *Keep in mind that the lining material is riveted or bonded to a metal backing plate and the metal portion is not included in this measurement.*

8   If it is difficult to determine the exact thickness of the remaining pad material by the above method, or if you are at all concerned about the condition of the pads, remove the caliper(s), then remove the pads from the calipers for further inspection (refer to Chapter 9).

9   Once the pads are removed from the calipers, clean them with brake cleaner and re-measure them with a ruler or a vernier caliper.

10   Measure the disc thickness with a micrometer to make sure that it still has service life remaining. If any disc is thinner than the specified minimum thickness, replace it (refer to Chapter 9). Even if the disc has service life remaining, check its condition. Look for scoring, gouging and burned spots. If these conditions exist, remove the disc and have it resurfaced (see Chapter 9).

11   Before installing the wheels, check all brake lines and hoses for damage, wear, deformation, cracks, corrosion, leakage, bends and twists, particularly in the vicinity of the rubber hoses at the calipers **(see illustra-**

tion). Check the clamps for tightness and the connections for leakage. Make sure that all hoses and lines are clear of sharp edges, moving parts and the exhaust system. If any of the above conditions are noted, repair, reroute or replace the lines and/or fittings as necessary (see Chapter 9).

Drum brakes

Refer to illustrations 19.15, 19.16 and 19.17

12   On rear drum brakes, make sure the parking brake is off then proceed to tap on the outside of the drum with a rubber mallet to loosen it.

13   Remove the brake drums.

14   With the drums removed, carefully clean the brake assembly with brake system cleaner. **Warning:** *Don't blow the dust out with compressed air and don't inhale any of it (it may contain asbestos, which is harmful to your health).*

15   Note the thickness of the lining material on both front and rear brake shoes. If the material has worn away to within 1/16-inch of

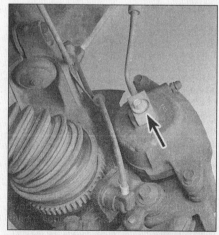

19.11 Check along the brake hoses and at each fitting (arrow) for deterioration and cracks

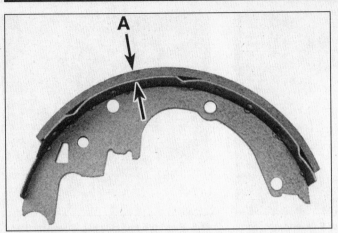

19.15 If the lining is bonded to the brake shoe, measure the lining thickness from the outer surface to the metal shoe, as shown here. If the lining is riveted to the shoe, measure from the lining outer surface to the rivet head

19.16 Typical assembled view of a rear drum brake (left side shown, right side is the exact opposite)

the recessed rivets or 1/8- inch of the metal backing on bonded type shoes, the shoes should be replaced **(see illustration)**. The shoes should also be replaced if they're cracked, glazed (shiny areas), or covered with brake fluid.

16   Make sure all the brake assembly springs are connected and in good condition **(see illustration)**.

17   Check the brake components for signs of fluid leakage. With your finger or a small screwdriver, carefully pry back the rubber cups on the wheel cylinder located at the top of the brake shoes **(see illustration)**. Any leakage here is an indication that the wheel cylinders should be overhauled immediately (see Chapter 9). Also, check all hoses and connections for signs of leakage.

18   Wipe the inside of the drum with a clean rag and denatured alcohol or brake cleaner. Again, be careful not to breathe the dangerous asbestos dust.

19   Check the inside of the drum for cracks, score marks, deep scratches and "hard spots" which will appear as small discolored areas. If imperfections cannot be removed

with fine emery cloth, the drum must be taken to an automotive machine shop for resurfacing.

20   Repeat the procedure for the remaining wheel. If the inspection reveals that all parts are in good condition, reinstall the brake drums, install the wheels and lower the vehicle to the ground.

Brake booster check

21   Sit in the driver's seat and perform the following sequence of tests.

22   With the engine stopped, depress the brake pedal several times - the travel distance should not change.

23   With the brake fully depressed, start the engine - the pedal should move down a little when the engine starts.

24   Depress the brake, stop the engine and hold the pedal in for about 30 seconds - the pedal should neither sink nor rise.

25   Restart the engine, run it for about a minute and turn it off. Then firmly depress the brake several times - the pedal travel should decrease with each application.

26   If your brakes do not operate as

described, the brake booster has failed. Refer to Chapter 9 for the replacement procedure.

Parking brake

27   Vehicles equipped with rear drum brakes utilize a self-adjusting parking brake mechanism and do not require regular scheduled maintenance. Only vehicles equipped rear disc brakes require regular scheduled maintenance. For more detailed information on the parking brake assembly see Chapter 9.

---

20   Air filter check and replacement (every 30,000 miles or 24 months)

Refer to illustrations 20.2a, 202b and 20.3

1   The air filter is located inside the housing on the left (driver's) side of the engine compartment on all vehicles.

2   To remove the air filter on 1995 through 1998 3.0L engines, release the spring clip that secures the two halves of the air cleaner housing together, the lift the cover up and remove the air filter element **(see illustration)**.

19.17 Check the wheel cylinder boots for leaking fluid indicating that the cylinder must be replaced or rebuilt

20.2a On 3.0L engines and 1995 3.8L engines, detach the clips and separate the cover from the air cleaner housing

20.2b Lift the cover up and slide the element out of the housing

20.3 On 1999 and later 3.0L engines and 1996 and later 3.8L engines, release the clamp (arrow) then separate the housing halves to access the air filter element

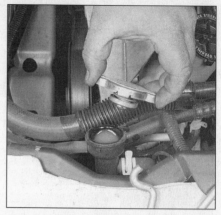

21.5 Push the radiator cap downward and rotate it counterclockwise - never remove it when the engine is hot!

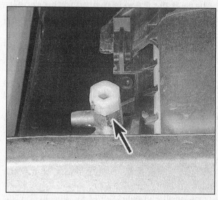

21.6 The radiator drain fitting (arrow) is located at the lower corner of the radiator

3   To remove the air filter on 1999 and later 3.0L engines and 1996 on 3.8L engines, release the clamp that secures the two halves of the air cleaner housing together, then separate the cover halves and remove the air filter element (see illustrations). **Note:** *The air cleaner housing on 1995 3.8L models is similar to the 1995 through 1998 3.OL engine except that the top cover is retained by bolts. Remove the bolts and follow Step 2 to remove the air filter element.*

4   Inspect the outer surface of the filter element. If it is dirty, replace it. If it is only moderately dusty, it can be reused by blowing it clean from the back to the front surface with compressed air. Because it is a pleated paper type filter, it cannot be washed or oiled. If it cannot be cleaned satisfactorily with compressed air, discard and replace it. While the cover is off, be careful not to drop anything down into the housing. **Caution:** *Never drive the vehicle with the air cleaner removed. Excessive engine wear could result and backfiring could even cause a fire under the hood.*

5   Wipe out the inside of the air cleaner housing.

6   Place the new filter into the air cleaner housing, making sure it seats properly.

7   Installation of the housing is the reverse of removal.

21 Cooling system servicing (draining, back flushing and refilling) (every 30,000 miles or 24 months)

Refer to illustrations 21.5 and 21.6

Warning 1: *Do not allow antifreeze to come in contact with your skin or painted surfaces of the vehicle. Rinse off spills immediately with plenty of water. Antifreeze is highly toxic if ingested. Never leave antifreeze lying around in an open container or in puddles on the floor; children and pets are attracted by it's sweet smell and may drink it. Check with local authorities about disposing of used antifreeze. Many communities have collection*

centers which will see that antifreeze is disposed of safely.

Warning 2: *Some models covered by this manual are equipped with air suspension systems. Always disconnect electrical power to the suspension system before lifting or towing the vehicle (see Chapter 10). Failure to perform this procedure may result in unexpected shifting or movement of the vehicle which could cause personal injury.*

1   Periodically, the cooling system should be drained, flushed and refilled to replenish the antifreeze mixture and prevent formation of rust and corrosion, which can impair the performance of the cooling system and cause engine damage.

2   At the same time the cooling system is serviced, all hoses and the radiator cap should be inspected and replaced if defective (see Section 15).

3   Since antifreeze is a corrosive and poisonous solution, be careful not to spill any of the coolant mixture on the vehicle's paint or your skin. If this happens, rinse it off immediately with plenty of clean water. Consult local authorities about the dumping of antifreeze before draining the cooling system. In many areas, reclamation centers have been set up to collect automobile oil and drained antifreeze/water mixtures, rather than allowing them to be added to the sewage system.

Draining

4   Apply the parking brake and block the wheels. If the vehicle has just been driven, wait several hours to allow the engine to cool down before beginning this procedure.

5   Once the engine is completely cool, remove the radiator cap and the reservoir cap **(see illustration)**.

6   Drain the radiator by opening the drain plug at the bottom of the radiator **(see illustration)**. If the drain plug is corroded and can't be turned easily, or if the radiator isn't equipped with a plug, disconnect the lower radiator hose to allow the coolant to drain. Be careful not to get antifreeze on your skin or in your eyes.

7   After the coolant stops flowing out of the radiator, remove the lower radiator hose

and allow the remaining fluid in the upper half of the engine block to drain.

8   While the coolant is draining from the engine block, disconnect the hose from the coolant reservoir and remove the reservoir (see Chapter 3 if necessary). Flush the reservoir out with water until it's clean, and if necessary, wash the inside with soapy water and a brush to make reading the fluid level easier.

9   While the coolant is draining, check the condition of the radiator hoses, heater hoses and clamps (refer to Section 15 if necessary).

10   Replace any damaged clamps or hoses (refer to Chapter 3 for detailed replacement procedures).

Back flushing

11   Once the system is completely drained, remove the thermostat from the engine (see Chapter 3). Then reinstall the thermostat housing without the thermostat. This will allow the system to be back flushed.

12   Reinstall the lower radiator hose and tighten the radiator drain plug.

13   Disconnect the upper radiator hose, then place a garden hose in the upper radiator inlet and flush the system until the water runs clear at the upper radiator hose.

14   In severe cases of contamination or clogging of the radiator, remove the radiator (see Chapter 3) and have a radiator repair

22.7 Pry the pan free of the gasket and allow the fluid to drain

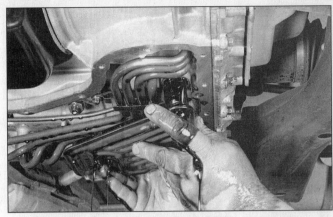

22.10a Pull straight down on the filter to remove it

facility clean and repair it if necessary.

15   Many deposits can be removed by the chemical action of a cleaner available at auto parts stores. Follow the procedure outlined in the manufacturer's instructions. **Note:** *When the coolant is regularly drained and the system refilled with the correct antifreeze/water mixture, there should be no need to use chemical cleaners or descalers.*

Refilling

16   To refill the system, install the thermostat, reconnect any radiator hoses and install the reservoir and the overflow hose.

17   Place the heater temperature control in the maximum heat position.

18   Make sure to use the proper coolant listed in this Chapter's Specifications. Slowly fill the radiator with the recommended mixture of antifreeze and water to the base of the filler neck. Then add coolant to the reservoir until it reaches the FULL COLD mark. Wait five minutes and recheck the coolant level in the radiator, adding if necessary.

19   Leave the radiator cap off and run the engine in a well-ventilated area until the thermostat opens (coolant will begin flowing through the radiator and the upper radiator hose will become hot).

20   Turn the engine off and let it cool. Add more coolant mixture to bring the level back up to the base of the filler neck.

21   Squeeze the upper radiator hose to expel air, then add more coolant mixture if necessary. Replace the radiator cap.

22   Place the heater temperature control and the blower motor speed control to their maximum setting.

23   Start the engine, allow it to reach normal operating temperature and check for leaks.

22   Automatic transaxle fluid and filter change (every 30,000 miles or 24 months)

Refer to illustrations 22.7, 22.10a, 22.10b, 22.11, 22.12a and 22.12b

Warning: *Some models covered by this manual are equipped with air suspension sys-*

tems. Always disconnect electrical power to the suspension system before lifting or towing the vehicle (see Chapter 10). Failure to perform this procedure may result in unexpected shifting or movement of the vehicle which could cause personal injury.

1   At the specified time intervals, the transaxle fluid should be drained and replaced. Since the fluid will remain hot long after driving, perform this procedure only after everything has cooled down completely.

2   Before beginning work, purchase the specified transaxle fluid (see *Recommended lubricants and fluids* at the front of this Chapter) and a new filter.

3   Other tools necessary for this job include jackstands to support the vehicle in a raised position, a drain pan capable of holding several quarts, newspapers and clean rags.

4   Raise the vehicle and support it securely on jackstands.

5   With a drain pan in place, remove the front and side transaxle pan mounting bolts.

6   Loosen the rear pan bolts one turn.

7   Carefully pry the transaxle pan loose with a screwdriver, allowing the fluid to drain **(see illustration)**.

8   Remove the remaining bolts, pan and gasket. Carefully clean the gasket surface of the transaxle to remove all traces of the old gasket and sealant.

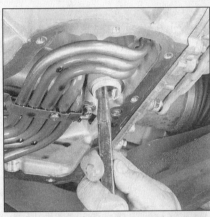

22.10b If the filter seal did not come out with the filter, remove it from the transaxle - be careful not to gouge the seal bore in anyway

9   Drain the fluid from the transaxle pan, clean the pan with solvent and dry it with compressed air. Be careful not to lose the magnet.

10   Remove the filter and pry out the seal **(see illustrations)**.

11   Install a new seal and filter **(see illustration)**.

12   Make sure the gasket surface on the transaxle pan is clean, then install the magnet

22.11 Install a new seal on the transaxle filter

22.12a Be sure to clean all traces of the old gasket from the pan

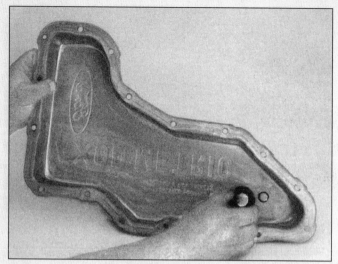

22.12b After cleaning the pan, place the magnet in position and install the new gasket

ACCEPTABLE

Cracks Running Across "V" Portions of Belt

Missing Two or More Adjacent Ribs 1/2" or longer

1/2"

UNACCEPTABLE

24.4 Small cracks in the underside of a V-ribbed belt are acceptable - lengthwise cracks, or missing pieces that cause the belt to make noise, are cause for replacement

Cracks Running Parallel to "V" Portions of Belt

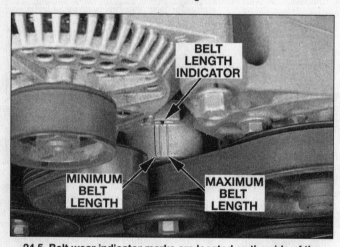

BELT LENGTH INDICATOR

MINIMUM BELT LENGTH

MAXIMUM BELT LENGTH

24.5 Belt wear indicator marks are located on the side of the tensioner body - when the belt reaches the maximum wear mark it must be replaced

24.7 Rotate the tensioner arm to relieve belt tension (3.0L engine shown, 3.8L engine similar)

and a new gasket **(see illustrations)**. Put the pan in place against the transaxle and install the bolts. Working around the pan, tighten each bolt a little at a time until the final torque figure listed in this Chapter's Specifications is reached. Don't overtighten the bolts!

13   Lower the vehicle and add the specified amount of automatic transmission fluid through the filler tube (see Section 7).

14   With the shift lever in Park and the parking brake set, run the engine at a fast idle, but don't race it.

15   Move the shift lever through each gear and back to Park. Check the fluid level.

16   Check under the vehicle for leaks during the first few trips.

23   Brake fluid change (every 30,000 miles or 24 months)

Warning: *Brake fluid can harm your eyes and damage painted surfaces, so use extreme caution when handling or pouring it. Do not*

use brake fluid that has been standing open or is more than one year old. Brake fluid absorbs moisture from the air. Excess moisture can cause a dangerous loss of braking effectiveness.

1   At the specified intervals, the brake fluid should be drained and replaced. Since the brake fluid may drip or splash when pouring it, place plenty of rags around the master cylinder to protect any surrounding painted surfaces.

2   Before beginning work, purchase the specified brake fluid (see *Recommended lubricants and fluids* at the beginning of this Chapter).

3   Remove the cap from the master cylinder reservoir.

4   Using a hand suction pump or similar device, withdraw the fluid from the master cylinder reservoir.

5   Add new fluid to the master cylinder until it rises to the base of the filler neck.

6   Bleed the brake system as described in Chapter 9 at all four brakes until new and uncontaminated fluid is expelled from the

bleeder screw. Be sure to maintain the fluid level in the master cylinder as you perform the bleeding process. If you allow the master cylinder to run dry, air will enter the system.

7   Refill the master cylinder with fluid and check the operation of the brakes. The pedal should feel solid when depressed, with no sponginess. **Warning:** *Do not operate the vehicle if you are in doubt about the effectiveness of the brake system.*

24   Drivebelt and drivebelt tensioner check and replacement (every 60,000 miles or 48 months)

Refer to illustrations 24.4, 24.5, 24.7, 24.9 and 24.10

1   The drivebelts are located at the front of the engine and play an important role in the overall operation of the vehicle and its components. Due to their function and material make-up, the drivebelts are prone to failure after a period of time and should be

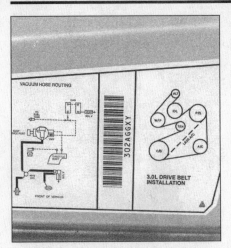

24.9 The routing schematic for the serpentine belt is usually found on the fan shroud (this one's for the 3.0L engine)

24.10 Remove the retaining bolt from the center of the tensioner body (3.0L engine shown, 3.8L engine similar)

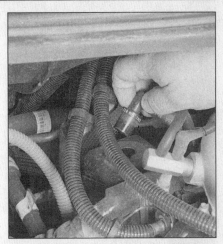

25.1 The PCV valve is located in the right valve cover adjacent to the firewall (3.0L engine shown, 3.8L similar)

inspected and adjusted periodically to prevent major engine damage.

2   The vehicles covered by this manual are equipped with a single self-adjusting serpentine drivebelt, which is used to drive all of the accessory components such as the alternator, power steering pump, water pump and air conditioning compressor.

Check

3   With the engine off, open the hood and locate the drivebelt at the front of the engine. Using your fingers (and a flashlight, if necessary), move along the belts checking for cracks and separation of the belt plies. Also check for fraying and glazing, which gives the belt a shiny appearance. Both sides of each belt should be inspected, which means you will have to twist the belt to check the underside.

4   Check the ribs on the underside of the belt. They should all be the same depth, with none of the surface uneven **(see illustration)**.

5   The tension of the belt is automatically adjusted by the belt tensioner and does not require any adjustments. Drivebelt wear can be checked visually by inspecting the wear indicator marks located on the side of the tensioner body. Locate the belt tensioner at the front of the engine on the right (passenger) side, adjacent to the lower crankshaft pulley, then find the tensioner operating marks **(see illustration)**. If the indicator mark is outside the operating range, the belt should be replaced.

6   To check the tensioner first remove the drivebelt (see Step 7). Then spin the pulley and listen for noise. Excessive noise from the pulley indicates a dry or faulty bearing which should be replaced. **Note:** *Often times the bearing and pulley can be replaced without replacing the entire tensioner assembly. Check your local auto parts store for these applications. Next rotate the tensioner body and check for a binding or frozen condition. if either conditions exist the tensioner should be replaced.*

Replacement

7   To replace the belt, place a wrench on the tensioner pulley bolt and rotate the tensioner body until tension on the belt is relieved **(see illustration)**.,

8   Remove the belt from the auxiliary components and carefully release the tensioner.

9   Route the new belt over the various pulleys, again rotating the tensioner to allow the belt to be installed, then release the belt tensioner. Make sure the belt fits properly into the pulley grooves - it must be completely engaged. **Note:** *Most models have a drivebelt routing decal on the upper radiator panel to help during drivebelt installation* **(see illustration)**.

10   To replace the drive belt tensioner, simply remove the retaining bolt from the center of the tensioner body and remove it from the engine **(see illustration)**.

11   Installation is the reverse of removal.

25 Positive Crankcase Ventilation (PCV) valve check (every 60,000 miles or 48 months)

Refer to illustrations 25.1 and 25.2

Note: *To maintain efficient operation of the PCV system, clean the hoses and check the PCV valve at the intervals recommended in the maintenance schedule. For additional information on the PCV system, refer to Chapter 6.*

1   The PCV valve on 3.0L and 3.8L engines is located in the right (rear) valve cover **(see illustration)**.

2   Start the engine and allow it to idle, then disconnect the PCV valve from the valve cover and feel for vacuum at the end of the valve **(see illustration)**. If vacuum is felt, the PCV valve/system is working properly (see Chapter 6 for additional PCV system information).

3   If no vacuum is felt, remove the valve and check for vacuum at the hose. If vacuum

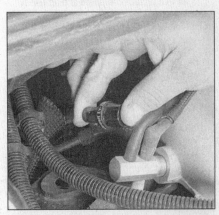

25.2 With the engine running at idle, remove the PCV valve and verify that vacuum can be felt at the end of the valve

is present at the hose but not at the valve, replace the valve. If no vacuum is felt at the hose, check for a plugged or cracked hose between the PCV valve and the intake plenum.

4   Check the rubber grommet in the valve cover for cracks and distortion. If it's damaged, replace it.

5   If the valve is clogged, the hose is also probably plugged. Remove the hose between the valve and the intake manifold and clean it with solvent.

6   After cleaning the hose, inspect it for damage, wear and deterioration. Make sure it fits snugly on the fittings.

7   If necessary, install a new PCV valve.

26 Spark plug check and replacement (every 60,000 miles or 48 months)

Refer to illustrations 26.2, 26.5a, 26.5b, 26.8, 26.9 and 26.10

Note 1: *Later model engines are equipped with two different type of platinum spark*

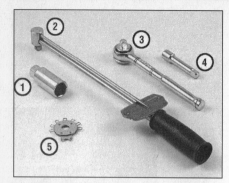

26.2 Tools required for changing spark plugs

1   *Spark plug socket - This will have special padding inside to protect the spark plug's porcelain insulator*

2   *Torque wrench - Although not mandatory, using this tool is the best way to ensure the plugs are tightened properly*

3   *Ratchet - Standard hand tool to fit the spark plug socket*

4   *Extension - Depending on model and accessories, you may need special extensions and universal joints to reach one or more of the plugs*

5   *Spark plug gap gauge - This gauge for checking the gap comes in a variety of styles. Make sure the gap for your engine is included*

plugs. If the original spark plugs are to be removed and reinstalled in the engine, the spark plugs must be marked and reinstalled in the original cylinder from which they were removed.

Note 2: *All spark plugs located in the right hand cylinder head (adjacent to the firewall) must be removed from below the vehicle. Special flexible sockets and extensions will be necessary. In some cases, it may be easier to remove the cowl cover (see Chapter 11) and remove the plugs from the top.*

1   All vehicles covered by this manual are equipped with transversely mounted V6 engines which locate the spark plugs on the side of the engine at the front and the rear of the engine compartment. The left side (front) spark plugs can be reached from the front of the vehicle while the right side (rear) spark plugs are located between the engine and the firewall which requires removal from beneath the vehicle.

2   In most cases, the tools necessary for spark plug replacement include a spark plug socket which fits onto a ratchet (spark plug sockets are padded inside to prevent damage to the porcelain insulators on the new plugs), various extensions and a gap gauge to check and adjust the gaps on the new plugs **(see illustration)**. A special plug wire removal tool is available for separating the wire boots from the spark plugs, and is a good idea on these models because the boots fit very tightly. A torque wrench should be used to tighten the new plugs. It is a good idea to allow the engine to cool before

26.5a Spark plug manufacturers recommend using a wire-type gauge when checking the gap - if the wire does not slide between the electrodes with a slight drag, adjustment is required

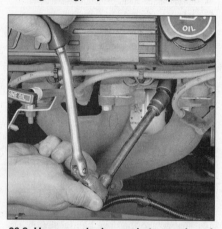

26.8 Use a spark plug socket wrench and extension to unscrew the spark plug

removing or installing the spark plugs.

3   The best approach when replacing the spark plugs is to purchase the new ones in advance, adjust them to the proper gap and replace the plugs one at a time. When buying the new spark plugs, be sure to obtain the correct plug type for your particular engine. The plug type can be found in the Specifications at the front of this Chapter and on the Emission Control Information label located under the hood. If these two sources list different plug types, consider the emission control label correct.

4   Allow the engine to cool completely before attempting to remove any of the plugs. While you are waiting for the engine to cool, check the new plugs for defects and adjust the gaps.

5   Check the gap by inserting the proper thickness gauge between the electrodes at the tip of the plug **(see illustration)**. The gap between the electrodes should be the same as the one specified on the Emissions Control Information label or in Chapter 5. The wire should slide between the electrodes with a slight amount of drag. If the gap is incorrect, use the adjuster on the gauge body to

26.5b To change the gap, bend the side electrode only, as indicated by the arrows, and be very careful not to crack or chip the porcelain insulator surrounding the center electrode

26.9 Apply a thin film of anti-seize compound to the spark plug threads to prevent damage to the cylinder head

bend the curved side electrode slightly until the proper gap is obtained **(see illustration)**. If the side electrode is not exactly over the center electrode, bend it with the adjuster until it is. Check for cracks in the porcelain insulator (if any are found, the plug should not be used).

6   With the engine cool, remove the spark plug wire as described in Section 27 from one spark plug. Pull only on the boot at the end of the wire - do not pull on the wire. A plug wire removal tool should be used if available.

7   If compressed air is available, use it to blow any dirt or foreign material away from the spark plug hole. A common bicycle pump will also work. The idea here is to eliminate the possibility of debris falling into the cylinder as the spark plug is removed.

8   The spark plugs on these models are, for the most part, difficult to reach so a spark plug socket incorporating a universal joint will be necessary. Place the spark plug socket over the plug and remove it from the engine by turning it in a counterclockwise direction **(see illustration)**. **Note:** *When removing the*

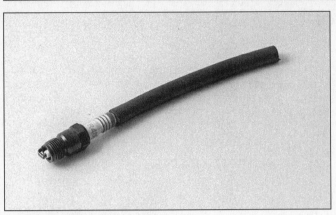

26.10 A length of rubber hose will save time and prevent damaged threads when installing the spark plugs

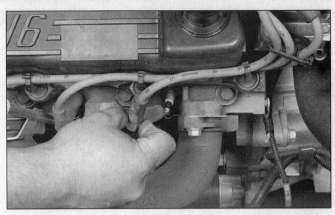

27.4 When removing the spark plug wires from the spark plug, pull only on the boot and use a twisting/pulling motion

firewall side spark plugs it is suggested (not mandatory) that you remove the cowl cover first (see Chapter 11).

9   Compare the spark plug with the chart shown on the inside back cover of this manual to get an indication of the general running condition of the engine. Before installing the new plugs, it is a good idea to apply a thin coat of anti-seize compound to the threads **(see illustration)**.

10   Thread one of the new plugs into the hole until you can no longer turn it with your fingers, then tighten it with a torque wrench (if available) or the ratchet. It's a good idea to slip a short length of rubber hose over the end of the plug to use as a tool to thread it into place **(see illustration)**. The hose will grip the plug well enough to turn it, but will start to slip if the plug begins to cross-thread in the hole - this will prevent damaged threads and the accompanying repair costs.

11   Before pushing the spark plug wire onto the end of the plug, inspect it following the procedures outlined in Section 27.

12   Attach the plug wire to the new spark plug, again using a twisting motion on the boot until it's seated on the spark plug.

13   Repeat the procedure for the remaining spark plugs, replacing them one at a time to prevent mixing up the spark plug wires.

27 Ignition system component check and replacement (every 60,000 miles or 48 months)

Refer to illustrations 27.4, 27.8 and 27.12

Spark plug wires

Note: *Every time a spark plug wire is detached from a spark plug or the coil, silicone dielectric compound (a white grease available at auto parts stores) must be applied to the inside of each boot before reconnection. Use a small standard screwdriver to coat the entire inside surface of each boot with a thin layer of the compound.*

1   The spark plug wires should be checked and, if necessary, replaced at the same time new spark plugs are installed.

2   The easiest way to identify bad wires is to make a visual check while the engine is running. In a dark, well-ventilated garage, start the engine and look at each plug wire. Be careful not to come into contact with any moving engine parts. If there is a break in the wire, you will see arcing or a small spark at the damaged area. If arcing is noticed, make a note to obtain new wires.

3   The spark plug wires should be inspected one at a time, beginning with the spark plug for the number one cylinder, (the cylinder closest to the engine drivebelt on the right bank), to prevent confusion. Clearly label each original plug wire with a piece of tape marked with the correct number. The plug wires must be reinstalled in the correct order to ensure proper engine operation.

4   Disconnect the plug wire from the first spark plug **(see illustration)**. A removal tool can be used, or you can grab the wire boot, twist it slightly and pull the wire free. Do not pull on the wire itself, only on the rubber boot.

5   Push the wire and boot back onto the end of the spark plug. It should fit snugly. If it doesn't, detach the wire and boot once more and use a pair of pliers to carefully crimp the metal connector inside the wire boot until it does.

6   Using a clean rag, wipe the entire length of the wire to remove built-up dirt and grease.

7   Once the wire is clean, check for burns, cracks and other damage. Do not bend the wire sharply or you might break the conductor.

8   Disconnect the wire from the coil pack. Pull only on the rubber boot. Check for corrosion and a tight fit **(see illustration)**. Reinstall the wire.

9   Inspect each of the remaining spark plug wires, making sure that each one is securely fastened on each end.

10   If new spark plug wires are required, purchase a set for your specific engine model. Pre-cut wire sets with the boots already installed are available. Remove and replace the wires one at a time to avoid mix-ups in the firing order. Should a mix up occur refer to the Specifications listed at the beginning of this Chapter.

Ignition coil pack

11   Clean the coil pack with a dampened cloth and dry them thoroughly.

12   Inspect each coil pack for cracks, damage and carbon tracking **(see illustration)**. If damage exists refer to Chapter 5 for the replacement procedure.

27.8 Remove each spark plug wire from the ignition coil pack and at the spark plug - check for corrosion and a tight fit

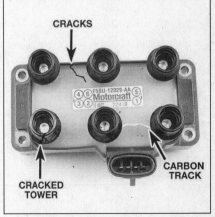

27.12 Shown here are some of the common defects to look for when inspecting the ignition coil pack

Notes

Chapter 2 Part A
Engines

Contents

2A

Specifications

General

Firing order	1-4-2-5-3-6

Torque specifications

	Ft-lbs
3.0L V6 engine	
Camshaft sprocket-to-camshaft bolt	41 to 51
Crankshaft pulley-to-damper bolt	30 to 44
Crankshaft damper-to-crankshaft bolt	93 to 121
Cylinder head bolts*	
1995	
Step 1	33 to 41
Step 2	63 to 73
1996 and later	
Step 1	59
Step 2	Loosen all bolts 1 turn (360-degrees)
Step 3	36
Step 4	67

*Use new bolts on installation.

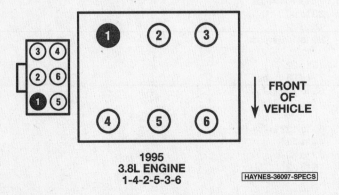

1995
3.8L ENGINE
1-4-2-5-3-6

HAYNES-36097-SPECS

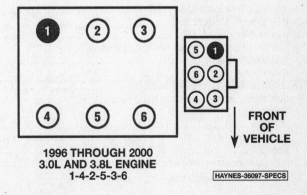

1996 THROUGH 2000
3.0L AND 3.8L ENGINE
1-4-2-5-3-6

HAYNES-36097-SPECS

Cylinder and coil terminal location

Torque specifications   Ft-lbs

3.0L V6 engine (continued)

Driveplate-to-crankshaft bolts	54 to 64
Engine mount upper through-bolt (front or rear)	30 to 40
Engine mount-to-frame nut (front or rear)	65 to 87
Exhaust manifold bolts/studs	15 to 22
Intake manifold-to-cylinder head bolts	
1995 through 1998	
Step 1	15 to 22
Step 2	20 to 23
1999 on	
Step 1	132 in-lbs
Step 2	24
Oil pan-to-block bolt	88 to 123 in-lbs
Oil pump-to-block bolt	30 to 40
Rocker arm-to-cylinder head bolts	
Step 1	62 to 132 in-lbs
Step 2	20 to 28
Valve cover-to-cylinder head bolt/stud	89 to 123 in-lbs
Timing chain cover-to-block bolts	15 to 22

3.8L V6 engine

Camshaft sprocket-to-camshaft bolt	30 to 37
Crankshaft pulley-to-damper bolt	
1995 to 1997	30 to 44
1998 and later	19 to 28
Crankshaft damper-to-crankshaft bolt	103 to 132
Cylinder head bolts*	
Step 1	15
Step 2	29
Step 3	37
Step 4	Loosen each bolt 2 or 3 turns (from this point on work on one bolt at a time, loosening and tightening each bolt to the final torque specification in sequence - DO NOT loosen all the bolts at the same time)
Step 5	
1995	
Long bolts	11 to 18
Short bolts	84 to 170 in-lbs
1996 and later	
Long bolts	30 to 36
Short bolts	15 to 22
Step 6	
1995	Tighten an additional 1/4 turn (90-degrees)
1996 and later	Tighten an additional 1/2 turn (180-degrees)
Exhaust manifold bolts/studs	15 to 22
Engine mount upper nut (front or rear)	50 to 68
Engine mount-to-frame nut (front or rear)	65 to 87
Intake manifold-to-cylinder head bolts	
1995	
Step 1	156 in-lbs
Step 2	182 in-lbs
1996 and later	71 to 106 in-lbs
Oil inlet tube-to-main bearing cap nut	30 to 40
Oil inlet tube-to-block bolt	15 to 22
Oil pan-to-block bolt	80 to 106 in-lbs
Oil pump/oil filter adapter-to-timing chain cover bolts	
Large bolts	18 to 22
Small bolts	72 to 96 in-lbs
Valve cover-to-cylinder head bolt/stud	80 to 106 in-lbs
Rocker arm fulcrum-to-cylinder head bolts	
Step 1	
1995	60 to 132 in-lbs
1996 and later	44 in-lbs
Step 2	23 to 30
Timing chain cover-to-block bolts	15 to 22

Use new bolts on installation.

1 General information

This Part of Chapter 2 is devoted to in-vehicle repair procedures for the 3.0L and 3.8L V6 engines. All information concerning engine removal and installation, repairs which require engine removal and engine block and cylinder head overhaul can be found in Part B of this Chapter.

The following repair procedures are based on the assumption that the engine is installed in the vehicle. If the engine has been removed from the vehicle and mounted on a stand, many of the steps outlined in this Part of Chapter 2 will not apply.

The specifications included in this Part of Chapter 2 apply only to the procedures contained in this Part. Part B of Chapter 2 contains the specifications necessary for cylinder head and engine block rebuilding.

2 Repair operations possible with the engine in the vehicle

Warning: *Some models covered by this manual are equipped with an air suspension system. Always disconnect the electrical power to the suspension system before lifting or towing (see Chapter 10). Failure to perform this procedure may result in unexpected shifting or movement of the vehicle, which could cause personal injury.*

Many major repair operations can be accomplished without removing the engine from the vehicle.

Clean the engine compartment and the exterior of the engine with some type of pressure washer before any work is done. A clean engine will make the job easier and will help keep dirt out of the internal areas of the engine.

Depending on the components involved, it may be a good idea to remove the hood to improve access to the engine as repairs are performed (refer to Chapter 11 if necessary).

If vacuum, exhaust, oil or coolant leaks develop, indicating a need for gasket or seal replacement, the repairs can generally be made with the engine in the vehicle. The intake and exhaust manifold gaskets, oil pan gasket and cylinder head gaskets are all accessible with the engine in place.

Exterior engine components such as the intake and exhaust manifolds, the oil pan (and the oil pump), the water pump, the starter motor, the alternator and the fuel injection system components can be removed for repair with the engine in place. Although the timing chain and sprockets can be replaced with the engine in-vehicle on the 3.0L engine, this procedure can't be performed in-vehicle on the 3.8L engine. Replacement of the camshaft or balance shaft (3.8L only) on either engine can be performed only with the engine removed from the vehicle. See Part B of this Chapter for camshaft and balance-shaft procedures.

3.6a Timing mark location - 3.0L engine

Since the cylinder heads can be removed without pulling the engine, valve component servicing can also be accomplished with the engine in the vehicle.

In extreme cases caused by a lack of necessary equipment, repair or replacement of piston rings, pistons, connecting rods and rod bearings is possible with the engine in the vehicle. However, this practice is not recommended because of the cleaning and preparation work that must be done to the components involved.

3 Top Dead Center (TDC) for number one piston - locating

Refer to illustrations 3.6a and 3.6b

1   Top Dead Center (TDC) is the highest point in the cylinder that each piston reaches as it travels up-and-down when the crankshaft turns. Each piston reaches TDC on the compression stroke and again on the exhaust stroke, but TDC generally refers to piston position on the compression stroke. The timing marks on the vibration damper installed on the front of the crankshaft are referenced to the number one piston at TDC on the compression stroke.

2   Positioning the piston(s) at TDC is an essential part of procedures such as timing chain and sprocket replacement.

3   In order to bring any piston to TDC, the crankshaft must be turned using one of the methods outlined below. When looking at the timing-chain end of the engine, normal crankshaft rotation is clockwise. **Warning:** *Before beginning this procedure, be sure to remove the ignition key.*

a)   *The preferred method is to turn the crankshaft with a large socket and breaker bar attached to the large bolt threaded into the center of the crankshaft pulley.*

b)   *A remote starter switch, which may save some time, can also be used. Attach the switch leads to the S (switch) and B (battery) terminals on the starter motor. Once the piston is close to TDC, use a*

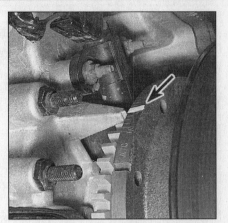

3.6b Timing mark location (arrow) - 3.8L engine

socket and breaker bar as described in the previous paragraph.

c)   *If an assistant is available to turn the ignition switch to the Start position in short bursts, you can get the piston close to TDC without a remote starter switch. Use a socket and breaker bar as described in Paragraph a) to complete the procedure.*

4   Disable the ignition system by disconnecting the primary electrical connectors at the ignition coil pack/modules (see Chapter 5).

5   Remove the spark plugs and install a compression gauge in the number one cylinder. Turn the crankshaft clockwise with a socket and breaker bar as described above.

6   When the piston approaches TDC, compression will be noted on the compression gauge. Continue turning the crankshaft until the notch in the crankshaft damper is aligned with the TDC mark on the front cover **(see illustrations)**. At this point number one cylinder is at TDC on the compression stroke. If the marks aligned but there was no compression, the piston was on the exhaust stroke. Continue rotating the engine once more.

7   After the number one piston has been positioned at TDC on the compression stroke, TDC for any of the remaining cylinders can be located by turning the crankshaft and following the firing order (refer to the Specifications). Divide the crankshaft pulley into three equal sections with chalk marks at three points, each indicating 120-degrees of crankshaft rotation. Rotating the engine 120 degrees past TDC for cylinder no. 1 will put the engine at TDC for cylinder no. 4.

4 Valve covers - removal and installation

Removal

Refer to illustrations 4.5, 4.10, 4.11a and 4.11b

1   Disconnect the negative cable from the battery.

2   Disconnect the ignition wires from the

4.5 Remove and position aside any wiring harnesses that are retained to the valve cover bolts (arrow)

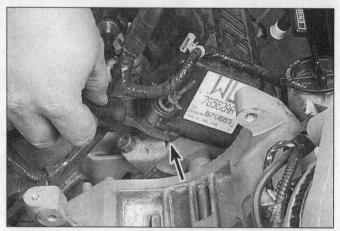

4.10 Cut carefully at these points (arrow) to separate the sealed areas of the gasket

4.11a Valve cover bolts (arrows) - 3.0L engine

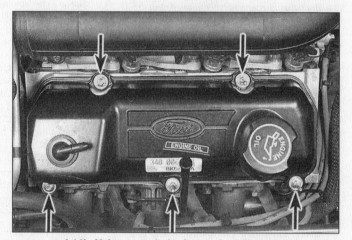

4.11b Valve cover bolts (arrows) - 3.8L engine

spark plugs. If they are not numbered, tag them for reassembly.

3   If removing the rear valve cover, refer to Chapter 11 and remove the cowl top vent panel.

4   If you're removing either valve cover on a 3.8L engine, remove the upper intake manifold (plenum) (see Chapter 4). If you're removing the right (rear) valve cover on a 3.0L engine, you'll have to remove the upper intake manifold, too.

5   Note the location of the wire routing clips and studs **(see illustration)** and pull the clips off the studs.

6   Remove the oil filler cap from the front valve cover and disconnect the PCV hose.

7   Remove the PCV valve (Chapter 6) and, on 3.8L models, position the air cleaner aside (right cover only).

8   On 3.0L models, remove the EGR tube from the exhaust manifold for access to the rear valve cover (see Chapter 6). Refer to Chapter 4 and remove the throttle body.

9   On 3.0L models, loosen, but do not remove, the valve cover bolts. **Note:** *The bolts do not come out of the cover, they are retained by the gasket.*

10   At the point (at each end of the valve cover) where the cylinder head and lower intake manifold meet on 3.0L engines, there is a section of the valve cover gasket sealed with RTV sealant. Carefully insert a thin putty knife or sharp blade and cut the silicone there without damaging the gasket **(see illustration)**.

11   Remove the bolts (3.8L) and remove the valve cover **(see illustrations)**.

Installation

Refer to illustration 4.14

12   Remove all traces of gasket material from the cylinder head and cover. Clean off any oil or dirt with acetone or lacquer thinner and a cloth. Use care when scraping aluminum components to prevent oil leaks.

13   Lightly oil all bolt threads prior to installation.

14   On 3.0L models, apply a 1/4-inch bead of RTV sealant at the cylinder head to intake manifold rail step (two places per cylinder head) **(see illustration)**.

15   Position a new gasket and install the

4.14 Apply a dab of RTV sealant at these points (arrows) on 3.0L engines

5.2 Loosen the rocker arm fulcrum bolt (arrow)

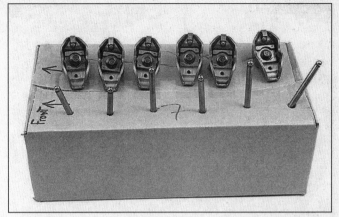

5.3 A perforated cardboard box can be used to store the components to ensure installation in their original locations

rocker cover. **Note:** *You can reuse the gaskets on 3.0L engines if they are in good condition.*

16   Tighten the bolts to the specified torque in several steps, starting with the center bolts and working towards the ends of the cover. Do not overtighten the bolts!

17   Reinstall the parts removed for access.

5   Rocker arms and pushrods - removal, inspection and installation

Removal

Refer to illustrations 5.2 and 5.3

1   Remove the valve cover(s) (refer to Section 4).

2   Loosen the rocker arm fulcrum bolts **(see illustration)**. **Note:** *If only removing the pushrods, loosen the fulcrum bolt until you can pivot the rocker arm to one side and withdraw the pushrod.*

3   Arrange to store the components in an organized manner so they can be returned to their original locations **(see illustration)**.

4   Remove the rocker arm, fulcrum and bolt as an assembly and withdraw the pushrod from the cylinder head.

Inspection

Refer to illustration 5.5

5   Clean and examine all components for wear and damage. Pushrods may be rolled over a flat surface such as a piece of glass to check for straightness. Check the fulcrums and rockers for galling and wear **(see illustration)**. Wear frequently occurs at the points where the pushrods contact the rockers. Replace any parts showing evidence of wear.

Installation

6   Prior to installation, apply moly-base grease or engine assembly lube to the ends of the pushrods, the fulcrums and the tips and pushrod seats in the rocker arms.

7   Install the pushrods, ensuring the ends are properly seated on the valve lifters.

8   Assemble the rocker arms, fulcrums and bolts and install each assembly onto the cylinder head in its proper location. Tighten the bolts hand-tight ONLY at this time. Make sure the pushrod end is properly seated in the rocker arm.

9   Position the number one cylinder at TDC on the compression stroke (see Section 3). Tighten the number one cylinder rocker arm bolts to the torque listed in this Chapter's Specifications.

10   Following the firing order, bring each of the remaining cylinders to TDC and tighten the respective rocker arm bolts to the specified torque. **Caution:** *Do not tighten a rocker arm bolt without the valve lifter positioned on the base circle of the camshaft lobe or damage to the valve train components may occur.*

11   Reinstall the valve cover(s) (Section 4).

6   Valve springs, retainers and seals - replacement

Refer to illustrations 6.4, 6.7, 6.8, 6.13 and 6.16

Note: *Broken valve springs and defective valve stem seals can be replaced without removing the cylinder head. Two special tools and a compressed air source are normally required to perform this operation, so read through this Section carefully and rent or buy the tools before beginning the job.*

1   Refer to Section 4 and remove the valve cover from the affected cylinder head. If all of the valve stem seals are being replaced, remove both valve covers.

2   Remove the spark plug from the cylinder which has the defective component. If all of the valve stem seals are being replaced, all of the spark plugs should be removed.

3   Turn the crankshaft until the piston in the affected cylinder is at top dead center on the compression stroke (see Section 3). If you are replacing all of the valve stem seals, begin with cylinder number one and work on the valves for one cylinder at a time. Move from cylinder-to-cylinder following the firing order sequence.

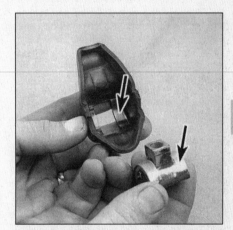

5.5 Check the rocker arm and fulcrum for wear and galling (arrows)

4   Thread an adapter into the spark plug hole and connect an air hose from a compressed air source to it. Most auto parts stores can supply the air hose adapter **(see illustration)**. **Note:** *Many cylinder compression gauges utilize a screw-in fitting that may work with your air hose quick-disconnect fitting.*

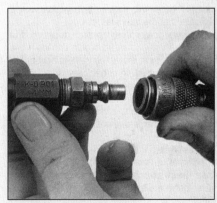

6.4 This is what the air-hose adapter that threads into the spark plug hole looks like - they're commonly available at auto parts stores

6.7 Compress the valve spring and remove the keepers with a magnet or needle-nose pliers

6.8 Once the valve spring assembly is removed, the seal (arrow) can be pulled off the valve guide

5   Remove the bolt, fulcrum and rocker arm for the valve with the defective part and pull out the pushrod. If all of the valve stem seals are being replaced, all of the rocker arms and pushrods should be removed (refer to Section 5).

6   Apply compressed air to the cylinder. The valves should be held in place by the air pressure.

7   Stuff shop rags into the cylinder head holes above and below the valves to prevent parts and tools from falling into the engine, then use a valve spring compressor to compress the valve spring. Remove the keepers with small needle-nose pliers or a magnet **(see illustration)**.

8   Remove the spring retainer shield and valve spring assembly, then remove the valve stem umbrella-type guide seal **(see illustration)**. **Note:** *If air pressure fails to hold the valve in the closed position during this operation, the valve face or seat is probably damaged. If so, the cylinder head will have to be removed for additional repair operations.*

9   Wrap a rubber band or tape around the top of the valve stem so the valve will not fall into the combustion chamber, then release the air pressure.

10   Inspect the valve stem for damage. Rotate the valve in the guide and check the end for eccentric movement, which would indicate that the valve is bent.

11   Move the valve up-and-down in the guide and make sure it doesn't bind. If the valve stem binds, either the valve is bent or the guide is damaged. In either case, the cylinder head will have to be removed for repair.

12   Reapply air pressure to the cylinder to retain the valve in the closed position, then remove the tape or rubber band from the valve stem.

13   Lubricate the valve stem with engine oil and install a new umbrella-type guide seal **(see illustration)**. 3.0L intake seals have a silver band while the exhaust seals have a red band.

14   Install the spring assembly in position over the valve.

15   Install the valve spring retainer. Compress the valve spring assembly.

16   Position the keepers in the grooves. Apply a small dab of grease to the inside of each keeper to hold it in place if necessary **(see illustration)**. Remove the pressure from the spring compressor and make sure the keepers are seated.

17   Disconnect the air hose and remove the adapter from the spark plug hole.

18   Refer to Section 5 and install the rocker arm(s) and pushrod(s).

19   Install the spark plug(s) and connect the wire(s).

20   Refer to Section 4 and install the valve cover(s).

21   Start and run the engine, then check for oil leaks and unusual sounds coming from the valve cover area. Once the lifters are fully primed (allow about five minutes at idle speed), there should be no valve train noise.

7   Intake manifold - removal and installation

Warning: *Wait until the engine is completely cool before beginning this procedure.*

Removal

Refer to illustrations 7.6, 7.7a and 7.7b

1   Disconnect the cable from the negative battery terminal. Drain the cooling system (see Chapter 1). Remove the upper intake manifold (plenum) and throttle body (see Chapter 4). **Note:** *Make sure to tag all of the electrical and vacuum connections.*

2   Disconnect the coolant hoses from the intake manifold. **Note:** *The bypass hose on the 3.8L engine is bolted to the front of the intake manifold. The rubber hose can be disconnected from the steel tube, but if the tube is pulled out, it should be reinstalled with a new O-ring.*

3   Disconnect the electrical connectors from the fuel injectors and the fuel lines from the fuel rails (see Chapter 4). **Note:** *The fuel injectors and fuel rails may be left in place on the lower manifold during removal. They only need to be removed if they are being transferred to a new manifold.*

4   On 3.0L engines, mark and remove the camshaft position sensor (see Chapter 6).

5   On 3.0L engines, remove the valve covers (see Section 4). On 1995 through 1998 models, loosen the number 3 intake rocker arm bolt enough to extract the pushrod. On 1999 and later models, remove loosen all rocker arm bolts and remove all push rods (see Section 5). The pushrod(s) go through the intake manifold and must be removed before removing the intake manifold.

6   Mark and disconnect the electrical connectors from the two temperature sensors and any other electrical connectors on the manifold. On 3.8L engines, disconnect the

6.13 Carefully seat the valve seal using a deep socket and hammer

6.16 Apply a small dab of grease to each keeper as shown here before installation - it'll hold them in place on the valve stem as the spring is released

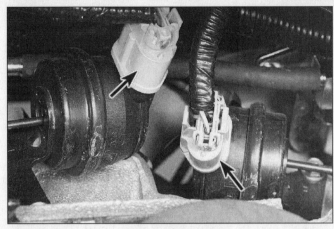

7.6 On 3.8L engines, disconnect the electrical connectors from the IMRC solenoids

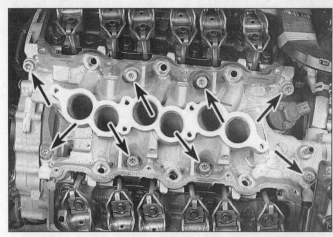

7.7a Lower intake manifold bolts (arrows) - 3.0L engine

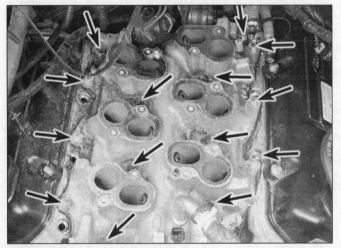

7.7b Lower intake manifold bolts (arrows) - 3.8L engine

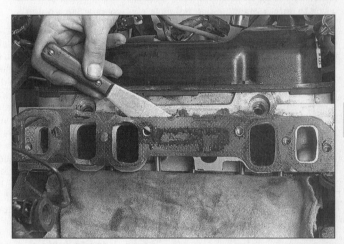

7.9 Cover the intake valley with clean shop rags, and use a scraper to remove the intake manifold gaskets

2A

electrical connectors and vacuum hoses from the IMRC (Intake Manifold Runner Control) motors at the front of the manifold **(see illustration)**.

7   Remove the lower intake manifold mounting bolts/studs (this requires a Torx-driver bit on 3.0L models), noting the locations of the studs (if equipped) for reinstallation **(see illustrations)**.

8   Remove the intake manifold. It may be necessary to pry on the end of the manifold with a small prybar to break the gasket seal. Use care to avoid damaging machined surfaces.

Installation

Refer to illustrations 7.9, 7.11, 7.12, 7.14a and 7.14b

9   Clean away all traces of old gasket material **(see illustration)**. Remove oil and dirt with a cloth and solvent, such as acetone or lacquer thinner.

10   Lightly oil all bolts and studs prior to assembly.

11   Install the rubber end seals to the block, using a thin layer of RTV sealant **(see illustration)**. **Note:** *The rear end seal on 3.0L engines must fit around the hole for the*

7.11 Install the rubber end seals over a thin bead of RTV sealant, then apply a thin bead on top (arrow)

7.12 Apply a dab of RTV sealant to the corners where the manifold gaskets meet the rubber end seals (3.8L shown)

camshaft position sensor. It only installs one way with proper clearance for the sensor.

12   Position the new manifold gaskets on the engine, engaging their notches to the tabs. Be sure the locating tabs engage properly with the tabs on the corners of the cylin-

der head gaskets, and apply a dab of RTV sealant where the manifold gaskets meet the rubber end seals **(see illustration)**. **Note:** *Assembly must be completed within several minutes. Don't allow the RTV sealant to dry.*

13   Carefully set the lower manifold into

7.14a Intake manifold bolt tightening sequence - 3.0L engine

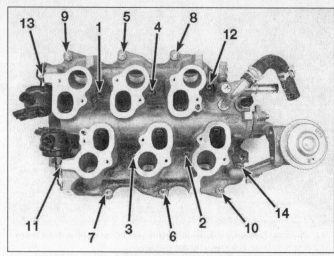

7.14b Intake manifold bolt tightening sequence - 3.8L engine

place. Be sure the gaskets don't shift out of place. Install the bolts and studs in their original locations.

14   Tighten the bolts/studs in the recommended sequence **(see illustrations)** to the torque listed in this Chapter's Specifications.

15   Reinstall all parts removed for access in the reverse order of removal.

16   Refill the cooling system and start the engine, checking for fuel, vacuum and coolant leaks.

8   Exhaust manifolds - removal and installation

Removal

Refer to illustrations 8.2, 8.4 and 8.6

1   Disconnect the negative battery cable.

2   From below, disconnect the exhaust pipe(s) from the exhaust manifold(s) **(see illustration)**.

3   Remove the spark plugs (Chapter 1).

4   If removing the left bank (front) manifold, unbolt the oil dipstick tube bracket **(see illustration)**.

5   If removing the left bank (front) manifold on a 3.8L engine or a right bank (rear) manifold, on a 3.0L engine, disconnect the EGR tube from the manifold (see Chapter 6).

6   If removing the right bank (rear) manifold **(see illustration)**, refer to Chapter 11 and remove the cowl top vent panel.

7   If removing the right bank (rear) manifold, on a 3.8L engine, remove the air cleaner and air cleaner duct (see Chapter 1).

8   Unbolt and remove the exhaust manifold from the vehicle. **Note:** *The bolts should be soaked with penetrating oil before removing, and keep track of where the studs are located.*

Installation

9   Clean all gasket surfaces thoroughly and inspect the manifold(s) for cracks and damage. Check the fasteners and bolt holes for stripped or damaged threads.

10   Lightly oil all bolts prior to installation.

11   On 3.8L engines, position a new gasket in place on the cylinder head (3.0L engines do not have an exhaust manifold gasket). Install the exhaust manifolds. On 3.8L engines, use a pilot bolt (left side - lower front

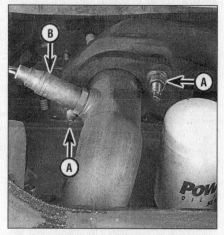

8.2 Remove the nuts (A) and separate the exhaust pipe from the exhaust manifold - disconnect the oxygen sensor (B)

bolt on number 5 cylinder, right side - lower rear bolt on number 2 cylinder) to aid in alignment. **Note:** *Slight warpage of the exhaust manifold may cause a misalignment between the bolt holes in the cylinder head and mani-*

8.4 Unbolt the oil dipstick tube bracket nut (A), and the wiring clip to the bracket (B) - 3.0L shown, 3.8L similar

8.6 Remove the bolts (arrows indicate the three top bolts) and the manifold (3.0L shown, 3.8L similar)

9.3 Mark the pulley and damper relationship, then remove the four pulley bolts (arrows)

9.5 Remove the vibration damper with a bolt-type puller

9.6 Carefully pry the seal out of the bore - DO NOT nick or scratch the crankshaft

9.8 The seal can be installed with a large socket and hammer

fold. *Elongate the holes in the manifold with a round file as necessary to correct the misalignment. Do not elongate the pilot hole.*

12   Install the remaining bolts and studs in their correct locations and tighten them to the torque listed in this Chapter's Specifications, starting with the two bolts for the center exhaust port, then each end port.

13   Reinstall the remaining parts in the reverse order of removal.

14   Start the engine and check for exhaust leaks.

9   Crankshaft front oil seal - replacement

Refer to illustrations 9.3, 9.5, 9.6 and 9.8

1   Disconnect the negative cable from the battery and remove the drivebelt(s) (Chapter 1).

2   Raise the front of the vehicle and support it securely on jackstands. **Warning:** *Some models covered by this manual are equipped with an air suspension system. Always disconnect the electrical power to the suspension system before lifting or towing (see Chapter 10). Failure to perform this procedure may*

result in unexpected shifting or movement of the vehicle, which could cause personal injury. Remove the right front wheel. Remove the plastic inner fender liner (Chapter 11).

3   Mark the vibration damper and pulley so the pulley can be reinstalled in the same relative position. Remove the four bolts attaching the lower pulley to the vibration damper **(see illustration)** and remove the pulley.

4   Remove the starter motor (see Chapter 5) and have an assistant wedge a large screwdriver in the teeth of the driveplate while you loosen the large damper bolt. **Caution:** *Considerable torque will be required to break this bolt loose, use a breaker bar and make sure your hands are protected in case the bolt loosens suddenly.*

5   Remove the vibration damper with a puller, available at most auto parts stores or tool rental facilities. **Caution:** *Don't use a gear-type puller as it will damage the damper. Use a puller that bolts into the hub* **(see illustration)**.

6   Carefully pry out the old seal with a screwdriver or seal puller **(see illustration)**.

7   Clean and inspect the seal bore and crankshaft surfaces for damage, nicks, burrs or other roughness which may cause a new seal to fail. Correct as necessary. Inspect the teeth of the crankshaft position sensor ring for any damage, and look for cracks in the front cover. **Note:** *If the damper on 3.8L engines is to be replaced with a new one, transfer any balancing pins from the old unit to the new one, making sure they are installed in the same relative positions as on the original damper to maintain proper balance.*

8   Lubricate the new seal lip with multipurpose grease and the outside edge of the seal with engine oil, and carefully tap the seal into place using a large socket and a hammer **(see illustration)**.

9   Apply a dab of RTV sealant to the keyway in the damper and position the damper on the crankshaft. Be sure the keyway is aligned with the crankshaft key. Install the damper using an installation tool. If unavailable, start the damper on with a soft-faced hammer and finish installation using the

damper retaining bolt. Tighten the bolt to the torque listed in this Chapter's Specifications.

10   Reinstall the remaining parts in the reverse order of removal.

11   Run the engine and check for oil leaks.

10   Timing chain and sprockets - check, removal and installation

2A

Timing chain deflection check

Refer to illustration 10.3
Note: *Timing chain deflection increases due to wear. The following check is a method of measuring wear without disassembling the engine.*

1   Remove the left (front) valve cover (Section 4).

2   Loosen the number 5 cylinder exhaust rocker arm bolt. This is the fourth rocker arm from the drivebelt end of the engine. Rotate the rocker arm aside.

3   Install a dial indicator on the end of the pushrod **(see illustration)**.

10.3 Install a dial indicator to measure timing chain deflection - use a short length of vacuum hose to hold the plunger over the pushrod end, if you encountered difficulty keeping the plunger on the pushrod

10.16a Timing chain cover bolts/studs (arrows) - 3.0L engines

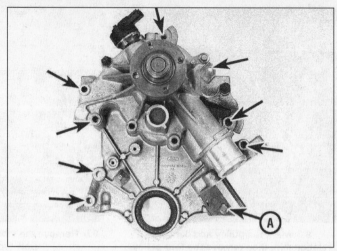

10.16b Timing chain bolt/stud locations (arrows) - 3.8L engines - (A) is the socket-head bolt hidden behind the oil pump, don't overlook it

10.18 Scrape all traces of gasket and sealant from the block and cover

10.19 With the cover off (3.8L), inspect the camshaft position sensor driven gear (A) and shaft (B)

10.20a Align the timing marks (arrows) on the camshaft and crankshaft sprockets

4   Rotating the crankshaft clockwise, position the engine with cylinder number one at TDC on the compression stroke (see Section 3).

5   Zero the dial indicator.

6   Slowly turn the crankshaft counterclockwise until the first movement is seen on the dial indicator. Stop and observe the timing marks to determine the number of degrees from TDC.

7   If the reading exceeds 6 degrees, replace the timing chain and sprockets.

Timing chain cover removal

Refer to illustrations 10.16a, 10.16b, 10.18 and 10.19

Note: *The timing chain cover can't be removed on the 3.8L engine with the engine in-vehicle. Refer to Part B of this Chapter for engine removal.*

8   Disconnect the negative cable from the battery (see Chapter 1).

9   Drain the cooling system and remove the drivebelt(s) (Chapter 1). **Note:** *Loosen the water pump pulley bolts before removing the drivebelt.*

10   Loosen the right front wheel lug nuts,

raise the vehicle and support it securely on jackstands. Remove the right front wheel and inner fender liner (Chapter 11).

11   Drain the engine oil (see Chapter 1), then refer to Section 14 and remove the oil pan.

12   Refer to Section 3 and position the engine with number one cylinder at TDC on the compression stroke, then refer to Section 9 and remove the crankshaft pulley and vibration damper.

13   Remove the drivebelt tensioner (see Chapter 1).

14   Remove the water pump pulley and disconnect the coolant hoses from the water pump (see Chapter 3).

15   Refer to Chapter 6 and disconnect the electrical connector from the crankshaft position sensor on 3.0L engines. On 3.8L engines, remove the camshaft position sensor (see Chapter 6).

16   Remove the timing chain cover bolts **(see illustrations)**. Unless the water pump is being replaced, it isn't necessary to remove all the water pump bolts. **Note:** *On 3.8L engines, remove the oil filter and oil cooler (see Chapter 3), setting the cooler aside with*

the hoses still connected.

17   Tap the cover loose with a soft face hammer and remove it from the engine. **Caution:** *Do not use excessive force or you may crack the cover. If the cover is difficult to remove, recheck for remaining bolts.*

18   Thoroughly clean and inspect all parts and remove all traces of gasket material **(see illustration)**. Remove oil film with a solvent such as lacquer thinner, acetone or brake system cleaner.

19   The oil pump is mounted in the timing chain cover on 3.8L engines. See Section 15 for further information. To remove the intermediate shaft from the front cover, remove the clip from the shaft and slide the shaft out of the cover **(see illustration)**.

Timing chain and sprockets removal

Refer to illustrations 10.20a, 10.20b, 10.21 and 10.24

20   Check that the upper and lower timing chain sprocket marks are aligned **(see illustrations)**. If they are not, install the vibration damper bolt and use it to turn the crankshaft clockwise until the two marks are adjacent to

10.20b On 3.8L engines, when the timing marks are aligned at TDC, the balance shaft keyway (arrow) will point straight up

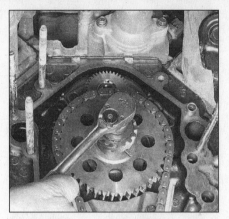

10.21 Remove the camshaft sprocket bolt without rotating the engine

10.24 Compress the timing chain tensioner with a screwdriver and insert an Allen wrench or small punch as shown to retain it (3.8L engine)

10.25 Assemble the chain and both sprockets with the marks aligned (arrows), then slip the assembly onto the camshaft and crankshaft

each other.

21 Remove the camshaft sprocket retaining bolt and (3.8L models only) the camshaft position sensor drive gear (see illustration).

22 Pull the camshaft and crankshaft sprockets away from the engine and remove the two sprockets with the chain. If the

crankshaft sprocket won't come off by hand, carefully pry it off with two screwdrivers.

23 3.8L engines are equipped with a balance shaft, driven by a gear behind the camshaft sprocket. It is not necessary to remove the balance shaft gears for chain replacement. See Part B of this Chapter for balance shaft timing, if necessary.

24 On 3.8L engines, retract the timing chain tensioner and install a pin in the tensioner to hold it in this position (see illustration).

Installation

Refer to illustrations 10.25 and 10.27

25 Assemble the new chain and sprockets with their marks aligned (see illustration). **Caution:** *Severe engine damage could result from improper timing.*

26 Slip the timing chain and sprocket assembly onto the camshaft and crankshaft. If the crankshaft sprocket goes on tight, tap it into place using a large socket and hammer.

27 On 3.8L engines, install the camshaft position sensor drive gear, aligning its keyway (see illustration).

28 Install the camshaft sprocket bolt and washer and tighten it to the torque listed in this Chapter's Specifications.

29 On 3.8L engines, remove the pin retain-

ing the timing chain tensioner (see illustration 10.24).

30 Position a new gasket on the block and install the timing chain cover.

31 On 3.8L engines, install the camshaft position sensor (see Chapter 6).

32 Tighten the bolts to the torque listed in this Chapter's Specifications.

33 Reinstall the remaining parts in the reverse order of removal.

34 Add oil and coolant as needed, start the engine and check for leaks.

11 Valve lifters - removal, inspection and installation

Removal

Refer to illustrations 11.4, 11.5 and 11.6

1 Refer to Section 3 and position the engine with number one cylinder at TDC on the compression stroke.

2 Remove the intake manifold (Section 7).

3 Remove the rocker arms and pushrods (Section 5).

4 Unbolt and remove the guide plate retainer and guide plates (see illustration).

2A

10.27 Align the camshaft position sensor gear's keyway (arrow) with the Woodruff key on the camshaft (3.8L engine)

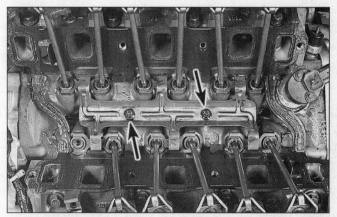

11.4 Remove the lifter guide retainer plate bolts (arrows) - 3.0L shown, 3.8L has two retainers, one on each bank

11.5 You may be able to remove the lifters with a magnet

11.6 Be sure to store the lifters in an organized manner to make sure they are reinstalled in their original locations

11.8a Inspect the pushrod seat (arrow) in the top of each lifter for wear

5   There are several ways to extract the lifters from the bores. Special tools designed to grip and remove lifters are available from most auto parts stores. but may not be needed in every case. On newer engines without a lot of varnish buildup, the lifters can often be removed with a small magnet **(see illustration)** or even with your fingers. A machinist's scribe with a bent end can be used to pull the lifters out by positioning the point under the retainer ring in the top of each lifter. **Caution:** *Do not use pliers to remove the lifters unless you intend to replace them with new ones (along with the camshaft). The pliers may damage the precision machined and hardened lifters, rendering them useless. On engines with considerable gum and varnish, work the lifters up and down, using carburetor cleaner spray to loosen the deposits.*

6   Before removing the lifters, arrange to store them in a clearly labeled box to ensure that they are reinstalled in their original locations **(see illustration). Note:** *Be sure to mark the front of each lifter before removal (if there's not already a notch or protrusion) so the rollers will turn in the same direction when reinstalled.*

Inspection

Refer to illustrations 11.8a and 11.8b

7   Clean the lifters with solvent and dry them thoroughly while still keeping them in order.

8   Check each lifter wall, pushrod seat and roller for scuffing, score marks and uneven wear **(see illustrations)**.

Installation

Refer to illustration 11.10

9   If new lifters are being installed, a new camshaft must also be installed. If a new camshaft is installed, then use new lifters as well. Never install lifters unless the original camshaft is used and the lifters can be installed in their original locations.

10   Lubricate the lifters with clean engine oil before installation. The remainder of the installation is the reverse of the removal process. **Note:** *When installing the guide plates,*

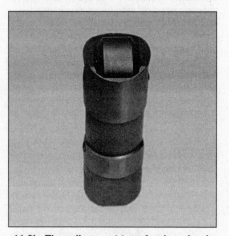

11.8b The roller must turn freely - check for wear and excessive play as well

make sure the word UP is showing **(see illustration)**.

---

12 Camshaft, balance shaft and bearings - removal, inspection and installation

The camshaft (and balance shaft on 3.8L engines) can't be removed with the engine in-vehicle. Refer to Part B of this Chapter for camshaft and balance shaft procedures.

---

13 Cylinder heads - removal and installation

Warning: *Wait until the engine is completely cool before beginning this procedure.*
Note: *Purchase new cylinder head bolts before beginning reassembly.*

Removal

Refer to illustrations 13.10, 13.11 and 13.15

1   Drain the cooling system (Chapter 1).

11.10 Install the guide plates so that the word UP is showing

2   Disconnect the negative cable from the battery.

3   Remove the air cleaner assembly.

4   Remove the drivebelt and drivebelt tensioner (Chapter 1).

5   Mark and remove the spark plug wires and the spark plugs (see Chapter 1).

6   Refer to Section 4 and remove the valve covers.

7   Refer to Section 7 and remove the intake manifold.

8   Refer to Section 8 and remove the exhaust manifold(s).

9   If removing the left bank (front) cylinder head, refer to Chapter 5 and remove the alternator and alternator mounting bracket.

10   If removing the left bank (front) cylinder head, remove the power steering pump and its bracket and set aside, without disconnecting the hoses **(see illustration)**.

11   If removing the right bank (rear) cylinder head on a 3.0L engine, unbolt the EGR vacuum regulator solenoid bracket and set aside, then move the speed sensor cable out of the way **(see illustration)**.

12   If removing the right bank (rear) cylinder head on a 3.8L engine, unbolt the power

13.10 Remove the bolts/nuts (arrows) from the power steering support bracket on the left bank (front) cylinder head

13.11 Remove the bolts (arrows) and set aside the EGR vacuum regulator solenoid and its mount - 3.0L engine

13.15 Once the bolts are removed, pry the cylinder head loose at a point where the gasket surfaces won't be damaged

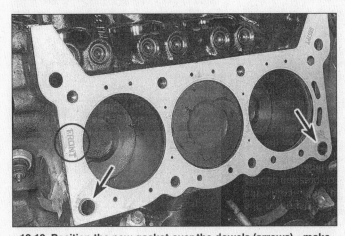

13.19 Position the new gasket over the dowels (arrows) - make sure the UP or FRONT mark is visible

13.21a Cylinder head bolt tightening sequence - 3.0L engine

2A

steering line mounting bracket and refer to Chapter 5 to remove and set aside the ignition coil-pack.

13 Loosen the rocker arm fulcrum bolts enough to allow the rocker arms to be lifted off the pushrods and rotate them to one side.

14 Remove the pushrods (Section 5). Store them so they can be reinstalled in the same location.

15 Remove and DISCARD the cylinder head bolts following the reverse of the tightening sequence (see illustrations 13.21a and 13.21b). Lift the cylinder head(s) off the engine (see illustration). Make sure there are no ground wires attached before removing the cylinder heads.

Installation

Refer to illustrations 13.19, 13.21a and 13.21b

16 Thoroughly remove all traces of gasket material with a gasket scraper and clean all parts with solvent. Use a rag and acetone, lacquer thinner or brake system cleaner to remove any traces of oil. See Chapter 2 Part B for cylinder head inspection procedures.

17 Use a tap of the correct size to chase the threads in the cylinder head bolt holes.

18 Recheck all cylinder head bolt holes and cylinder bores for any traces of coolant, oil or other foreign matter. Remove as needed.

19 Position the new gasket over the dowel pins on the block. The top of the gasket should be stamped TOP or THIS SIDE UP to ensure correct installation (see illustration). Don't use sealant on the gaskets.

20 Install the NEW cylinder head bolts finger tight. Caution: New cylinder head bolts must be installed whenever the cylinder heads are removed.

21 Following the sequence shown (see illustrations), tighten the cylinder head bolts in several stages to the torque listed in this Chapter's Specifications.

22 Reinstall the parts removed in the reverse order of removal. Lubricate the rocker arm components with moly-based assembly lube or high viscosity engine oil.

23 Install the pushrods and rocker arms in their original locations. Caution: Be sure to follow the recommended rocker arm tightening procedure (see Section 5).

13.21b Cylinder head bolt tightening sequence - 3.8L engine

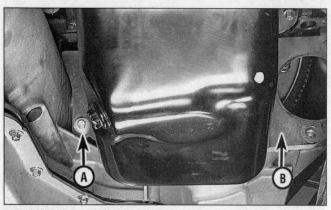

14.7 After the starter is removed, remove the bolt (A) and remove the converter housing rear cover (B)

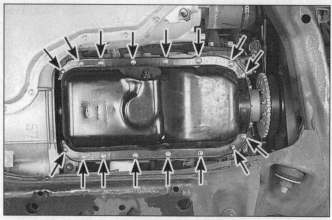

14.8 Remove the oil pan bolts (arrows) - 3.0L shown, 3.8L similar

24   Refill the cooling system, change the oil and filter (Chapter 1). Start the engine and inspect for leaks.

14 Oil pan - removal and installation

Removal

Refer to illustrations 14.7 and 14.8

1   Disconnect the negative cable from the battery.
2   Remove the oil dipstick.
3   Raise the vehicle and support it securely on jackstands. **Warning:** *Some models covered by this manual are equipped with an air suspension system. Always disconnect the electrical power to the suspension system before lifting or towing (see Chapter 10). Failure to perform this procedure may result in unexpected shifting or movement of the vehicle, which could cause personal injury.*
4   Drain the oil and remove the oil filter (see Chapter 1). If equipped with a low oil level sensor, remove the retainer clip from the oil pan sensor and unplug the wire harness from the sensor.
5   Disconnect the oxygen sensors (see Chapter 6), then remove the catalytic converter Y-pipe assembly (see Chapter 4).

6   Remove the starter motor (see Chapter 5).
7   Remove the rear cover from the converter housing on the transaxle **(see illustration)**.
8   Unbolt the oil pan and remove it from the vehicle **(see illustration)**. If the pan is difficult to break loose, tap on it with a rubber mallet.

Installation

Refer to illustration 14.11

9   Remove all traces of gasket material from the mating surfaces and clean the oil pan with solvent.
10   On 3.0L engines, install a new, one-piece gasket on the oil pan using gasket adhesive to hold it in place.
11   On 3.8L engines, install a new rubber end seal on the rear main bearing cap **(see illustration)** and a bead of RTV sealant around the perimeter of the oil pan flange.
12   On 3.0L engines, apply a 1/4-inch bead of RTV sealant to the junctions of the engine block/rear main bearing cap and the timing chain cover/engine block for a total of four places. **Note:** *Install the oil pan within five minutes of applying the sealant.*
13   Position the oil pan on the engine block and install the bolts finger-tight. When all

bolts are in place, tighten them to the torque listed in this Chapter's Specifications. **Note:** *On 3.0L engines, loosen all the bolts and then tighten a second time to the specified torque.*
14   Reinstall the remaining parts in the reverse order of removal.
15   Install a new oil filter, add oil and start the engine. Check for oil leaks.

15 Oil pump and pickup - removal and installation

3.0L engine

Refer to illustration 15.2

Removal

1   Raise the vehicle and support it securely on jackstands. **Warning:** *Some models covered by this manual are equipped with an air suspension system. Always disconnect the electrical power to the suspension system before lifting or towing (see Chapter 10). Failure to perform this procedure may result in unexpected shifting or movement of the vehicle, which could cause personal injury.*
2   Remove the oil pan (see Section 14). Remove the oil pump mounting bolt **(see illustration)**.
3   Lower the oil pump assembly from the

14.11 On 3.8L engines, install a new rubber rear seal on the rear main cap, with a dab of RTV on each side (arrows)

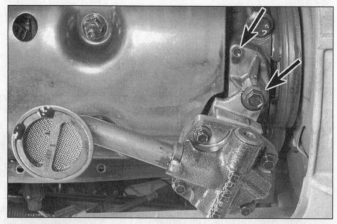

15.2 Oil pump mounting bolt and locating dowel (arrows) - 3.0L engine

15.11 Remove the oil pump cover/oil filter adapter mounting bolts (arrows)

15.12 To detach the oil pickup tube, remove the nut and bolts (arrows)

block. The oil pump intermediate shaft will come down with the pump.

Installation

4   Prime the pump by pouring oil into the oil pickup and turning the pump shaft by hand.

5   Fit the oil pump driveshaft into the pump until the shaft seats completely in the pump.

6   Install the oil pump assembly, inserting the intermediate shaft through the hole in the rear main bearing cap and engage the shaft with the camshaft position sensor synchronizer. If the shaft does not engage, turn the pump slightly and try again, do not try to force it.

7   Position the pump over the two locating dowels and tighten the bolt to the torque listed in this Chapter's Specifications.

8   Reinstall the oil pan (Section 14). Install a new filter and add engine oil. Start the engine and check for leaks.

3.8L engine

Refer to illustrations 15.11, 15.12 and 15.14

Removal

9   The oil pump and intermediate (pump drive) shaft is mounted in the timing chain cover. Intermediate shaft removal is included in Section 10.

10   Remove the oil filter (Chapter 1) and remove the oil cooler assembly (see Chapter 3).

11   Remove the oil pump/oil filter adapter from the front cover **(see illustration)**. Clean and inspect the oil pump cavity. If the oil pump gear pocket in the timing chain cover is damaged or worn or scored, replace the timing chain cover.

12   The oil pump pickup is mounted within the oil pan. For access, remove the oil pan (see Section 14). Remove the pickup tube bracket nut and the two mounting bolts **(see illustration)**. Lower the pickup from the engine.

Installation

13   Install the pump gears in the housing so they are seated flush with the cover surface.

If the gears don't seat fully, check the drive gear where it connects to the intermediate shaft. It probably wasn't engaged properly.

14   Pack the pump cavity with petroleum jelly, filling the space completely around the gears. Install the pump/adapter to the timing cover, using a new O-ring **(see illustration)**. Tighten the bolts to the torque listed in this Chapter's Specifications.

15   Reinstall the engine oil cooler (if equipped). Install a new oil filter and add engine oil. Start the engine and check for leaks.

---

16   Driveplate - removal and installation

Removal

Refer to illustration 16.3

1   Raise the vehicle and support it securely on jackstands, then refer to Chapter 7 and remove the transaxle. **Warning:** *Some models covered by this manual are equipped with an air suspension system. Always disconnect the electrical power to the suspension system before lifting or towing (see Chapter 10). Failure to perform this procedure may result in unexpected shifting or movement of the vehicle, which could cause personal injury.*

2   Look for factory paint marks that indicate driveplate-to-crankshaft alignment. If they aren't there, use a center-punch or paint to make alignment marks on the driveplate and crankshaft to ensure correct alignment during reinstallation.

3   Remove the bolts that secure the driveplate to the crankshaft **(see illustration)**. If the crankshaft turns, use a driveplate-holding tool or wedge a screwdriver through one of the driveplate openings.

4   Remove the driveplate from the crankshaft. **Warning:** *The ring gear teeth may be sharp, wear gloves to protect your hands.*

5   Clean the driveplate with brake cleaner to remove grease and oil. Inspect the surface for cracks. Check for cracked and broken ring gear teeth. Lay the driveplate on a flat

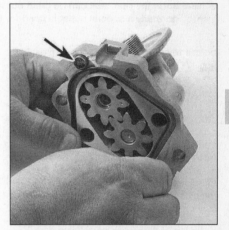

15.14 Install a new O-ring (arrow) lubricated with clean engine oil before reinstalling the adapter

surface and use a straightedge to check for warpage.

6   Clean and inspect the mating surfaces of the driveplate and the crankshaft. If the crankshaft rear seal is leaking, replace it

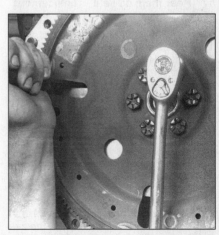

16.3 Insert a prybar through one of the holes in the driveplate to keep it from turning while removing the driveplate-to-crankshaft bolts

2A

17.2 If you're very careful not to damage the crankshaft or the seal bore, the rear seal can be pried out with a screwdriver - normally a special puller is used for this procedure

18.6 The three-bar engine support mounts above the engine, supporting it against the body so that engine mounts or subframe work can be performed safely

before reinstalling the driveplate (see Section 17).

Installation

7　Position the driveplate against the crankshaft. Be sure to align the marks made during removal. Before installing the bolts, apply Teflon thread sealant to the threads. **Note:** *On 3.8L engines only, if the driveplate is being replaced with a new one, check for balancing pins on the old driveplate. Transfer the pins to the same relative positions on the new driveplate to maintain proper engine balance.*

8　Use a driveplate-holding tool or wedge a screwdriver through the starter motor opening to keep the driveplate from turning as you tighten the bolts to the torque listed in this Chapter's Specifications.

9　The remainder of installation is the reverse of the removal procedure.

17　Rear main oil seal - replacement

Refer to illustration 17.2

1　Raise the vehicle and support it securely on jackstands. **Warning:** *Some models covered by this manual are equipped with an air suspension system. Always disconnect the electrical power to the suspension system before lifting or towing (see Chapter 10). Failure to perform this procedure may result in unexpected shifting or movement of the vehicle, which could cause personal injury.* Refer to Chapter 7 and remove the transaxle, then detach the driveplate from the engine (Section 16).

2　The old seal can be removed by prying it out with a screwdriver **(see illustration)** or by making one or two small holes in the seal flange with a sharp pick, then using a screw-in type slide-hammer puller. Be sure to note how far the seal is recessed into the bore before removing it; the new seal will have to

be recessed an equal amount. **Caution:** *Be very careful not to scratch or otherwise damage the crankshaft or the bore in the housing or oil leaks could develop!*

3　Clean the crankshaft and seal bore with lacquer thinner or acetone. Check the seal contact surface on the crankshaft very carefully for scratches and nicks that could damage the new seal lip and cause oil leaks. If the crankshaft is damaged, the only alternative is a new or different crankshaft.

4　Make sure the bore is clean, then apply a thin coat of engine oil to the outer edge of the new seal. Apply multi-purpose grease to the seal lips. The seal must be pressed squarely into the bore; a special seal installation tool available at auto parts stores is highly recommended. Hammering it into place is not recommended. If you don't have access to the special tool, you may be able to tap the seal in with a large section of pipe and a hammer. If you must use this method, be very careful not to damage the seal or crankshaft! And work the seal lip carefully over the end of the crankshaft with a blunt tool such as the rounded end of a socket extension.

5　Reinstall the driveplate and the transaxle.

18　Engine mounts - check and replacement

Warning: *Do not place any part of your body under the engine when it is supported only by a jack. Jack failure could result in severe injury or death!*

Note: *Transaxle mount replacement is covered in Chapter 7.*

Check

1　The engine mounts may be inspected with the engine in the vehicle.

2　Disconnect the negative cable at the battery.

3　Raise the vehicle and support it securely on jackstands. **Warning:** *Some models covered by this manual are equipped with an air suspension system. Always disconnect the electrical power to the suspension system before lifting or towing (see Chapter 10). Failure to perform this procedure may result in unexpected shifting or movement of the vehicle, which could cause personal injury.*

4　Position a jack under the engine oil pan, using a block of wood to protect the pan.

5　Raise the engine slightly to take the weight off the engine mounts. Inspect the mounts for cracks and separation. Sometimes the rubber will split right down the center. Replace as needed. **Note:** *Whenever self-locking fasteners are removed, replace them with new self-locking fasteners.*

Replacement

Refer to illustrations 18.6, 18.10, 18.13 and 18.15

Front engine mount

6　When engine mounts are replaced, the engine should be supported from above by a three-bar support fixture **(see illustration)**. These can be rented at most tool rental yards. Attach a chain to the front and rear of the engine from the support.

7　Raise the vehicle and support it securely on jackstands.

8　On models equipped with air conditioning, unbolt the compressor (Chapter 3) without disconnecting the refrigerant lines and set it aside.

9　Remove the nut or through-bolt attaching the engine mount to the compressor bracket **(see illustration 18.10)**.

10　Remove the lower nut from the subframe **(see illustration)**.

11　Raise the engine one inch with the three-bar support fixture and remove the old mount insulator and install the new one. Tighten the nuts and/or through-bolt to the torque listed in this Chapter's Specifications.

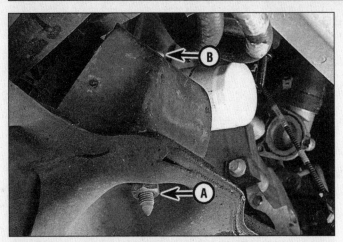

18.10 Remove the lower nut (A) from the subframe and the upper nut or through-bolt (B) from the bracket

18.13 Remove the nut (arrow) for the rear mount at the subframe

Rear engine mount

12   Raise the vehicle and support it securely on jackstands.

13   Remove the nut attaching the engine mount to the subframe **(see illustration)**.

14   Loosen the attaching nut on the rear engine mount and heat shield assembly, and install a three-bar engine support **(see illustration 18.6)**.

15   Use the threaded hooks (on the support fixture) to raise the engine by the alternator brace one inch, and remove the rear mount insulator **(see illustration)**. **Note:** *Some models may have a heat shield that also must be removed.*

16   Install the new insulator with the heat shield and lower the engine. **Note:** *On 3.8L models, the insulator has a stud at top and bottom and a locating pin at the top. Make sure the locating pin fits into the hole in the upper bracket before tightening the upper nut.*

17   Tighten the nuts to the torque listed in this Chapter's Specifications.

2A

18.15 Remove the through-bolt (A), then the insulator (B)

Notes

Chapter 2 Part B
General engine overhaul procedures

Contents

Specifications

General

3.0L V6 engine

General

Cylinder compression pressure	101 psi minimum (see text)
Oil pressure (hot at 2500 rpm)	40 to 60 psi

Cylinder head and valves

Warpage limit	0.003 inch in any 6 inches (0.007 overall)
Minimum valve margin	1/32 inch
Intake valve	
Stem diameter	
Standard	0.3134 to 0.3126 inch
0.015 oversize	0.3283 to 0.3276 inch
0.030 oversize	0.3433 to 0.3425 inch
Stem-to-guide clearance	0.0010 to 0.0028 inch
Exhaust valve	
Stem diameter	
Standard	0.3129 to 0.3121 inch
0.015 oversize	0.3279 to 0.3271 inch
0.030 oversize	0.3428 to 0.3420 inch
Stem-to-guide clearance	0.0015 to 0.0032 inch
Valve spring pressure (not including dampener)	
Valve open	180 lbs at 1.16 inches
Valve closed	65 lbs at 1.58 inches
Valve spring installed height	1.736 to 1.650 inches
Valve lifter	
Diameter (standard)	0.8740 to 0.8745 inch
Lifter-to-bore clearance	0.0007 to 0.0027 inch
Collapsed lifter clearance	0.09 to 0.19 inch

3.0L V6 engine (continued)

Crankshaft and connecting rods

Connecting rod journal
Diameter..	2.1253 to 2.1261 inches
Out-of-round limit...	0.0003 inch
Taper limit..	0.0006 inch

Bearing oil clearance
Desired..	0.0010 to 0.0014 inch
Allowable..	0.0008 to 0.0027 inch

Connecting rod side clearance (endplay)
Standard..	0.006 to 0.014 inch
Service limit..	0.014 inch

Main bearing journal
Diameter..	2.5190 to 2.5198 inches
Out-of-round limit...	0.0003 inch
Taper limit..	0.0006 inch
Runout limit..	0.002 inch

Bearing oil clearance
Desired..	0.0001 to 0.0014 inch
Allowable..	0.0005 to 0.0023 inch

Crankshaft endplay
Standard..	0.004 to 0.008 inch
Service limit..	0.012 inch maximum

Pistons and rings

Piston-to-bore clearance limits
Standard..	0.0012 to 0.0022 inch
Service limit..	0.0031 inch

Piston ring end gap
Compression rings	0.010 to 0.020 inch
Oil ring..	0.010 to 0.049 inch

Side clearance
Compression rings	0.0016 to 0.0037 inch
Oil rings	Snug fit

Camshaft

Lobe lift
Intake..	0.260 inch
Exhaust..	0.260 inch
Allowable lobe lift loss	0.005 inch
Endplay, service limit...	0.007 inch
Journal-to-bearing clearance	0.001 to 0.003 inch
Journal diameter (all)	2.0074 to 2.0084 inches
Cam bearing inside diameter...	2.0094 to 2.0104 inches
Runout limit..	0.005 inch (runout of no. 2 or no. 3 relative to no. 1 and no. 4)
Out-of-round limit...	0.004 inch

Bearing bore inside diameter
No. 1...	2.1531 to 2.1541 inches
No. 2...	2.1334 to 2.1344 inches
No. 3...	2.1334 to 2.1344 inches
No. 4...	2.1531 to 2.1541 inches
Bearing oil clearance, standard...	0.001 0.003 inch

Torque specifications*

Ft-lbs (unless otherwise indicated)
Camshaft sprocket-to-camshaft bolt	37 to 51
Camshaft thrust plate bolts	88 in-lbs
Connecting rod cap nuts	20 to 28
Driveplate bolts..	54 to 64
Main bearing cap bolts...	56 to 62
Timing chain cover bolts..	15 to 22

3.8L V6 engine

General

Cylinder compression pressure..	101 psi minimum (see text)
Oil pressure (hot at 2500 rpm)...	40 to 60 psi

Cylinder head and valves

Warpage limit...	0.007 inch
Minimum valve margin...	1/32 inch

Intake valve
    Stem diameter .... 0.3423 to 0.3415 inch
    Valve stem-to-guide clearance .... 0.001 to 0.0028 inch
Exhaust valve
    Stem diameter .... 0.3418 to 0.3410 inch
    Valve stem-to-guide clearance .... 0.0015 to 0.0033 inch
Valve spring pressure (not including dampener)
    Valve open .... 220 lbs at 1.18 inches
    Valve closed .... 85 lbs at 1.65 inches
Valve lifter
    Diameter (standard) .... 0.8740 to 0.8745 inch
    Lifter-to-bore clearance .... 0.0007 to 0.0027 inch

Crankshaft and connecting rods

Connecting rod journal
    Diameter .... 2.3103 to 2.3111 inches
    Bearing oil clearance
        Desired .... 0.001 to 0.0014 inch
        Allowable .... 0.00086 to 0.0027 inch
Connecting rod side clearance (endplay)
    Standard .... 0.0047 to 0.0114 inch
    Service limit .... 0.014 inch maximum
Main bearing journal
    Diameter .... 2.5190 to 2.5198 inches
    Bearing oil clearance
        Desired .... 0.001 to 0.0014 inch
        Allowable .... 0.0005 to 0.0023 inch
Crankshaft endplay (at thrust bearing) .... 0.004 to 0.008 inch

Pistons and rings

Piston-to-bore clearance limit .... 0.0014 to 0.0032 inch
Piston ring end gap
    1995
        Top compression ring .... 0.011 to 0.012 inch
        Second compression ring .... 0.010 to 0.020 inch
        Oil ring .... 0.015 to 0.058 inch
    1996 and later
        Top compression ring .... 0.010 to 0.016 inch
        Second compression ring .... 0.010 to 0.020 inch
        Oil ring .... 0.006 to 0.065 inch

Camshaft

Lobe lift
    Intake .... 0.245 inch
    Exhaust .... 0.259 inch
Allowable lobe lift loss .... 0.005 inch
Endplay, service limit .... 0.001 to 0.006 inch
Bearing journal diameter (all) .... 2.0515 to 2.0505 inches
Bearing oil clearance (all) .... 0.001 to 0.003 inch

Balance shaft

Journal diameter .... 2.0515 to 2.0505 inches
Endplay .... 0.003 to 0.008 inch

Torque specifications*

Ft-lbs (unless otherwise indicated)

Main bearing cap bolts
    1995 .... 65 to 81
    1996 and later .... 81 to 88
Connecting rod cap nuts
    1995 .... 31 to 36
    1996 and later
        Step 1 .... 29 to 34
        Step 2 .... Rotate an additional 90 to 120 degrees
Camshaft sprocket-to-camshaft bolt .... 30 to 37
Camshaft thrust plate bolts .... 71 to 124 in-lbs
Balance shaft thrust plate bolts .... 71 to 124 in-lbs
Driveplate-to-crankshaft bolts .... 54 to 64

*Note: *Refer to Chapter 2, Part A for additional torque specifications.*

2B

2.2 3.0L oil pressure sender (arrow) - near the transmission dipstick

2.3 Connect a mechanical oil pressure gauge (A) to the block oil port (B) and check your engine's pressure compared to Specifications

1   General information - engine overhaul

Included in this portion of Chapter 2 are the general overhaul procedures for the cylinder head and internal engine components.

The information ranges from advice concerning preparation for an overhaul and the purchase of replacement parts to detailed, step-by-step procedures covering Removal and installation of internal engine components and the inspection of parts.

The following Sections have been written based on the assumption that the engine has been removed from the vehicle. For information concerning in-vehicle engine repair, as well as removal and installation of the external components necessary for the overhaul, see Chapter 2A.

The Specifications included in this Part are only those necessary for the inspection and overhaul procedures which follow. Refer to Chapter 2, Part A for additional Specifications.

It's not always easy to determine when, or if, an engine should be completely overhauled, as a number of factors must be considered.

High mileage is not necessarily an indication that an overhaul is needed, while low mileage doesn't preclude the need for an overhaul. Frequency of servicing is probably the most important consideration. An engine that's had regular and frequent oil and filter changes, as well as other required maintenance, will most likely give many thousands of miles of reliable service. Conversely, a neglected engine may require an overhaul very early in its life.

Excessive oil consumption is an indication that piston rings, valve seals and/or valve guides are in need of attention. Make sure that oil leaks aren't responsible before deciding that the rings and/or guides are bad. Perform a cylinder compression check to determine the extent of the work required (see Section 4). Also check the vacuum readings under various conditions (see Section 3).

Loss of power, rough running, knocking or metallic engine noises, excessive valve train noise and high fuel consumption rates may also point to the need for an overhaul, especially if they're all present at the same time. If a complete tune-up doesn't remedy the situation, major mechanical work is the only solution.

An engine overhaul involves restoring the internal parts to the specifications of a new engine. During an overhaul, the piston rings are replaced and the cylinder walls are reconditioned (re-bored and/or honed). If a re-bore is done by an automotive machine shop, new oversize pistons will also be installed. The main bearings, connecting rod bearings and camshaft bearings are generally replaced with new ones and, if necessary, the crankshaft may be reground to restore the journals. Generally, the valves are serviced as well, since they're usually in less-than-perfect condition at this point. While the engine is being overhauled, other components, such as the distributor, starter and alternator, can be rebuilt as well. The end result should be a like new engine that will give many trouble free miles. **Note:** *Critical cooling system components such as the hoses, drivebelts, thermostat and water pump should be replaced with new parts when an engine is overhauled. The radiator should be checked carefully to ensure that it isn't clogged or leaking (see Chapter 3). If you purchase a rebuilt engine or short block, some rebuilders will not warranty their engines unless the radiator has been professionally flushed. Also, we don't recommend overhauling the oil pump - always install a new one when an engine is rebuilt.*

Before beginning the engine overhaul, read through the entire procedure to familiarize yourself with the scope and requirements of the job. Overhauling an engine isn't difficult, but it is time-consuming. Plan on the vehicle being tied up for a minimum of two weeks, especially if parts must be taken to an automotive machine shop for repair or reconditioning. Check on availability of parts and make sure that any necessary special tools and equipment are obtained in advance. Most work can be done with typical hand tools, although a number of precision measuring tools are required for inspecting parts to determine if they must be replaced. Often an automotive machine shop will handle the inspection of parts and offer advice concerning reconditioning and replacement. **Note:** *Always wait until the engine has been completely disassembled and all components, especially the engine block, have been inspected before deciding what service and repair operations must be performed by an automotive machine shop. Since the block's condition will be the major factor to consider when determining whether to overhaul the original engine or buy a rebuilt one, never purchase parts or have machine work done on other components until the block has been thoroughly inspected. As a general rule, time is the primary cost of an overhaul, so it doesn't pay to install worn or substandard parts.*

As a final note, to ensure maximum life and minimum trouble from a rebuilt engine, everything must be assembled with care in a spotlessly-clean environment.

2   Oil pressure check

Refer to illustrations 2.2 and 2.3

1   Low engine oil pressure can be a sign of an engine in need of rebuilding. A "low oil pressure" indicator (often called an "idiot light") is not a test of the oiling system. Such indicators only come on when the oil pressure is dangerously low. Even a factory oil pressure gauge in the instrument panel is only a relative indication, although much better for driver information than a warning light. A better test is with a mechanical (not electrical) oil pressure gauge. When used in conjunction with an accurate tachometer, an engine's oil pressure performance can be compared to factory Specifications for that year and model.

2   Find the oil pressure indicator sending unit **(see illustration)**. **Note:** *The 3.8 sender is located at the lower left (passenger side) corner of the engine, near the oil filter. It isn't easy to see.*

3   Remove the oil pressure sending unit and install a fitting which will allow you to directly connect your hand-held, mechanical oil pressure gauge **(see illustration)**. Use Teflon tape or sealant on the threads of the adapter and the fitting on the end of your gauge's hose.

4   Connect an accurate tachometer to the engine, according to the tachometer manufacturer's instructions.

5   Check the oil pressure with the engine running (full operating temperature) at the specified engine speed, and compare it to this Chapter's Specifications. If it's extremely low, the bearings and/or oil pump are probably worn out.

3   Vacuum gauge diagnostic checks

Refer to illustrations 3.4 and 3.6

A vacuum gauge provides valuable information about what is going on in the engine at a low-cost. You can check for worn rings or cylinder walls, leaking head or intake manifold gaskets, incorrect carburetor adjustments, restricted exhaust, stuck or burned valves, weak valve springs, improper ignition or valve timing and ignition problems.

Unfortunately, vacuum gauge readings are easy to misinterpret, so they should be used in conjunction with other tests to confirm the diagnosis.

Both the absolute readings and the rate of needle movement are important for accurate interpretation. Most gauges measure vacuum in inches of mercury (in-Hg). The following references to vacuum assume the diagnosis is being performed at sea level. As elevation increases (or atmospheric pressure decreases), the reading will decrease. For every 1,000 foot increase in elevation above approximately 2000 feet, the gauge readings will decrease about one inch of mercury.

Connect the vacuum gauge directly to intake manifold vacuum, not to ported (throttle body) vacuum **(see illustration)**. Be sure no hoses are left disconnected during the test or false readings will result.

Before you begin the test, allow the engine to warm up completely. Block the wheels and set the parking brake. With the transmission in Park, start the engine and allow it to run at normal idle speed. **Warning:** *Keep your hands and the vacuum gauge clear of the fans.*

3.4 A simple vacuum gauge, attached to a manifold vacuum port (arrow) can make a very useful diagnosis of an engine's condition

Read the vacuum gauge; an average, healthy engine should normally produce about 17 to 22 inches of vacuum with a fairly steady needle **(see illustration)**. Refer to the following vacuum gauge readings and what they indicate about the engine's condition:

1   A low steady reading usually indicates a leaking gasket between the intake manifold and cylinder head(s) or throttle body, a leaky

2B

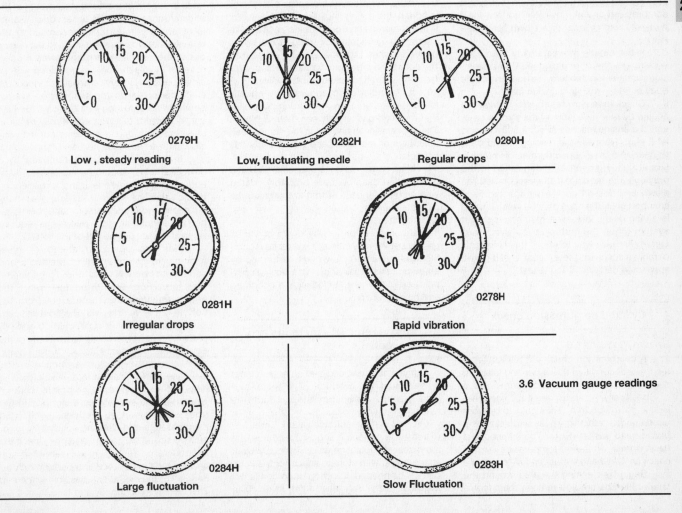

Low , steady reading   0279H

Low, fluctuating needle   0282H

Regular drops   0280H

Irregular drops   0281H

Rapid vibration   0278H

Large fluctuation   0284H

Slow Fluctuation   0283H

3.6 Vacuum gauge readings

vacuum hose, late ignition timing or incorrect camshaft timing. Check ignition timing with a timing light and eliminate all other possible causes, utilizing the tests provided in this Chapter before you remove the timing chain cover to check the timing marks.

2   If the reading is three to eight inches below normal and it fluctuates at that low reading, suspect an intake manifold gasket leak at an intake port or a faulty fuel injector.

3   If the needle has regular drops of about two-to-four inches at a steady rate, the valves are probably leaking. Perform a compression check or leak-down test to confirm this.

4   An irregular drop or down-flick of the needle can be caused by a sticking valve or an ignition misfire. Perform a compression check or leak-down test and read the spark plugs.

5   A rapid vibration of about four in.-Hg vibration at idle combined with exhaust smoke indicates worn valve guides. Perform a leak-down test to confirm this. If the rapid vibration occurs with an increase in engine speed, check for a leaking intake manifold gasket or head gasket, weak valve springs, burned valves or ignition misfire.

6   A slight fluctuation, say one inch up and down, may mean ignition problems. Check all the usual tune-up items and, if necessary, run the engine on an ignition analyzer.

7   If there is a large fluctuation, perform a compression or leak-down test to look for a weak or dead cylinder or a blown head gasket.

8   If the needle moves slowly through a wide range, check for a clogged PCV system, incorrect idle fuel mixture, carburetor/throttle body or intake manifold gasket leaks.

9   Check for a slow return after revving the engine by quickly snapping the throttle open until the engine reaches about 2,500 rpm and let it shut. Normally the reading should drop to near zero, rise above normal idle reading (about 5 in.-Hg over) and then return to the previous idle reading. If the vacuum returns slowly and doesn't peak when the throttle is snapped shut, the rings may be worn. If there is a long delay, look for a restricted exhaust system (often the muffler or catalytic converter). An easy way to check this is to temporarily disconnect the exhaust ahead of the suspected part and redo the test.

4   Cylinder compression check

Refer to illustration 4.6

1   A compression check will tell you what mechanical condition the upper end (pistons, rings, valves, head gaskets) of your engine is in. Specifically, it can tell you if the compression is down due to leakage caused by worn piston rings, defective valves and seats or a blown head gasket. **Note:** *The engine must be at normal operating temperature for this check and the battery must be fully charged.*

2   Begin by cleaning the area around the spark plugs before you remove them (com-pressed air works best for this). This will prevent dirt from getting into the cylinders as the compression check is being done.

3   Remove all of the spark plugs from the engine (see Chapter 1).

4   Block the throttle wide open.

5   Disable the ignition system by disconnecting the primary (low voltage) wires from the coil(s). Also, disable the fuel pump (see Chapter 4, Section 2).

6   With the compression gauge in the number one spark plug hole, crank the engine over at least four compression strokes and watch the gauge **(see illustration)**. The compression should build up quickly in a healthy engine. Low compression on the first stroke, followed by gradually increasing pressure on successive strokes, indicates worn piston rings. A low compression reading on the first stroke, which does not build up during successive strokes, indicates leaking valves or a blown head gasket (a cracked head could also be the cause). Record the highest gauge reading obtained.

7   Repeat the procedure for the remaining cylinders and compare the results to the Specifications.

8   Add some engine oil (about three squirts from a plunger-type oil can) to each cylinder, through the spark plug hole, and repeat the test.

9   If the compression increases after the oil is added, the piston rings are definitely worn. If the compression does not increase significantly, the leakage is occurring at the valves or head gasket. Leakage past the valves may be caused by burned valve seats and/or faces or warped, cracked or bent valves.

10   If two adjacent cylinders have equally low compression, there is a strong possibility that the head gasket between them is blown. The appearance of coolant in the combustion chambers or the crankcase would verify this condition.

11   If the compression is unusually high, the combustion chambers are probably coated with carbon deposits. If that is the case, the cylinder heads should be removed and decarbonized.

12   If compression is way down or varies greatly between cylinders, it would be a good idea to have a leak-down test performed by an automotive repair shop. This test will pinpoint exactly where the leakage is occurring and how severe it is.

5   Engine removal - methods and precautions

If you have decided that an engine must be removed for overhaul or major repair work, several preliminary steps should be taken.

Locating a suitable work area is extremely important. A shop is, of course, the most desirable place to work. Adequate work space, along with storage space for the vehicle, will be needed. If a shop or garage is not available, at the very least a flat, level, clean

4.6 A compression gauge with a threaded fitting for the spark plug hole is preferred over the type that requires hand pressure to maintain the seal - be sure to open the throttle valve as far as possible during the compression check

work surface made of concrete or asphalt is required.

Cleaning the engine compartment and engine before beginning the removal procedure will help keep tools clean and organized.

An engine hoist or A-frame will be needed. Make sure that the equipment is rated in excess of the combined weight of the engine/transaxle and its accessories. Safety is of primary importance, considering the potential hazards involved in lifting the engine out of the vehicle. **Warning:** *Some models covered by this manual are equipped with an air suspension system. Always disconnect the electrical power to the suspension system before lifting or towing (see Chapter 10). Failure to perform this procedure may result in unexpected shifting or movement of the vehicle, which could cause personal injury.*

If the engine is being removed by a novice, a helper should be available. Advice and aid from someone more experienced would also be helpful. There are many instances when one person cannot simultaneously perform all of the operations required when lifting the engine out of the vehicle.

Plan the operation ahead of time. Arrange for or obtain all of the tools and equipment you will need prior to beginning the job. For the Windstar van, the engine/transaxle must come out from the bottom of the vehicle, with the front suspension subframe. This is a difficult job and will require the use of a vehicle hoist. Some of the equipment necessary to perform engine removal and installation safely and with relative ease are (in addition to a vehicle hoist) a heavy duty floor jack, complete sets of wrenches and sockets as described in the front of this manual, wooden blocks and plenty of rags and cleaning solvent for mopping up spilled oil, coolant and gasoline. If the hoist is to be rented, make sure that you arrange for it in advance and perform beforehand all of the operations possible without it. This will save you money and time.

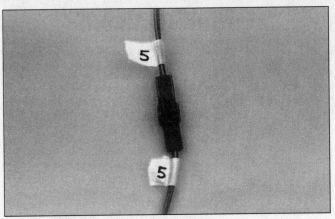

6.6a Label each wire before unplugging the connector

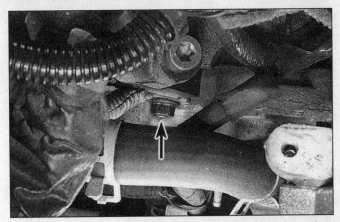

6.6b Don't overlook ground straps, such as this one (arrow) attached to the rear cylinder head

Plan for the vehicle to be out of use for a considerable amount of time. Once the engine/transaxle and subframe are removed from your van, the vehicle will be difficult to even move around without a large shop-type bumper jack.

A machine shop will be required to perform some of the work which the do-it-yourselfer cannot accomplish due to a lack of special equipment. These shops often have a busy schedule, so it would be wise to consult them before removing the engine in order to accurately estimate the amount of time required to rebuild or repair components that may need work.

Always use extreme caution when removing and installing the engine. Serious injury can result from careless actions. Plan ahead, take your time and a job of this nature, although major, can be accomplished successfully.

6   Engine - removal and installation

Note: *Engine/transaxle removal on the Windstar van is a difficult job, especially for the do-it-yourself mechanic working at home in the driveway. Because of the vehicle's design, the engine and transaxle must be removed from the bottom of the vehicle, not the top, and the front suspension subframe must come down with them. With just a floor jack and jackstands, the vehicle can't be raised high enough for the engine/transaxle/subframe to slide out the front. Also a special jack that supports the subframe should be used to guide the drivetrain out without the engine tipping over. The factory recommends that the engine removal be performed only when there is access to a vehicle hoist. Once the front wheels are removed, the vehicle can be lowered until the subframe rests on a wooden pallet, preferably with caster wheels. When everything is disconnected from the vehicle, it can be raised up, leaving the engine/transaxle/subframe on a moveable, working platform.*

Warning 1: *The air conditioning system is under high pressure. DO NOT loosen any fittings or remove any components until after*

the system has been discharged. Air conditioning refrigerant should be properly discharged into an EPA-approved container at a dealer service department or an automotive air conditioning repair facility. Always wear eye protection when disconnecting air conditioning system fittings.
Warning 2: *Gasoline is extremely flammable, so take extra precautions when you work on any part of the fuel system. Don't smoke or allow open flames or bare light bulbs near the work area, and don't work in a garage where a natural gas-type appliance (such as a water heater or a clothes dryer) with a pilot light is present. Since gasoline is carcinogenic, wear latex gloves when there's a possibility of being exposed to fuel, and, if you spill any fuel on your skin, rinse it off immediately with soap and water. Mop up any spills immediately and do not store fuel-soaked rags where they could ignite. The fuel system is under constant pressure, so, if any fuel lines are to be disconnected, the fuel pressure in the system must be relieved first (see Chapter 4 for more information). When you perform any kind of work on the fuel system, wear safety glasses and have a Class B type fire extinguisher on hand.*
Warning 3: *Some models covered by this manual are equipped with an air suspension system. Always disconnect the electrical power to the suspension system before lifting or towing (see Chapter 10). Failure to perform this procedure may result in unexpected shifting or movement of the vehicle, which could cause personal injury.*

Removal

Refer to illustrations 6.6a, 6.6b and 6.28
1   Relieve the fuel system pressure (see Chapter 4).
2   Disconnect the negative cable from the battery.
3   Cover the fenders and cowl and remove the hood and cowl top vent panel (see Chapter 11). Special pads are available to protect the fenders, but an old bedspread or blanket will also work.
4   Remove the air cleaner assembly.
5   Drain the cooling system (see Chapter 1).

6   Label the vacuum lines, emissions system hoses, electrical connectors, ground straps and fuel lines that would interfere with engine removal, to ensure correct reinstallation, then detach them. Pieces of masking tape with numbers or letters written on them work well **(see illustration)**. If there's any possibility of confusion, make a sketch of the engine compartment and clearly label the lines, hoses and wires. Don't overlook ground wires that may be bolted to the engine **(see illustration)**.
7   Label and detach all coolant hoses from the engine.
8   Remove the cooling fans assembly and radiator (see Chapter 3).
9   Remove the drivebelt(s) (see Chapter 1).
10   Disconnect the accelerator cable and cruise control cable (if equipped) from the engine (see Chapter 4).
11   Unbolt the power steering pump (see Chapter 10). Leave the lines/hoses attached and make sure the pump is kept in an upright position in the engine compartment (use wire or rope to restrain it out of the way).
12   On air-conditioned models, unbolt the compressor (see Chapter 3) and set it aside. Do not disconnect the hoses.
13   Refer to Chapter 7 and disconnect the shift linkage, fluid cooler lines and the electrical connectors at the transaxle.
14   Disconnect the main wiring harness plugs in the engine compartment and position the harness out of the way.
15   Drain the engine oil (see Chapter 1) and remove the filter. On 3.8L models with oil coolers, disconnect the oil cooler hoses and remove the oil cooler (see Chapter 3).
16   Remove the starter motor (see Chapter 5).
17   Remove the alternator (see Chapter 5).
18   Unbolt the exhaust system from the engine (see Part A and Chapter 4).
19   Turn the steering wheel to the straight-ahead position and lock the column by removing the key.
20   Raise the vehicle on the hoist and remove the front wheels. Refer to Chapter 10 and disconnect the steering column from the rack at the pinch-bolt and remove the intermediate shaft.

2B

21   Remove the links from the front stabilizer bar and separate the tie-rods and lower suspension arms from the steering knuckles (see Chapter 10).

22   Refer to Chapter 8 and remove the left and right driveaxles.

23   Either use a subframe jack to support the subframe/engine/transaxle, or lower the vehicle hoist until the subframe is supported on something sturdy like a wooden pallet.

24   Remove the four subframe-to-body bolts and raise the vehicle up, leaving the subframe and drivetrain on the wooden pallet. Move the pallet and drivetrain assembly out from under the vehicle.

25   Attach a standard engine-lifting hoist (available at most tool rental facilities) to lifting hooks on the engine and raise the engine/transaxle slightly, enough to remove the engine mounts.

26   Raise the engine slightly. Carefully work it forward to separate it from the transmission. Be sure the torque converter stays in the transmission (clamp a pair of vise-grips to the transmission housing to keep the converter from sliding out). Slowly raise the engine out of the subframe. Check carefully to make sure nothing is hanging up.

27   Remove the driveplate (see Part A or B of this Chapter).

28   Mount the engine on an engine stand **(see illustration)**.

29   Once the engine is removed, support the transmission with blocks of wood on the subframe.

Installation

30   Check the engine and transmission mounts. If they're worn or damaged, replace them.

31   Carefully lower the engine into the subframe - make sure the engine mounts line up.

32   Guide the torque converter into the crankshaft following the procedure outlined in Chapter 7. Don't pull the converter away from the transaxle; let the engine move back against the transmission.

33   Install the transaxle-to-engine bolts and tighten them securely. **Caution:** DO NOT use the bolts to force the transaxle and engine together!

34   Install the engine mounts between the engine and the subframe.

35   Lower the vehicle hoist until the subframe can be raised close enough to the body to insert the four subframe bolts. Reinstall the remaining components in the reverse order of removal. Don't forget to install the torque converter-to-driveplate nuts.

36   Add coolant and oil as needed. Run the engine and check for leaks and proper operation of all accessories, then install the hood and test drive the vehicle.

7   Engine rebuilding alternatives

The do-it-yourselfer is faced with a number of options when performing an engine

6.28 Use long, high-strength bolts (arrows) to hold the engine block on the engine stand - make sure they are tight before lowering the hoist and placing the entire weight of the engine on the stand (lower bolts not shown)

overhaul. The decision to replace the engine block, piston/connecting rod assemblies and crankshaft depends on a number of factors, with the number one consideration being the condition of the block. Other considerations are cost, access to machine shop facilities, parts availability, time required to complete the project and the extent of prior mechanical experience on the part of the do-it-yourselfer.

Some of the rebuilding alternatives include:

Individual parts - If the inspection procedures reveal that the engine block and most engine components are in reusable condition, purchasing individual parts may be the most economical alternative. The block, crankshaft and piston/connecting rod assemblies should all be inspected carefully. Even if the block shows little wear, the cylinder bores should be surface honed.

Crankshaft kit - This rebuild package consists of a reground crankshaft and a matched set of pistons and connecting rods. The pistons will already be installed on the connecting rods. Piston rings and the necessary bearings will be included in the kit. These kits are commonly available for standard cylinder bores, as well as for engine blocks which have been bored to a regular oversize.

Short block - A short block consists of an engine block with renewed crankshaft and piston/connecting rod assemblies already installed. All new bearings are incorporated and all clearances will be correct. The existing cylinder head(s), camshaft, valve train components and external parts can be bolted to the short block with little or no machine shop work necessary.

Long block - A long block consists of a short block plus an oil pump, oil pan, cylinder heads, valve covers, camshaft and valve train components, timing sprockets, timing chain and timing cover. All components are installed with new bearings, seals and gaskets incorporated throughout. The installation of manifolds and external parts is all that is necessary.

Used Engine - While overhaul provides the best assurance of a like-new engine, used engines available from wrecking yards and importers are often a very simple and economical solution. Many used engines come with warranties, but always give any used engine a thorough diagnostic check-out before purchase. Check compression and also for signs of oil leakage. If possible, have the seller run the engine either in the vehicle or on a test stand so you can be sure it runs smoothly with no knocking or other noises. Give careful thought to which alternative is best for you and discuss the situation with local automotive machine shops, auto parts dealers or parts store countermen before ordering or purchasing replacement parts.

8   Engine overhaul - disassembly sequence

Caution: The cylinder head bolts on all engines in the vehicles covered by this manual are "torque-to-yield" bolts and are NOT reusable. A predetermined stretch of the bolt gives the even clamping load needed to seal the cylinders properly. Once removed they must be replaced.

1   It's much easier to disassemble and work on the engine if it's mounted on a portable engine stand. A stand can often be rented quite cheaply from an equipment rental yard. Before the engine is mounted on a stand, the driveplate and engine rear (sheet metal) plate should be removed from the engine.

2   If a stand isn't available, it's possible to disassemble the engine with it blocked up on the floor. Be extra careful not to tip or drop the engine when working without a stand.

3   If you're going to obtain a rebuilt engine, all external components must come off first, to be transferred to the replacement engine, just as they will if you're doing a complete engine overhaul yourself. These include:

Alternator and brackets
Emissions control components
Camshaft position sensor
Crankshaft position sensor
Spark plug wires and spark plugs
Thermostat and housing cover
Water pump
EFI components
Intake/exhaust manifolds
Oil filter (replace)
Engine mounts
Driveplate
Engine rear plate
Crankshaft damper

Note: When removing the external components from the engine, pay close attention to details that may be helpful or important during installation. Note the installed position of gaskets, seals, spacers, pins, brackets, washers, bolts and other small items.

4   If you're obtaining a short block, which consists of the engine block, crankshaft, pistons and connecting rods all assembled, then the cylinder head(s), oil pan and oil pump will

9.1 A small plastic bag, with an appropriate label, can be used to store the valve train components so they can be kept together and reinstalled in the original position

9.2 Use a valve spring compressor to compress the spring, then remove the keepers from the valve stem with needle-nose pliers or a magnet

9.3 If the valve won't pull through the guide, deburr the edge of the stem end and the area around the top of the keeper groove with a file or whetstone

have to be removed as well. See *Engine rebuilding alternatives* for additional information regarding the different possibilities to be considered.

5   If you're planning a complete overhaul, the engine must be disassembled and the components removed in the following order:

Driveplate
Valve covers
Intake manifold
Exhaust manifolds
Rocker arms and pushrods
Valve lifters
Vibration damper
Timing chain cover
*Timing chain(s), sprockets, guides and
    tensioners*
Camshaft
Balance shaft (3.8L only)
Cylinder heads
Oil pan
Oil pump (replace)
Piston/connecting rod assemblies
Crankshaft and main bearings ((replace)

6   Before beginning the disassembly and overhaul procedures, make sure the following items are available. Also, refer to *Engine overhaul - reassembly sequence* for a list of tools and materials needed for engine reassembly.

Common hand tools
*Small cardboard boxes and plastic bags
    for storing parts*
Gasket scraper
Ridge reamer
Vibration damper puller
Micrometers
Telescoping gauges
Dial indicator set
Valve spring compressor
Cylinder surfacing hone
Piston ring groove cleaning tool
Electric drill motor
Tap and die set
Wire brushes
Oil gallery brushes
Cleaning solvent

9   Cylinder head - disassembly

Refer to illustrations 9.1, 9.2 and 9.3
Note: *New and rebuilt cylinder heads are commonly available for most engines at dealerships and auto parts stores. Due to the fact that some specialized tools are necessary for the disassembly and inspection procedures, and replacement parts may not be readily available, it may be more practical and economical for the home mechanic to purchase replacement head(s) rather than taking the time to disassemble, inspect and recondition the original(s).*

1   Cylinder head disassembly involves removal of the intake and exhaust valves and related components. If they're still in place, remove the rocker arm nuts, fulcrums and rocker arms from the cylinder head studs. Label the parts or store them separately **(see illustration)** so they can be reinstalled in their original locations and in the same valve guides they are removed from.

2   Compress the springs on the first valve with a spring compressor and remove the keepers **(see illustration)**. Carefully release the valve spring compressor and remove the retainer, the spring and the spring seat (if used).

3   Pull the valve out of the head, then remove the oil seal from the guide. If the valve binds in the guide (won't pull through), push it back into the head and deburr the area around the keeper groove with a fine file or whetstone **(see illustration)**.

4   Repeat the procedure for the remaining valves. Remember to keep all the parts for each valve together so they can be reinstalled in the same locations.

5   Once the valves and related components have been removed and stored in an organized manner, the head should be thoroughly cleaned and inspected. If a complete engine overhaul is being done, finish the engine disassembly procedures before beginning the cylinder head cleaning and inspection process.

10   Cylinder head - cleaning and inspection

1   Thorough cleaning of the cylinder head(s) and related valve train components, followed by a detailed inspection, will enable you to decide how much valve service work must be done during the engine overhaul. **Note:** *If the engine was severely overheated, the cylinder head is probably warped (see Step 12).*

Cleaning

2   Scrape all traces of old gasket material and sealing compound off the head gasket, intake manifold and exhaust manifold sealing surfaces. Be very careful not to gouge the cylinder head. Special gasket removal solvents that soften gaskets and make removal much easier are available at auto parts stores.

3   Remove all built up scale from the coolant passages.

4   Run a stiff wire brush through the various holes to remove deposits that may have formed in them.

5   Run an appropriate-size tap into each of the threaded holes to remove corrosion and thread sealant that may be present. If compressed air is available, use it to clear the holes of debris produced by this operation. **Warning:** *Wear eye protection when using compressed air!*

6   Clean the exhaust manifold stud threads, if equipped.

7   Clean the cylinder head with solvent and dry it thoroughly. Compressed air will speed the drying process and ensure that all holes and recessed areas are clean. **Note:** *Decarbonizing chemicals are available and may prove very useful when cleaning cylinder heads and valve train components. They are very caustic and should be used with caution. Be sure to follow the instructions on the container.*

2B

10.12 Check the cylinder head gasket surface for warpage by trying to slip a feeler gauge under the straightedge (see this Chapter's Specifications for the maximum warpage allowed and use a feeler gauge of that thickness)

10.14 Lay the head on its edge, pull each valve out about 1/16 inch, set up a dial indicator with the probe touching the valve, wiggle the valve and measure its movement

8   Clean the valvetrain components with solvent and dry them thoroughly (don't mix them up during the cleaning process). **Note:** *Compressed air will speed the drying process and can be used to clean out the oil passages.*

9   Clean all the valve springs, spring seats, keepers and retainers with solvent and dry them thoroughly. Work the components from one valve at a time to avoid mixing up the parts.

10   Scrape off any heavy deposits that may have formed on the valves, then use a motorized wire brush to remove deposits from the valve heads and stems. Again, make sure the valves don't get mixed up.

Inspection

Note: *Be sure to perform all of the following inspection procedures before concluding that machine shop work is required. Make a list of the items that need attention.*

Cylinder heads

Refer to illustrations 10.12 and 10.14

11   Inspect the heads very carefully for cracks, evidence of coolant leakage and other damage. **Note:** *Sometimes cracks are not visible. The cylinder heads should be Magnafluxed (or checked by some other professional method) at an automotive machine shop.* If cracks are found, check with an automotive machine shop concerning repair. If repair isn't possible, a new cylinder head should be obtained.

12   Using a straightedge and feeler gauge, check the head gasket mating surface for warpage **(see illustration)**. Check the head both straight across and corner-to-corner. If the warpage exceeds the limit listed in this Chapter's Specifications, it can be resurfaced at an automotive machine shop. **Note:** *If the cylinder heads are resurfaced more than .010-inch, the intake manifold flanges will also require machining.*

13   Examine the valve seats in each of the combustion chambers. If they're pitted, cracked or burned, the head will require valve

service that's beyond the scope of the home mechanic.

14   Check the valve stem-to-guide clearance by measuring the lateral movement of the valve stem with a dial indicator attached securely to the head **(see illustration)**. The valve must be in the guide and approximately 1/16-inch off the seat. The total valve stem movement indicated by the gauge needle must be divided by two to obtain the actual clearance. After this is done, if there's still some doubt regarding the condition of the valve guides they should be checked by an automotive machine shop (the cost should be minimal).

Valves

Refer to illustrations 10.15 and 10.16

15   Carefully inspect each valve face for uneven wear, deformation, cracks, pits and burned areas **(see illustration)**. Check the valve stem for scuffing and galling and the neck for cracks. Rotate the valve and check for any obvious indication that it's bent. Look for pits and excessive wear on the end of the stem. The presence of any of these conditions indicates the need for valve service by an automotive machine shop.

16   Measure the margin width on each valve **(see illustration)**. Any valve with a margin narrower than 1/32-inch will have to be replaced with a new one.

Valve components

Refer to illustrations 10.17 and 10.18

17   Check each valve spring for wear (on the ends) and pits. Measure the free length and compare it to the Specifications **(see illustration)**. Any springs that are shorter than specified have sagged and should not be reused. The tension of all springs should be checked with a special fixture before deciding that they're suitable for use in a rebuilt engine (take the springs to an automotive machine shop for this check).

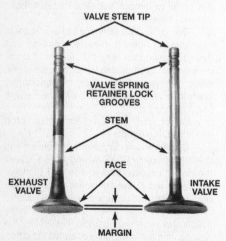

10.15 Check for valve wear at the points shown here

10.16 The margin width on each valve must be as specified (if no margin exists, the valve cannot be reused)

18   Stand each spring on a flat surface and check it for squareness **(see illustration)**. If any of the springs are distorted or sagged, replace all of them with new parts.

19   Check the spring retainers and keepers for obvious wear and cracks. Any questionable parts should be replaced with new ones, as extensive damage will occur if they fail during engine operation.

Rocker arm components

Refer to illustrations 10.22a, 10.22b and 10.22c

20   Clean all the parts thoroughly. Make sure all oil passages are open.

21   Check the rocker arm faces (the areas that contact the pushrod ends and valve stems) for pits, wear, galling, score marks and rough spots.

22   Check the rocker arm pivot contact areas and fulcrums. Look for cracks in each rocker arm and bolt **(see illustrations)**.

23   Inspect the pushrod ends for scuffing and excessive wear. Roll each pushrod on a flat surface, like a piece of plate glass, to determine if it's bent.

24   Check the rocker arm pedestals in the cylinder heads for damaged threads and secure installation.

25   Any damaged or excessively worn parts must be replaced with new ones.

26   If the inspection process indicates that the valve components are in generally poor condition and worn beyond the limits specified, which is usually the case in an engine that's being overhauled, reassemble the valves in the cylinder head and refer to Section 11 for valve servicing recommendations.

27   Clean all the parts thoroughly. Make sure all oil passages are open.

28   Any damaged or excessively worn parts must be replaced with new ones.

29   If the inspection process indicates that the valve components are in generally poor condition and worn beyond the limits specified, which is usually the case in an engine that's being overhauled, reassemble the valves in the cylinder head and refer to Section 11 for valve servicing recommendations.

10.17 Measure the free length of each valve spring with a dial or vernier caliper

11   Valves - servicing

1   Because of the complex nature of the job and the special tools and equipment needed, servicing of the valves, the valve seats and the valve guides, commonly known as a valve job, should be done by a professional.

2   The home mechanic can remove and disassemble the head, do the initial cleaning and inspection, then reassemble and deliver it to a dealer service department or an automotive machine shop for the actual service work. Doing the inspection will enable you to see what condition the head and valve train components are in and will ensure that you know what work and new parts are required when dealing with an automotive machine shop.

3   The dealer service department or automotive machine shop will remove the valves and springs, recondition or replace the valves and valve seats, recondition the valve guides, check and replace the valve springs, spring retainers or rotators and keepers (as necessary), replace the valve seals with new ones, reassemble the valve components and make sure the installed spring height is correct. The

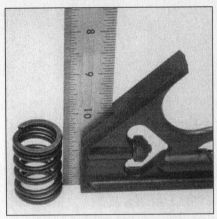

10.18 Check each valve spring for squareness; if it is bent it should be replaced

cylinder head gasket surface will also be resurfaced if it's warped.

4   After the valve job has been performed by a professional, the head will be in like-new condition. When the head is returned, be sure to clean it again before installation on the engine to remove any metal particles and abrasive grit that may still be present from the valve service or head resurfacing operations. Use compressed air, if available, to blow out all the oil holes and passages.

12   Cylinder head - reassembly

Refer to illustrations 12.3, 12.6 and 12.8

1   Regardless of whether or not the head was sent to an automotive repair shop for valve servicing, make sure it's clean before beginning reassembly.

2   If the head was sent out for valve servicing, the valves and related components will already be in place. Begin the reassembly procedure with Step 9.

3   On all engines, lubricate and install the valves, then install new seals on each of the valve guides. Using a hammer and deep socket, gently tap each seal into place until

2B

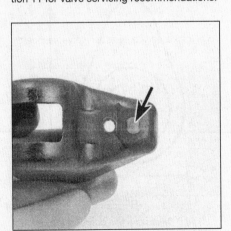

10.22a Check the rocker arms where they contact the valves (arrow) and where the pushrod rides . . .

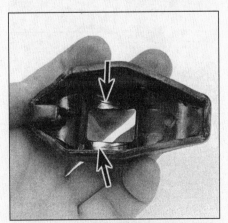

10.22b . . . the fulcrum seats (arrows) in the top of the rocker arm . . .

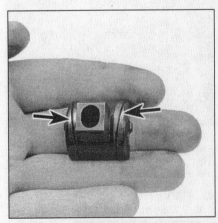

10.22c . . . and the fulcrums themselves for wear and galling (arrows)

12.3 Use a hammer and a seal installer (or a deep socket, as shown here) to drive the seal onto the valve guide/head casting boss (make sure the valve spring seat is in place first)

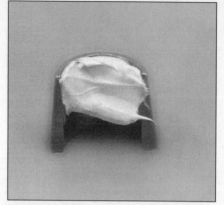

12.6 Apply a small dab of grease to each keeper as shown here before installation - it'll hold them in place on the valve stem as the spring is released

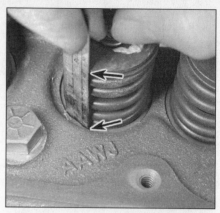

12.8 Valve spring installed height is the distance from the spring seat on the head to the bottom of the spring retainer

it's seated on the guide **(see illustration)**. Don't twist or cock the seals during installation or they will not seat properly on the valve stems. **Note:** *If it was equipped originally, the spring seat must be in place before installing the seal.*

4   Beginning at one end of the head, lubricate and install the first valve. Apply moly-base grease or clean engine oil to the valve stem.

5   Set the valve spring and retainer in place.

6   Apply a small dab of grease to each keeper to hold it in place **(see illustration)**.

7   Compress the springs with a valve spring compressor and carefully install the keepers in the upper groove **(see illustration 9.2)**, then slowly release the compressor and make sure the keepers seat properly. Repeat the procedure for the remaining valves. Be sure to return the components to their original locations - don't mix them up!

8   Check the installed valve spring height with a ruler graduated in 1/32-inch increments or a dial caliper. If the head was sent out for service work, the installed height

should be correct (but don't automatically assume that it is). The measurement is taken from the top of each spring seat or shim(s) to the bottom of the retainer **(see illustration)**. If the height is greater than the figure listed in this Chapter's Specifications, shims can be added under the springs to correct it. **Caution:** *Don't, under any circumstances, shim the springs to the point where the installed height is less than specified.*

9   Apply moly-base grease to the rocker arm faces and the fulcrums, then install the rocker arms and fulcrums on the cylinder head studs.

13   Camshaft, balance shaft and bearings - removal and inspection

Removal

Refer to illustrations 13.3 and 13.4

1   The timing chain, sprockets and lifters must be removed before extracting the camshaft (see Part A of this Chapter).

2   Check the camshaft endplay by attach-

ing a dial indicator to the front of the block, with the indicator plunger touching the camshaft. Pry the camshaft back and forth and compare the endplay to Specifications.

3   On 3.8L engines, remove the balance shaft drive gear from the front of the camshaft **(see illustration)**.

4   Remove the two camshaft thrust plate bolts (Torx in most models), and any spacer that may be in place **(see illustration)**. Installing a long bolt in the front of the camshaft to use as a handle, guide the camshaft gently out of the block, without nicking the lobes or journals on the camshaft bearings in the block.

5   On 3.8L models only, remove the two bolts holding the balance shaft retaining plate **(see illustration 13.4)** and slide the balance shaft out of the block.

Inspection

Camshaft

Refer to illustration 13.6

6   Using a micrometer, measure the lobe at its highest point **(see illustration)**. Then

13.3 Slide the balance shaft drive gear (arrow) from the camshaft (3.8L only)

13.4 Remove the camshaft thrust plate bolts (arrows) - 3.8L shown, with balance shaft driven gear and balance shaft retaining plate above

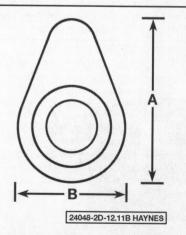

24048-2D-12.11B HAYNES

13.6 Measure the camshaft lobe maximum diameter (A) and the minimum (B) - the difference is the lobe lift

14.1 A ridge reamer is required to remove the ridge from the top of each cylinder - do this before removing the pistons!

14.3 Check the connecting rod side clearance (endplay) with a dial indicator or a feeler gauge

14.4 Mark the rod bearing caps in order from the front of the engine to the rear (numbers can be used or use one mark for the front cap, two for the second one and so on)

2B

measure the base circle perpendicular (90-degrees) to the lobe. Do this for each lobe and record the results.

7 Subtract the base circle measurement from the lobe height. The difference is the lobe lift. Compare your measurements to Specifications.

8 Inspect the camshaft journals for wear, pits or scoring. Measure the journals with a micrometer and compare the measurements to Specifications.

Balance shaft (3.8L only)

9 The balance shaft endplay can be checked as in Step 2 with a dial indicator before the balance shaft is removed from the block.

10 The front and rear journals (there are only two) of the balance shaft can be measured with a micrometer and compared to Specifications as with a camshaft.

11 The balance shaft rides in a bearing at the front of the block and a bushing at the rear. If the runout exceeds Specifications, replace the balance shaft, bearing and bushing as a set.

Camshaft/balance shaft bearing inspection

12 Check the camshaft/balance shaft bearings in the block for wear and damage. Look for galling, pitting and discolored areas.

13 The inside diameter of each bearing can be determined with a small hole gauge and outside micrometer or an inside micrometer. Subtract the camshaft/balance shaft bearing journal diameter(s) from the corresponding bearing inside diameter(s) to obtain the bearing oil clearance. If it's excessive, new bearings will be required regardless of the condition of the originals.

14 Balance shaft and camshaft bearing replacement requires special tools and expertise that place it outside the scope of the home mechanic. Take the block to an automotive machine shop to ensure the job is done correctly.

14 Pistons/connecting rods - removal

Refer to illustrations 14.1, 14.3, 14.4, 14.7 and 14.8

Note: *Prior to removing the piston/connecting rod assemblies, remove the cylinder head(s), the oil pan and the oil pump (on 5.0L engines) by referring to the appropriate Sections in Chapter 2, Part A or Part B, depending on which engine is being overhauled. On some models, there will be a sheet metal oil baffle plate attached with nuts to the main cap studs. Remove the nuts and the baffle.*

1 Use your fingernail to feel if a ridge has formed at the upper limit of ring travel (about 1/4-inch down from the top of each cylinder). If carbon deposits or cylinder wear have produced ridges, they must be completely removed with a special tool (see illustration). Follow the manufacturer's instructions provided with the tool. Failure to remove the ridges before attempting to remove the piston/connecting rod assemblies may result in piston breakage. **Note:** *Do not let the tool cut into the ring travel area more than 1/32-inch.*

2 After the cylinder ridges have been removed, turn the engine upside-down so the crankshaft is facing up.

3 Before the connecting rods are removed, check the endplay with a dial indicator or with feeler gauges (see illustration). Slide them between the first connecting rod and the crankshaft throw until the play is removed. The endplay is equal to the thickness of the feeler gauge(s). If the endplay exceeds the service limit, new connecting rods will be required. If new rods (or a new crankshaft) are installed, the endplay may fall under the specified minimum (if it does, the rods will have to be machined to restore it - consult an automotive machine shop for advice if necessary). Repeat the procedure for the remaining connecting rods.

4 Check the connecting rods and caps for identification marks. If they aren't plainly

14.7 To prevent damage to the crankshaft journals and cylinder walls, slip sections of rubber hose over the rod bolts during disassembly and assembly

marked, use a small center-punch, number stamping die (see illustration), or scribe, to make the appropriate number of indentations, or marks, on each rod and cap (1, 2, 3, etc., depending on the engine type and cylinder they're associated with).

5 Loosen each of the connecting rod cap bolts 1/2-turn at a time until they can be removed by hand.

6 Remove the connecting rod cap and bearing insert. Don't drop the bearing insert out of the cap.

7 To protect the rod journals of the crankshaft during disassembly or assembly, cover the rod bolts with rubber tubing, so that nothing is scratched as the piston/rod assembly is withdrawn or installed (see illustration). **Note:** *On 3.8L engines, the rod caps are retained by bolts that stay with the cap. On these engines, make studs by cutting off the heads of two rod bolts. Thread these in the rod after removing the cap, and cover these studs with rubber tubing. These rod bolts on 3.8L engines are not reusable. New ones must be installed on assembly.*

14.8 Use a hammer handle to drive the piston and connecting rod assembly down and out of the cylinder block, being very careful not to nick the crankshaft on the way out

15.1 Checking crankshaft endplay with a dial indicator

8   Remove the bearing insert and push the connecting rod/piston assembly out through the top of the engine. Use a wooden or plastic hammer handle to push on the upper bearing surface in the connecting rod **(see illustration)**. If resistance is felt, double-check to make sure that all of the ridge was removed from the cylinder.

9   Repeat the procedure for the remaining cylinders.

10   After removal, reassemble the connecting rod caps and bearing inserts in their respective connecting rods and install the cap nuts finger tight. Leaving the old bearing inserts in place until reassembly will help prevent the connecting rod bearing surfaces from being accidentally nicked or gouged.

11   Don't separate the pistons from the connecting rods (see Section 19 for additional information).

main bearings should correct the endplay.

3   If a dial indicator isn't available, feeler gauges can be used. Gently pry or push the crankshaft all the way to the front of the engine. Slip feeler gauges between the crankshaft and the front face of the thrust main bearing to determine the clearance **(see illustration)**.

4   Check the main bearing caps to see if they're marked to indicate their locations **(see illustration)**. They should be numbered consecutively from the front of the engine to the rear. If they aren't, mark them with number stamping dies or a center-punch. Main bearing caps generally have a cast-in arrow, which points to the front of the engine.

Note: *The thrust bearing is at the number three main bearing cap location. It has an upper and lower thrust bearing shell.*

5   Loosen the main bearing cap bolts/studs 1/4-turn at a time each, until they

can be removed by hand. Note if any stud bolts are used and make sure they're returned to their original locations when the crankshaft is reinstalled.

6   Gently tap the caps with a soft-face hammer, then separate them from the engine block. If necessary, use the bolts as levers to remove the caps. Try not to drop the bearing inserts if they come out with the caps. All main caps should have an arrow cast in to indicate the front of the engine, and a number to indicate which position they have on the block.

7   Carefully lift the crankshaft out of the engine. It may be a good idea to have an assistant available, since the crankshaft is quite heavy. With the bearing inserts in place in the engine block and main bearing caps, return the caps to their respective locations on the engine block and tighten the bolts finger tight.

15   Crankshaft - removal

Refer to illustrations 15.1, 15.3 and 15.4

Note: *The crankshaft can be removed only after the engine has been removed from the vehicle. It's assumed that the driveplate, vibration damper, timing chain(s) or gears, oil pan, oil pump and piston/connecting rod assemblies have already been removed.*

1   Before the crankshaft removal procedure is started, check the endplay. Mount a dial indicator with the stem in line with the crankshaft and just touching the end of the crankshaft **(see illustration)**.

2   Push the crankshaft all the way to the rear and zero the dial indicator. Next, pry the crankshaft to the front as far as possible and check the reading on the dial indicator. The distance that it moves is the endplay. If it's greater than limit listed in this Chapter's Specifications, check the crankshaft thrust surfaces for wear. If no wear is evident, new

15.3 Checking crankshaft endplay with a feeler gauge

15.4 The main bearing caps are usually marked to indicate their locations and direction - they should be numbered consecutively from the front of the engine to the rear

16.1a A hammer and a large punch can be used to knock the core plugs sideways in their bores

16.1b Pull the core plugs from the block with pliers

16.4 Remove the oil gallery plugs at the front and rear of the block (arrows) indicate two at the rear of a typical block

16 Engine block - cleaning

Refer to illustrations 16.1a, 16.1b, 16.4, 16.8 and 16.10

Caution: *The core plugs (also known as freeze plugs or soft plugs) may be difficult or impossible to retrieve if they're driven into the block coolant passages.*

1   Using the wide end of a punch **(see illustration)** tap in on the outer edge of the core plug to turn the plug sideways in the bore. Then, using a pair of pliers, pull the core plug from the engine block **(see illustration)**. Don't worry about the condition of the old core plugs as they are being removed because they will be replaced on reassembly with new plugs.

2   Using a gasket scraper, remove all traces of gasket material from the engine block. Be very careful not to nick or gouge the gasket sealing surfaces.

3   Remove the main bearing caps and separate the bearing inserts from the caps and the engine block (see Section 15). Tag the bearings, indicating which main cap/saddle they were removed from and whether they were in

the cap or the block, then set them aside.

4   Remove all of the threaded oil gallery plugs from the block **(see illustration)**. The plugs are usually very tight - they may have to be drilled out and the holes re-tapped. Use new plugs when the engine is reassembled.

5   If the engine is extremely dirty it should be taken to an automotive machine shop to be steam cleaned or hot tanked.

6   After the block is returned, clean all oil holes and oil galleries one more time. Brushes specifically designed for this purpose are available at most auto parts stores. Flush the passages with warm water until the water runs clear, dry the block thoroughly and wipe all machined surfaces with a light, rust preventive oil. If you have access to compressed air, use it to speed the drying process and to blow out all the oil holes and galleries. **Warning:** *Wear eye protection when using compressed air!*

7   If the block isn't extremely dirty or sludged up, you can do an adequate cleaning job with hot soapy water and a stiff brush. Take plenty of time and do a thorough job. Regardless of the cleaning method used, be sure to clean all oil holes and galleries very thoroughly, dry the block completely and

coat all machined surfaces with light oil.

8   The threaded holes in the block must be clean to ensure accurate torque readings during reassembly. Run the proper size tap into each of the holes to remove rust, corrosion, thread sealant or sludge and restore damaged threads **(see illustration)**. If possible, use compressed air to clear the holes of debris produced by this operation.

9   Reinstall the main bearing caps and tighten all bolts finger tight.

10   After coating the sealing surfaces of the new core plugs with Permatex no. 2 sealant (or equivalent), install them in the engine block **(see illustration)**. Make sure they're driven in straight and seated properly or leakage could result. Special tools are available for this purpose, but a large socket, with an outside diameter that will just slip into the core plug, a 1/2-inch drive extension and a hammer will work just as well.

11   Apply non-hardening sealant (such as Permatex no. 2 or Teflon pipe sealant) to the new oil gallery plugs and thread them into the holes in the block. Make sure they're tightened securely.

12   If the engine isn't going to be reassembled right away, cover it with a large plastic trash bag to keep it clean.

17 Engine block - inspection

Refer to illustrations 17.4a, 17.4b and 17.4c

1   Before the block is inspected, it should be cleaned as described in Section 16.

2   Visually check the block for cracks, rust and corrosion. Look for stripped threads in the threaded holes. It's also a good idea to have the block checked for hidden cracks by an automotive machine shop that has the special equipment to do this type of work. If defects are found, have the block repaired, if possible, or replaced.

3   Check the cylinder bores for scuffing and scoring.

4   Check the cylinders for taper and out-of-round conditions as follows **(see illustrations)**.

16.8 All bolt holes in the block - particularly the main bearing cap and head bolt holes - should be cleaned and restored with a tap (be sure to remove debris from the holes after this is done)

16.10 A large socket on an extension can be used to drive the new core plugs into the bores

2B

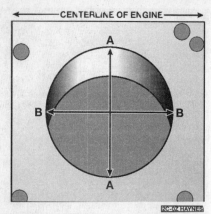

17.4a Measure the diameter of each cylinder at a right angle to the engine centerline (A), and parallel to engine centerline (B) - out-of-round is the difference between A and B; taper is the difference between A and B at the top of the cylinder and A and B at the bottom of the cylinder

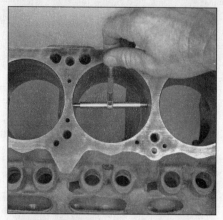

17.4b The ability to "feel" when the telescoping gauge is at the correct point will be developed over time, so work slowly and repeat the check until you're satisfied the bore measurement is accurate

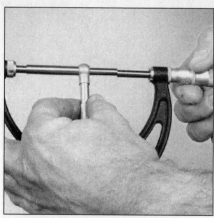

17.4c The gauge is then measured with a micrometer to determine the bore size

18.3a A "bottle brush" hone is much easier to use than a spring-loaded, stone type hone

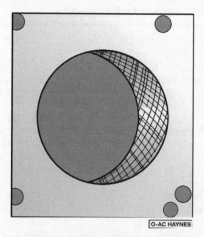

18.3b The cylinder hone should leave a smooth, crosshatch pattern with the lines intersecting at approximately a 60-degree angle

5   Measure the diameter of each cylinder at the top (just under the ridge area), center and bottom of the cylinder bore, parallel to the crankshaft axis.

6   Next measure each cylinder's diameter at the same three locations perpendicular to the crankshaft axis.

7   The taper of the cylinder is the difference between the bore diameter at the top of the cylinder and the diameter at the bottom. The out-of-round specification of the cylinder bore is the difference between the parallel and perpendicular readings. Compare your results to those listed in this Chapter's Specifications.

8   Repeat the procedure for the remaining pistons and cylinders.

9   If the cylinder walls are badly scuffed or scored, or if they're out-of-round or tapered beyond the limits given in this Chapter's Specifications, have the engine block rebored and honed at an automotive machine shop. If a rebore is done, oversize pistons and rings will be required.

10   If the cylinders are in reasonably good condition and not worn to the outside of the limits, and if the piston-to-cylinder clearances can be maintained properly, then they don't have to be rebored. Honing is all that's necessary (see Section 18).

18   Cylinder honing

Refer to illustrations 18.3a and 18.3b

1   Prior to engine reassembly, the cylinder bores must be honed so the new piston rings will seat correctly and provide the best possible combustion chamber seal. **Note:** *If you don't have the tools or don't want to tackle the honing operation, most automotive machine shops will do it for a reasonable fee.*

2   Before honing the cylinders, install the main bearing caps and tighten the bolts to the torque listed in this Chapter's Specifications. Make sure you use only the original main cap bolts, not the new ones for final assembly.

3   Two types of cylinder hones are commonly available - the flex hone or "bottle brush" type and the more traditional surfacing hone with spring-loaded stones. Both will do the job, but for the less experienced mechanic the "bottle brush" hone will probably be easier to use. You'll also need some kerosene or honing oil, rags and a 1/2-inch electric drill motor. Proceed as follows:

a) *Mount the hone in the drill motor, compress the stones and slip it into the first cylinder* **(see illustration)**. *Be sure to wear safety goggles or a face shield!*

b) *Lubricate the cylinder with plenty of honing oil, turn on the drill and move the hone up-and-down in the cylinder at a pace that will produce a fine crosshatch pattern on the cylinder walls. Ideally, the crosshatch lines should intersect at approximately a 60-degree angle* **(see illustration)**. *Be sure to use plenty of lubricant and don't take off any more*

material than is absolutely necessary to produce the desired finish. **Note:** *Piston ring manufacturers may specify a smaller crosshatch angle than the traditional 60-degrees - read and follow any instructions included with the new rings.*

c) *Don't withdraw the hone from the cylinder while it's running. Instead, shut off the drill and continue moving the hone up-and-down in the cylinder until it comes to a complete stop, then compress the stones and withdraw the hone. If you're using a "bottle brush" type hone, stop the drill motor, then turn the chuck in the normal direction of rotation while withdrawing the hone from the cylinder.*

d) *Wipe the oil out of the cylinder and repeat the procedure for the remaining cylinders.*

4   After the honing job is complete, chamfer the top edges of the cylinder bores with a small file so the rings won't catch when the pistons are installed. Be very careful not to nick the cylinder walls with the end of the file.

19.4a The piston ring grooves can be cleaned with a special tool, as shown here . . .

19.4b . . . or a section of a broken ring

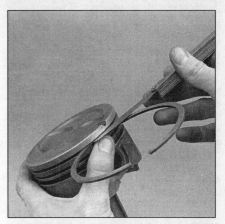

19.10 Check the ring side clearance with a feeler gauge at several points around the groove

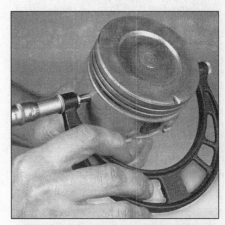

19.11 Measure the piston diameter at a 90-degree angle to the piston pin and in line with it

5   The entire engine block must be washed again very thoroughly with warm, soapy water to remove all traces of the abrasive grit produced during the honing operation. **Note:** *The bores can be considered clean when a lint-free white cloth - dampened with clean engine oil - used to wipe them out doesn't pick-up any more honing residue, which will show up as gray areas on the cloth. Be sure to run a brush through all oil holes and galleries and flush them with running water.*
6   After rinsing, dry the block and apply a coat of light rust preventive oil to all machined surfaces. Wrap the block in a plastic trash bag to keep it clean and set it aside until reassembly.

19 Pistons/connecting rods - inspection

Refer to illustrations 19.4a, 19.4b, 19.10 and 19.11
1   Before the inspection process can be carried out, the piston/connecting rod assemblies must be cleaned and the original piston rings removed from the pistons. **Note:** *Always use new piston rings when the engine is reassembled.*
2   Using a piston ring installation tool, carefully remove the rings from the pistons **(see illustration 23.11)**. Be careful not to nick or gouge the pistons in the process.
3   Scrape all traces of carbon from the top of the piston. A hand-held wire brush or a piece of fine emery cloth can be used once the majority of the deposits have been scraped away. Do not, under any circumstances, use a wire brush mounted in a drill motor or caustic chemicals to remove deposits from the pistons. The piston material is soft and may be eroded away.
4   Use a piston ring groove cleaning tool to remove carbon deposits from the ring grooves. If a tool isn't available, a piece broken off the old ring will do the job **(see illustrations)**. Be very careful to remove only the carbon deposits - don't remove any metal

and do not nick or scratch the sides of the ring grooves.
5   Once the deposits have been removed, clean the piston/rod assemblies with solvent and dry them with compressed air (if available). Make sure the oil return holes in the back sides of the ring grooves are clear.
6   If the pistons and cylinder walls aren't damaged or worn excessively, and if the engine block is not rebored, new pistons won't be necessary. Normal piston wear appears as even vertical wear on the piston thrust surfaces and slight looseness of the top ring in its groove. New piston rings, however, should always be used when an engine is rebuilt.
7   Carefully inspect each piston for cracks around the skirt, at the pin bosses and at the ring lands.
8   Look for scoring and scuffing on the thrust faces of the skirt, holes in the piston crown and burned areas at the edge of the crown. If the skirt is scored or scuffed, the engine may have been suffering from overheating and/or abnormal combustion, which caused excessively high operating temperatures. The cooling and lubrication systems should be checked thoroughly. A hole in the piston crown is an indication that abnormal combustion (pre-ignition) was occurring. Burned areas at the edge of the piston crown are usually evidence of spark knock (detonation). If any of the above problems exist, the causes must be corrected or the damage will occur again. The causes may include intake air leaks, incorrect fuel/air mixture, incorrect ignition timing and EGR system malfunctions.
9   Corrosion of the piston, in the form of small pits, indicates that coolant is leaking into the combustion chamber and/or the crankcase. Again, the cause must be corrected or the problem may persist in the rebuilt engine.
10   Measure the piston ring side clearance by laying a new piston ring in each ring groove and slipping a feeler gauge in beside it **(see illustration)**. Check the clearance at three or four locations around each groove. Be sure to use the correct ring for each groove - they are different. If the side clear-

ance is greater than the figure listed in this Chapter's Specifications, new pistons will have to be used.
11   Check the piston-to-bore clearance by measuring the bore (see Section 16) and the piston diameter. Make sure the pistons and bores are correctly matched. Measure the piston across the skirt, at a 90-degree angle to, and in line with, the piston pin **(see illustration)**. Subtract the piston diameter from the bore diameter to obtain the clearance. If it's greater than specified, the block will have to be rebored and new pistons and rings installed.
12   Check the piston-to-rod clearance by twisting the piston and rod in opposite directions. Any noticeable play indicates excessive wear, which must be corrected. The big-end of the rod should be checked for taper and out-of-round conditions, too. The piston/connecting rod assemblies should be taken to an automotive machine shop to have the pistons and rods resized and new pins installed.
13   If the pistons must be removed from the connecting rods for any reason, they should be taken to an automotive machine shop. While they are there have the connecting rods checked for bend and twist, since auto-

2B

motive machine shops have special equipment for this purpose. **Note:** *Unless new pistons and/or connecting rods must be installed, do not disassemble the pistons and connecting rods.*

14   Check the connecting rods for cracks and other damage. Temporarily remove the rod caps, lift out the old bearing inserts, wipe the rod and cap bearing surfaces clean and inspect them for nicks, gouges and scratches. After checking the rods, replace the old bearings, slip the caps into place and tighten the nuts finger tight. **Note:** *If the engine is being rebuilt because of a connecting rod knock, be sure to install new rods.*

20   Crankshaft - inspection

Refer to illustrations 20.1, 20.2, 20.5 and 20.7

1   Remove all burrs from the crankshaft oil holes with a stone, file or scraper (see illustration).
2   Clean the crankshaft with solvent and dry it with compressed air (if available). Be sure to clean the oil holes with a stiff brush (see illustration) and flush them with solvent.
3   Check the main and connecting rod bearing journals for uneven wear, scoring, pits and cracks.
4   Check the rest of the crankshaft for cracks and other damage. It should be Magnafluxed to reveal hidden cracks - an automotive machine shop will handle the procedure.
5   Using a micrometer, measure the diameter of the main and connecting rod journals and compare the results to this Chapter's Specifications (see illustration). By measuring the diameter at a number of points around each journal's circumference, you'll be able to determine whether or not the journal is out-of-round. Take the measurement at each end of the journal, near the crank throws, to determine if the journal is tapered.
6   If the crankshaft journals are damaged, tapered, out-of-round or worn beyond the limits given in the Specifications, have the

20.1 The oil holes should be chamfered so sharp edges don't gouge or scratch the new bearings

crankshaft reground by an automotive machine shop. Be sure to use the correct size bearing inserts if the crankshaft is reconditioned.
7   Check the oil seal journals at each end of the crankshaft for wear and damage. If the seal has worn a groove in the journal, or if it's nicked or scratched (see illustration), the new seal may leak when the engine is reassembled. In some cases, an automotive machine shop may be able to repair the journal by pressing on a thin sleeve. If repair isn't feasible, a new or different crankshaft should be installed.
8   Refer to Section 21 and examine the main and rod bearing inserts.

21   Main and connecting rod bearings - inspection

Refer to illustration 21.1

1   Even though the main and connecting rod bearings should be replaced with new ones during the engine overhaul, the old bearings should be retained for close examination, as they may reveal valuable informa-

20.2 Use a wire or stiff plastic bristle brush to clean the oil passages in the crankshaft

tion about the condition of the engine (see illustration).
2   Bearing failure occurs because of lack of lubrication, the presence of dirt or other foreign particles, overloading the engine and corrosion. Regardless of the cause of bearing failure, it must be corrected before the engine is reassembled to prevent it from happening again.
3   When examining the bearings, remove them from the engine block, the main bearing caps, the connecting rods and the rod caps and lay them out on a clean surface in the same general position as their location in the engine. This will enable you to match any bearing problems with the corresponding crankshaft journal.
4   Dirt and other foreign particles get into the engine in a variety of ways. It may be left in the engine during assembly, or it may pass through filters or the PCV system. It may get into the oil, and from there into the bearings. Metal chips from machining operations and normal engine wear are often present. Abrasives are sometimes left in engine components after reconditioning, especially when parts are not thoroughly cleaned using the

20.5 Measure the diameter of each crankshaft journal at several points to detect taper and out-of-round conditions

20.7 If the seals have worn grooves in the crankshaft journals, or if the seal contact surfaces are nicked or scratched, the new seals will leak

proper cleaning methods. Whatever the source, these foreign objects often end up embedded in the soft bearing material and are easily recognized. Large particles will not embed in the bearing and will score or gouge the bearing and journal. The best prevention for this cause of bearing failure is to clean all parts thoroughly and keep everything spotlessly clean during engine assembly. Frequent and regular engine oil and filter changes are also recommended.

5   Lack of lubrication (or lubrication breakdown) has a number of interrelated causes. Excessive heat (which thins the oil), overloading (which squeezes the oil from the bearing face) and oil leakage or throw off (from excessive bearing clearances, worn oil pump or high engine speeds) all contribute to lubrication breakdown. Blocked oil passages, which usually are the result of misaligned oil holes in a bearing shell, will also oil starve a bearing and destroy it. When lack of lubrication is the cause of bearing failure, the bearing material is wiped or extruded from the steel backing of the bearing. Temperatures may increase to the point where the steel backing turns blue from overheating.

6   Driving habits can have a definite effect on bearing life. Low speed operation in too high a gear (lugging the engine) puts very high loads on bearings, which tends to squeeze out the oil film. These loads cause the bearings to flex, which produces fine cracks in the bearing face (fatigue failure). Eventually the bearing material will loosen in pieces and tear away from the steel backing. Short trip driving leads to corrosion of bearings because insufficient engine heat is pro-

duced to drive off the condensed water and corrosive gases. These products collect in the engine oil, forming acid and sludge. As the oil is carried to the engine bearings, the acid attacks and corrodes the bearing material.

7   Incorrect bearing installation during engine assembly will lead to bearing failure as well. Tight fitting bearings leave insufficient bearing oil clearance and will result in oil starvation. Dirt or foreign particles trapped behind a bearing insert result in high spots on the bearing which lead to failure.

22 Engine overhaul - reassembly sequence

1   Before beginning engine reassembly, make sure you have all the necessary new parts (including new head bolts, and new rod bolts on 3.8L engines), gaskets and seals as well as the following items on hand:

Common hand tools
A 1/2-inch drive torque wrench
A 3/8-inch drive torque wrench (inch-lb. measurement)
Piston ring installation tool
Piston ring compressor
Vibration damper installation tool
Connecting rod guide bolts
Plastigage
Feeler gauges
A fine-tooth file
New engine oil
Engine assembly lube
Gasket sealant
Thread-locking compound

2   In order to save time and avoid problems, engine reassembly must be done in the following general order:

New camshaft bearings to be installed by an automotive machine shop)
New balance shaft bearings (to be installed by an automotive machine shop)
Piston rings
Crankshaft and main bearings
Piston/connecting rod assemblies
Oil pump
Oil pan
Camshaft
Valve lifters
Timing chain and sprockets
Timing chain cover
Cylinder heads
Rocker arms and pushrods
Intake and exhaust manifolds
Valve covers
Rear main seal
Driveplate

23 Piston rings - installation

Refer to illustrations 23.3, 23.4, 23.8a, 23.8b and 23.11

1   Before installing the new piston rings, the ring end gaps must be checked. It's assumed that the piston ring side clearance has been checked and verified correct (see Section 19).

2   Lay out the piston/connecting rod assemblies and the new ring sets so the ring sets will be matched with the same piston and cylinder during the end gap measurement and engine assembly.

3   Insert the top (number one) ring into the first cylinder and square it up with the cylinder walls by pushing it in with the top of the piston **(see illustration)**. The ring should be near the bottom of the cylinder, at the lower limit of ring travel.

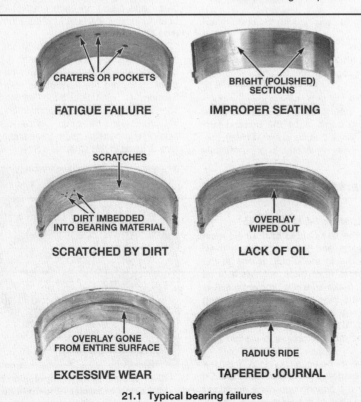

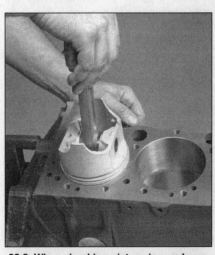

23.3 When checking piston ring end gap, the ring must be square in the cylinder bore (this is done by pushing the ring down with the top of a piston as shown)

FATIGUE FAILURE — CRATERS OR POCKETS

IMPROPER SEATING — BRIGHT (POLISHED) SECTIONS

SCRATCHED BY DIRT — SCRATCHES / DIRT IMBEDDED INTO BEARING MATERIAL

LACK OF OIL — OVERLAY WIPED OUT

EXCESSIVE WEAR — OVERLAY GONE FROM ENTIRE SURFACE

TAPERED JOURNAL — RADIUS RIDE

21.1 Typical bearing failures

23.4 With the ring square in the cylinder, measure the end gap with a feeler gauge

23.8a Installing the spacer/expander in the oil control ring groove . . .

23.8b . . . followed by the side rails - DO NOT use a piston ring installation tool when installing the oil ring side rails

4   To measure the end gap, slip feeler gauges between the ends of the ring until a gauge equal to the gap width is found **(see illustration)**. The feeler gauge should slide between the ring ends with a slight amount of drag. Compare the measurement to this Chapter's Specifications. If the gap is larger or smaller than specified, double-check to make sure you have the correct rings before proceeding. If there is any doubt contact the parts store where the rings were purchased, to verify that the correct ring set is being used.

5   Excess end gap isn't critical unless it's greater than 0.040-inch. Again, double-check to make sure you have the correct rings for your engine.

6   Repeat the procedure for each ring that will be installed in the first cylinder and for each ring in the remaining cylinders. Remember to keep rings, pistons and cylinders matched up.

7   Once the ring end gaps have been checked/corrected, the rings can be installed on the pistons.

8   The oil control ring (lowest one on the piston) is usually installed first. It's composed of three separate components. Slip the spacer/expander into the groove **(see illustration)**. Next, install the lower side rail. Don't use a piston ring installation tool on the oil ring side rails, as they may be damaged. Instead, place one end of the side rail into the groove between the spacer/expander and the ring land, hold it firmly in place and slide a finger around the piston while pushing the rail into the groove **(see illustration)**. Next, install the upper side rail in the same manner.

9   After the three oil ring components have been installed, check to make sure that both the upper and lower side rails can be turned smoothly in the ring groove.

10   The number two (middle) ring is installed next. It's usually stamped with a mark which must face up, toward the top of the piston. **Note:** *Always follow the instructions printed on the ring package or box - different manufacturers may require different approaches. Do not mix up the top and middle rings, as they have different cross sections.*

11   Use a piston ring installation tool and make sure the identification mark is facing

the top of the piston, then slip the ring into the middle groove on the piston **(see illustration)**. Don't expand the ring any more than necessary to slide it over the piston.

12   Install the number one (top) ring in the same manner. Make sure the mark is facing up. Be careful not to confuse the number one and number two rings.

13   Repeat the procedure for the remaining pistons and rings.

24 Crankshaft - installation and main bearing oil clearance check

1   Crankshaft installation is the first step in engine reassembly. It's assumed at this point that the engine block and crankshaft have been cleaned, inspected and repaired or reconditioned.

2   Position the engine with the bottom facing up.

3   Remove the main bearing cap bolts/studs and lift out the caps. Lay them out in the proper order to ensure correct installation.

4   If they're still in place, remove the original bearing inserts from the block and the main bearing caps. Wipe the bearing surfaces of the block and caps with a clean, lint-free cloth. They must be kept spotlessly clean.

Main bearing oil clearance check

Refer to illustrations 24.11 and 24.15

5   Clean the back sides of the new main bearing inserts and lay one in each main bearing saddle in the block. If one of the bearing inserts from each set has a large groove in it, make sure the grooved insert is installed in the block. Lay the other bearing from each set in the corresponding main bearing cap. Make sure the tab on the bearing insert fits into the recess in the block or cap.

Caution: *The oil holes in the block must line up with the oil holes in the bearing insert. Do not hammer the bearing into place and don't nick or gouge the bearing faces. No lubrica-*

tion should be used at this time.

6   The flanged thrust bearing must be installed in the third cap and saddle.

7   Clean the faces of the bearings in the block and the crankshaft main bearing journals with a clean, lint-free cloth.

8   Check or clean the oil holes in the crankshaft, as any dirt here can go only one way - straight through the new bearings.

9   Once you're certain the crankshaft is clean, carefully lay it in position in the main bearings.

10   Before the crankshaft can be permanently installed, the main bearing oil clearance must be checked.

11   Cut several pieces of the appropriate-size Plastigage (they must be slightly shorter than the width of the main bearings) and place one piece on each crankshaft main bearing journal, parallel with the journal axis **(see illustration)**.

12   Clean the faces of the bearings in the caps and install the caps in their respective positions (don't mix them up) with the arrows pointing toward the front of the engine (see Section 14). Don't disturb the Plastigage.

13   Starting with the center main and working out toward the ends, tighten the main bearing cap bolts/studs, in three steps, to the torque listed in this Chapter's Specifications.

23.11 Installing the compression rings with a ring expander - the mark (arrow) must face up

24.11 Lay the Plastigage strips on the main bearing journals, parallel to the crankshaft centerline

24.15 Compare the width of the crushed Plastigage to the scale on the envelope to determine the main bearing oil clearance (always take the measurement at the widest point of the Plastigage); be sure to use the correct scale - standard and metric ones are included

25.1 Coat the camshaft lobes and journals with assembly lube before installation

14   Remove the bolts/studs and carefully lift off the main bearing caps. Keep them in order. Don't disturb the Plastigage or rotate the crankshaft. If any of the main bearing caps are difficult to remove, tap them gently from side-to-side with a soft-face hammer to loosen them.

15   Compare the width of the crushed Plastigage on each journal to the scale printed on the Plastigage envelope to obtain the main bearing oil clearance **(see illustration)**. Check the Specifications to make sure it's correct.

16   If the clearance is not as specified, the bearing inserts may be the wrong size (which means different ones will be required). Before deciding that different inserts are needed, make sure that no dirt or oil was between the bearing inserts and the caps or block when the clearance was measured. If the Plastigage was wider at one end than the other, the journal may be tapered (refer to Section 20).

17   Carefully scrape all traces of the Plastigage material off the main bearing journals and/or the bearing faces. Use your fingernail or the edge of a credit card - don't nick or scratch the bearing faces.

Final crankshaft installation

18   Carefully lift the crankshaft out of the engine.

19   Clean the bearing faces in the block, then apply a thin, uniform layer of moly-base grease or engine assembly lube to each of the bearing surfaces. Be sure to coat the thrust faces as well as the journal face of the thrust bearing.

20   Make sure the crankshaft journals are clean, then lay the crankshaft back in place in the block.

21   Clean the faces of the bearings in the caps, then apply lubricant to them.

22   Install the caps in their respective positions with the arrows pointing toward the front of the engine. **Note:** *Apply a thin bead of RTV sealant at the side joints between the rear main cap and the block. Install the cap and tighten the bolts within four minutes of applying the sealant.*

23   Install the main cap bolts.

24   Tighten all, except the thrust bearing cap bolts (number 3), to the torque listed in this Chapter's Specifications (work from the center out and approach the final torque in three steps).

25   Tighten the thrust bearing cap bolts finger tight.

26   Pry the crankshaft forward and while holding pressure on the crankshaft, pry the thrust bearing cap backward. Forcing these two in opposite directions, against each other, will align the thrust bearing surfaces.

27   While keeping forward pressure on the crankshaft, re-tighten ALL main bearing cap bolts to the torque listed in this Chapter's Specifications, starting with cap #3, then #2, #4 and #1.

28   Rotate the crankshaft a number of times by hand to check for any obvious binding.

29   The final step is to check the crankshaft endplay with a feeler gauge or a dial indicator (see Section 14). The endplay should be correct if the crankshaft thrust faces aren't worn or damaged and new bearings have been installed.

30   Install the rear main oil seal (see Chapter 2, Part A).

25   Camshaft and balance shaft - installation

Camshaft

Refer to illustration 25.1

1   Lubricate the camshaft bearing journals and cam lobes with a special camshaft installation lubricant **(see illustration)**.

2   Slide the camshaft into the engine, using a long bolt screwed into the front of the camshaft as a "handle". Support the cam near the block and be careful not to scrape or nick the bearings. Install the camshaft retainer plate and tighten the bolts to the torque listed in the Specifications.

3   On 3.0L engines, complete the installation of the timing chain and sprockets (see Part A). On 3.8L engines, perform the following procedure on balance shaft installation before installing the timing chain and sprockets.

Balance shaft (3.8L only)

Refer to illustration 25.6

4   Lubricate the front bearing and rear journal of the balance shaft with engine oil and insert the balance shaft, with its driven gear and retainer plate, carefully into the block.

5   Install and tighten the balance shaft retainer plate bolts.

6   Point the mark on the balance shaft driven gear (on balance shaft) straight down, then install the balance shaft drive gear over the front of the camshaft, aligning the marks on the two gears **(see illustration)**.

7   Tighten the balance shaft retainer bolts, and after the camshaft sprocket, crankshaft sprocket and timing chain have been installed (see Part A), tighten the balance shaft sprocket bolt.

2B

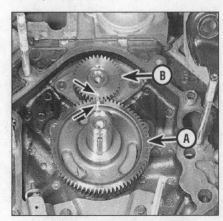

25.6 Align the marks (arrows) on the balance shaft drive gear on the camshaft (A) and the balance shaft driven gear (B) above it

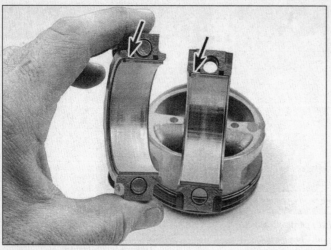

26.3 Insert the connecting rod bearing halves, making sure the bearing tab (arrows) are in the notches in the rod and cap

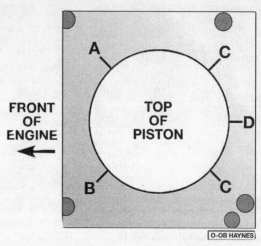

26.5 Ring end gap positions - align the oil ring spacer gap at D, the oil ring side rails at C (one inch either side of the pin centerline), and the compression rings at A and B, one inch either side of the pin centerline

26   Pistons/connecting rods - installation and rod bearing oil clearance check

Caution: *The connecting rod bolts on 3.8L engines are "torque-to-yield" bolts and are NOT reusable. A pre-determined stretch of the bolt, calculated by the manufacturer, gives the added rigidity required with this cylinder block. Once removed they must be replaced with new bolts. During clearance checks using Plastigage, use the old bolts and torque to Specifications, but use only new bolts for final assembly.*

1   Before installing the piston/connecting rod assemblies, the cylinder walls must be perfectly clean, the top edge of each cylinder must be chamfered, and the crankshaft must be in place.

2   Remove the cap from the end of the number one connecting rod (refer to the marks made during removal). Remove the original bearing inserts and wipe the bearing surfaces of the connecting rod and cap with a clean, lint-free cloth. They must be kept spotlessly clean.

Connecting rod bearing oil clearance check

Refer to illustrations 26.3, 26.5, 26.8, 26.9a, 26.9b, 26.11, 26.13 and 26.17

3   Clean the back side of the new upper bearing insert, then lay it in place in the connecting rod **(see illustration)**. Make sure the tab on the bearing fits into the recess in the rod. Don't hammer the bearing insert into place and be very careful not to nick or gouge the bearing face. Don't lubricate the bearing at this time.

4   Clean the back side of the other bearing insert and install it in the rod cap. Again, make sure the tab on the bearing fits into the recess in the cap, and don't apply any lubricant. It's critically important that the mating surfaces of the bearing and connecting rod are perfectly clean and oil free when they're assembled.

5   Position the piston ring gaps at intervals around the piston **(see illustration)**.

6   Lubricate the piston and rings with clean engine oil and attach a piston ring compressor to the piston. Leave the skirt protruding about 1/4-inch to guide the piston into the cylinder. The rings must be compressed until they're flush with the piston.

7   Rotate the crankshaft until the number one connecting rod journal is at BDC (bottom dead center) and apply a coat of engine oil to the cylinder walls.

8   With the arrow or notch on top of the piston **(see illustration)** facing the front of the engine, gently insert the piston/connecting rod assembly into the number one cylinder bore and rest the bottom edge of the ring compressor on the engine block.

9   The rod bolts on 3.8L engines are pressed lightly into the rod caps. They must

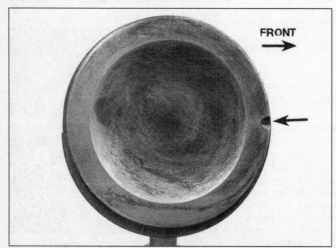

26.8 Turn the piston when installing it to make sure the mark/notch in the piston faces the front of the engine as they are installed

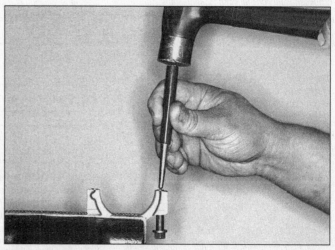

26.9a On 3.8L engines only, hold the rod cap lightly in a vise while driving out the old rod bolts - it takes only a little tap to drive them out; then drive new bolts in

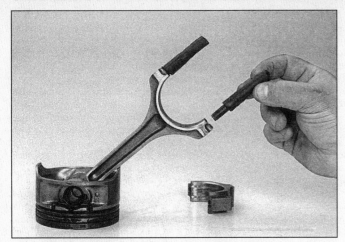

26.9b Two old rod bolts can be used (with the heads cut off) as rod installation guides with rubber hose over them

26.11 The piston can be driven gently into the cylinder bore with the end of a wooden or plastic hammer handle

be replaced with new bolts, so tap out the old ones **(see illustration)**. Take two of these and cut the heads off, slip rubber hose over them, and screw them into the connecting rod to use as alignment devices during piston/rod installation **(see illustration)**.

10   Tap the top edge of the ring compressor to make sure it's contacting the block around its entire circumference.

11   Gently tap on the top of the piston with the end of a wooden hammer handle **(see illustration)** while guiding the end of the connecting rod into place on the crankshaft journal. The piston rings may try to pop out of the ring compressor just before entering the cylinder bore, so keep some downward pressure on the ring compressor. Work slowly, and if any resistance is felt as the piston enters the cylinder, stop immediately. Find out what's hanging up and fix it before proceeding. Do not, for any reason, force the piston into the cylinder - you might break a ring and/or the piston.

12   Once the piston/connecting rod assembly is installed, the connecting rod bearing oil clearance must be checked before the rod

cap is permanently bolted in place.

13   Cut a piece of the appropriate-size Plastigage slightly shorter than the width of the connecting rod bearing and lay it in place on the number one connecting rod journal, parallel with the journal axis **(see illustration)**.

14   Clean the connecting rod cap bearing face, remove the protective "studs" from the connecting rod and install the rod cap (with the **old** bolts on 3.8L engines). Make sure the mating mark on the cap is on the same side as the mark on the connecting rod.

15   Tighten the bolts to the torque listed in this Chapter's Specifications, working up to it in three steps. **Note:** *Use a thin-wall socket to avoid erroneous torque readings that can result if the socket is wedged between the rod cap and bolt. If the socket tends to wedge itself between the bolt and the cap, lift up on it slightly until it no longer contacts the cap. Do not rotate the crankshaft at any time during this operation.*

16   Loosen the bolts/nuts and detach the rod cap, being very careful not to disturb the Plastigage.

17   Compare the width of the crushed Plas-

tigage to the scale printed on the Plastigage envelope to obtain the oil clearance **(see illustration)**. Compare it to the Specifications to make sure the clearance is correct.

18   If the clearance is not as specified, the bearing inserts may be the wrong size (which means different ones will be required). Before deciding that different inserts are needed, make sure that no dirt or oil was between the bearing inserts and the connecting rod or cap when the clearance was measured. Also, recheck the journal diameter. If the Plastigage was wider at one end than the other, the journal may be tapered (refer to Section 20).

Final connecting rod installation

19   Carefully scrape all traces of the Plastigage material off the rod journal and/or bearing face. Be very careful not to scratch the bearing - use your fingernail or the edge of a credit card.

20   Make sure the bearing faces are perfectly clean, then apply a uniform layer of

2B

26.13 Lay the Plastigage strips on each rod bearing journal, parallel to the crankshaft centerline

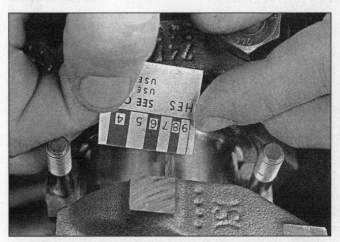

26.17 Measuring the width of the crushed Plastigage to determine the rod bearing oil clearance (be sure to use the correct scale - standard and metric ones are included)

clean moly-base grease or engine assembly lube to both of them. You'll have to push the piston up into the cylinder to expose the face of the bearing insert in the connecting rod - be sure to slip the protective hoses over the rod bolts first.

21   On 3.8L engines, slide the connecting rod back into place on the journal, remove the protective "studs" you installed, install the rod cap (this time with the **new** bolts) and tighten the bolts to the torque listed in this Chapter's Specifications. Again, work up to the torque in three steps. On 3.0L engines, tighten the rod cap nuts to Specifications in three steps also.

22   Repeat the entire procedure for the remaining pistons/connecting rods.

23   The important points to remember are:
  a) *Keep the back sides of the bearing inserts and the insides of the connecting rods and caps perfectly clean when assembling them.*
  b) *Make sure you have the correct piston/rod assembly for each cylinder.*
  c) *The notches or mark on the piston must face the FRONT of the engine.*
  d) *Lubricate the cylinder walls with clean oil.*
  e) *Lubricate the bearing faces when installing the rod caps after the oil clearance has been checked.*

24   After all the piston/connecting rod assemblies have been properly installed, rotate the crankshaft a number of times by hand to check for any obvious binding.

25   As a final step, the connecting rod side clearance must be checked. Refer to Section 14 for this procedure.

26   Compare the measured side clearance to the Specifications to make sure it's correct. If it was correct before disassembly and the original crankshaft and rods were reinstalled, it should still be right. If new rods or a new crankshaft were installed, the endplay may be inadequate. If so, the rods will have to be removed and taken to an automotive machine shop for resizing.

27   Initial start-up and break-in after overhaul

Warning: *Have a fire extinguisher handy when starting the engine for the first time.*

1   Once the engine has been installed in the vehicle, double-check the engine oil and coolant levels.

2   With the spark plugs out of the engine and the ignition and fuel systems disabled (see Section 4, Step 5), crank the engine until oil pressure registers on the gauge or the light goes out.

3   Install the spark plugs, hook up the plug wires and restore the ignition and fuel system functions.

4   Start the engine. It may take a few moments for the fuel system to build up pressure, but the engine should start without a great deal of effort. **Note:** *If backfiring occurs through the throttle body, recheck the valve timing and ignition timing.*

5   After the engine starts, it should be allowed to warm up to normal operating temperature. While the engine is warming up, make a thorough check for fuel, oil and coolant leaks.

6   Shut the engine off and recheck the engine oil and coolant levels.

7   Drive the vehicle to an area with no traffic, accelerate from 30 to 50 mph, then allow the vehicle to slow to 30 mph with the throttle closed. Repeat the procedure 10 or 12 times. This will load the piston rings and cause them to seat properly against the cylinder walls. Check again for oil and coolant leaks.

8   Drive the vehicle gently for the first 500 miles (no sustained high speeds) and keep a constant check on the oil level. It is not unusual for an engine to use oil during the break-in period.

9   At approximately 500 to 600 miles, change the oil and filter.

10   For the next few hundred miles, drive the vehicle normally. Do not pamper it or abuse it.

11   After 2000 miles, change the oil and filter again and consider the engine broken in.

Chapter 3
Cooling, heating and air conditioning systems

Contents

Specifications

Cooling system capacity
- Standard heater ... 12 quarts
- Auxiliary rear heater ... 14 quarts

Coolant type ... 50/50 mixture of non-phosphate ethylene glycol antifreeze and demineralized water

Thermostat
- Opening temperature ... 188 to 195-degrees
- Fully open temperature ... 208 to 215-degrees

Radiator pressure cap rating ... 16 psi
Refrigerant type ... R-134a
Refrigerant capacity
- Without auxiliary unit ... 44.0 ounces
- With auxiliary unit ... 56.0 ounces

Torque specifications
Ft-lbs (unless otherwise indicated)

Thermostat housing bolts
- 3.0L ... 89 to 123 in-lbs
- 3.8L ... 71 to 97 in-lbs

Water pump bolts
- 8mm bolts ... 15 to 22
- 6mm bolts ... 71 to 106 in-lbs
- Nuts ... 71 to 106 in-lbs

Water pump pulley-to-hub ... 15 to 22
Engine oil cooler adapter fitting (3.8L) ... 22 to 29
Fan shroud-to-radiator ... 71 to 106 in-lbs

1   General information

The cooling system consists of a radiator and coolant reserve system, a radiator pressure cap, a thermostat, two temperature-controlled electric cooling fans, and a pulley/belt-driven water pump.

The radiator cooling fans are mounted in a housing/shroud at the engine side of the radiator. They are designed to come on when the engine reaches a certain temperature, and shut off again when the engine cools down some, thereby keeping the engine in the desired operating-temperature range. The fans are two-speed design, with the higher speeds activated during certain load conditions and when the air conditioning is on.

The system is pressurized by a spring-loaded pressure relief cap, which, by maintaining pressure, increases the boiling point of the coolant. If the coolant temperature goes above this increased boiling point, the extra pressure in the system forces the radiator cap valve off its seat and exposes the overflow pipe or hose. The overflow pipe/hose leads to a coolant recovery system. This consists of a plastic reservoir, mounted to the right side of the radiator, into which the coolant that normally escapes due to expansion is retained. When the engine cools, the excess coolant is drawn back into the radiator by the vacuum created as the system cools, maintaining the system at full capacity. This is a continuous process and provided the level in the reservoir is correctly maintained, it is not necessary to add coolant to the radiator.

Coolant in the right side of the radiator circulates through the lower radiator hose to the water pump, where it is forced through the water passages in the cylinder block. The coolant then travels up into the cylinder head, circulates around the combustion chambers and valve seats, travels out of the cylinder head past the open thermostat into the upper radiator hose and back into the radiator.

When the engine is cold, the thermostat restricts the circulation of coolant to the engine. When the minimum operating temperature is reached, the thermostat begins to open, allowing coolant to return to the radiator.

An automatic transmission fluid cooler element is incorporated into one of the radiator end-tanks to cool the transmission fluid.

The heating system works by directing air through the heater core, which is like a small radiator mounted behind the dash. Hot engine coolant heats the core, over which air passes to the interior of the vehicle by a system of ducts. Temperature is controlled by mixing heated air with fresh air, using a system of flapper doors in the ducts, and a heater bower motor. On some models, a second blower motor is located in the center console, to direct airflow to the rear seat area.

Air conditioning is an optional acces-

2.4 An inexpensive hydrometer can be used to test the condition of your coolant

sory, consisting of an evaporator core located under the dash, a condenser in front of the radiator, an accumulator/drier in the engine compartment and a belt-driven compressor mounted at the front of the engine.

Available as an option is a rear-mounted auxiliary heating/air conditioning system. In this system, the driver has controls on the dashboard for both the front and rear heating/air conditioning systems. A second heater core, air conditioning evaporator core and blower motor are located at the rear of the vehicle behind the left rear interior panel.

2   Antifreeze - general information

Refer to illustration 2.4
Warning: *Do not allow antifreeze to come in contact with your skin or painted surfaces of the vehicle. Rinse off spills immediately with plenty of water. Antifreeze is highly toxic if ingested. Never leave antifreeze lying around in an open container or in puddles on the floor; children and pets are attracted by it's sweet smell and may drink it. Check with local authorities about disposing of used antifreeze. Many communities have collection centers which will see that antifreeze is disposed of safely. Never dump used antifreeze on the ground or pour it into drains.*
Note: *Non-toxic antifreeze is now manufactured and available at local auto parts stores, but even this type should be disposed of properly.*

The cooling system should be filled with a water/ethylene glycol based antifreeze solution which will prevent freezing down to at least -20-degrees F (even lower in cold climates). It also provides protection against corrosion and increases the coolant boiling point. The manufacturer recommends that only coolant designated as safe for aluminum engine components be used.

The cooling system should be drained, flushed and refilled at least every other year (see Chapter 1). The use of antifreeze solutions for periods of longer than two years is

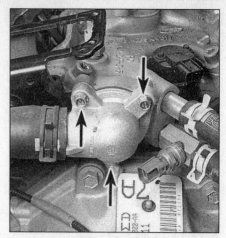

3.8a Thermostat cover bolts (arrows) - 3.0L engine

likely to cause damage and encourage the formation of rust and scale in the system.

Before adding antifreeze to the system, check all hose connections. Antifreeze can leak through very minute openings.

The exact mixture of antifreeze to water which you should use depends on the relative weather conditions. The mixture should contain at least 50-percent antifreeze, but should never contain more than 70-percent antifreeze. Consult the mixture ratio chart on the container before adding coolant. Hydrometers are available at most auto parts stores to test the coolant **(see illustration)**. Use antifreeze which meets Ford specifications for use with aluminum engine components.

3   Thermostat - check and replacement

Warning: *The engine must be completely cool when this procedure is performed.*
Note: *Don't drive the vehicle without a thermostat! The computer may stay in open loop and emissions and fuel economy will suffer.*

Check

1   Before condemning the thermostat, check the coolant level, drivebelt tension and temperature gauge (or light) operation.
2   If the engine takes a long time to warm up, the thermostat is probably stuck open. Replace the thermostat.
3   If the engine runs hot, check the temperature of the upper radiator hose. If the hose isn't hot, the thermostat is probably stuck shut. Replace the thermostat.
4   If the upper radiator hose is hot, it means the coolant is circulating and the thermostat is open. Refer to the *Troubleshooting* section at the front of this manual for the cause of overheating.
5   If an engine has been overheated, you may find damage such as leaking head gaskets, scuffed pistons and warped or cracked cylinder heads.

3.8b Thermostat cover bolts (arrows) - 3.8L engine

3.10 Once the cover is removed, scrape off the old gasket material and sealant

3.11 When installing the thermostat pay special attention to the direction in which it's placed in the engine; the spring goes into the intake manifold

Replacement

Refer to illustrations 3.8a, 3.8b, 3.10 and 3.11

6   Drain coolant from the radiator, until the coolant level is below the thermostat housing (see Chapter 1).

7   Disconnect the upper radiator hose from the thermostat housing cover.

8   Remove the bolts/studs and lift the cover off **(see illustrations)**.

9   Note how it's installed, then remove the thermostat. Be sure to use a replacement thermostat with the correct opening temperature (see this Chapter's Specifications).

10   Use a scraper or putty knife to remove all traces of old gasket material and sealant from the mating surfaces **(see illustration)**. Make sure no gasket material falls into the coolant passages; it is a good idea to stuff a rag in the passage. Wipe the mating surfaces with a rag saturated with lacquer thinner or acetone.

11   Install the thermostat with the spring directed toward the engine **(see illustration)**.

12   Apply a thin coat of RTV sealant to both sides of the new gasket and position it on the engine side, over the thermostat, and make

sure the gasket holes line up with the bolt holes in the housing. **Note:** *On 3.8L models, the gasket has a tang that aligns with a notch on the thermostat cover for alignment.*

13   Carefully position the cover and install the bolts. Tighten them to the torque listed in this Chapter's Specifications - do not over-tighten them or the cover may be cracked or distorted.

14   Reattach the radiator hose to the cover and tighten the clamp - now may be a good time to check and replace the hoses and clamps (see Chapter 1).

15   Refer to Chapter 1 and refill the system, then run the engine and check carefully for leaks.

16   Repeat steps 1 through 5 to be sure the repairs corrected the previous problem(s).

---

4   Engine cooling fan and circuit - check and fan/motor removal and installation

Warning 1: *The models covered by this manual are equipped with Supplemental Restraint Systems (SRS), more commonly known as airbags. Always disconnect the negative battery cable, then the positive battery cable and wait two minutes before working in the vicinity of the impact sensors, steering column or instrument panel to avoid the possibility of accidental deployment of the airbag, which could cause personal injury (see Chapter 12). Do not use any electrical test equipment on any of the airbag system wires or tamper with them in any way.*

Warning 2: *Do not work with your hands near the fans at any time that the engine is running or the key is ON. With the key ON, (even with the engine not running) the fan can start at any time, since it is controlled by coolant temperature.*

Check

Refer to illustrations 4.2, 4.6 ,4.9a and 4.9b

1   All models have two-speed electric fans

mounted in a plastic shroud attached to the back of the radiator.

2   Fan operation is controlled both by the Powertrain Control Module (PCM) and a pair of high and low-speed fan relays and the air conditioning relay. On 1997 and earlier models, the relays are incorporated in the Constant Control Relay Module (CCRM) (see Chapter 12 for more information regarding the CCRM). On 1998 and later models, a CCRM is not used and the relays are separate units located at the engine compartment relay panel **(see illustration)**.

3   The coolant temperature sensor signals the PCM of engine temperature. When the coolant temperature reaches a predetermined temperature, or when the air conditioning is switched on, the PCM completes the ground circuit of the CCRM/relay coil, energizing the cooling fan motors. The PCM commands the CCRM or relays to turn the fan on LOW speed when coolant temperature reaches 210-degrees F, then turn the fan off when temperature lowers to 204-degrees F. The fans will start running at high speed if the coolant temperature reaches 225-degrees F and stop running when the coolant temperature reaches 208-degrees F. Low speed fan operation is accomplished with the use of a dropping resistor located in the wiring harness leading to the cooling fans.

4   If the fans do not operate properly, the fault could be with the PCM, the CCRM/relays, the coolant temperature sensor or the related wiring. Refer to Chapter 6 for diagnosis of the sensor, and Chapter 12 for diagnosis of the CCRM.

5   Warm the engine up until the gauge on the instrument panel indicates the high side of NORMAL. The fan should come on. If not, check the cooling fan fuse in the fuse junction panel (see Chapter 12 for fuse locations).

6   If the fuse checked OK, check that the fans themselves are operational by disconnecting the electrical connector at each fan motor. Attach a fused jumper wire connected to battery voltage to the red/orange wire ter-

4.2 Location of low-speed and high-speed cooling fan relays (arrows) - 1998 models

3

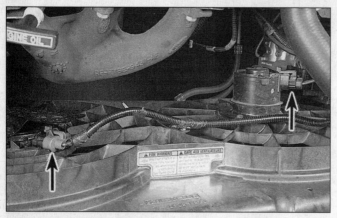

4.6 Cooling fan electrical connectors (arrows)

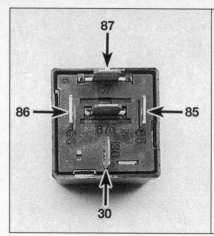

4.9a Terminal identification for the low-speed cooling fan relay - 1998 and later models

30   Relay output
85   Power (hot in Start or Run)
86   Relay solenoid control
87   Power (hot at all times)

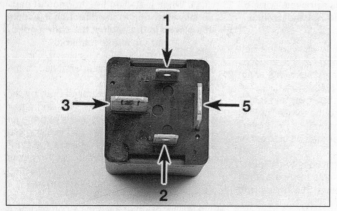

4.9b Terminal identification for the high-speed cooling fan relay - 1998 and later models

1   Power (hot in Start or Run)   3   Power (hot at all times)
2   Relay solenoid control   5   Relay output

5.3 Remove the two screws (arrows) and pull the reservoir straight up and out

minal and a jumper wire to the black wire terminal and a good chassis ground **(see illustration)**. If the fans do not operate, they should be replaced.

7   If they do operate with jumper wires, but don't operate under normal driving conditions, connect a voltmeter to a chassis ground and probe the connector at the red/orange wire. If the engine is hot and the temperature gauge shows above NORMAL, there should be battery voltage at this wire.

8   Check the ground of the circuit by switching your meter to the ohms scale. Connect the negative lead of the meter to a good chassis ground point and connect the positive probe to the black wire terminal of the fan connector. Resistance should be no more than 5 ohms. If resistance is high, trace the ground wire from the fan motor to the chassis ground point.

9   If battery power was not present at the terminals in Step 6, check that power is being supplied to the fan relays (1998 and later models) **(see illustrations)** or CCRM (see Chapter 12). The black/orange wire terminals (terminal 87 on the low-speed relay, terminal 3 on the high-speed relay or terminals 1, 2, 6, 7, 12 and 24 on the CCRM) should be hot at all times.

10   The red wire pin socket (for pin 85 on

the low-speed relay and pin 1 on the high-speed relay) on 1998 and later models should have battery power when the key is in the On or Run positions.

11   If these terminals check OK, refer to Chapter 12 for more information on testing relays and the CCRM.

Fan/motor removal and installation

12   Disconnect the electrical connectors from the fan motors. Remove the two fan shroud bolts and lift out the fan/shroud as an assembly. **Note:** *The bottom of the fan shroud has two projections that fit into tabs on the radiator.*

13   The manufacturer suggests that the fans, motors and shroud not be disassembled or repaired. A new shroud must be purchased with the fans/motors in place.

14   Installation is the reverse of removal.

5   Radiator and coolant reservoir - removal and installation

Warning 1: *The engine must be completely cool when this procedure is performed.*
Warning 2: *The models covered by this man-*

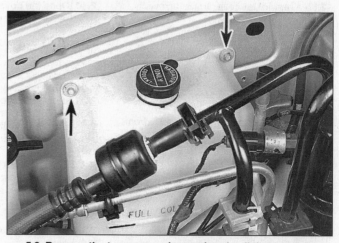

5.7a To disconnect a transmission cooling line, pry off the retaining clip (arrow) . . .

ual are equipped with Supplemental Restraint Systems (SRS), more commonly known as airbags. Always disconnect the negative battery cable, then the positive battery cable and wait two minutes before working in the vicinity of the impact sensors, steering column or instrument panel to avoid the possibility of accidental deployment of the airbag, which could cause personal injury (see Chapter 12). Do not use any electrical test equipment on any of the airbag system wires or tamper with them in any way.

5.7b . . . insert the special quick-connect coupling tool (arrow), then pull the tool and line out of radiator

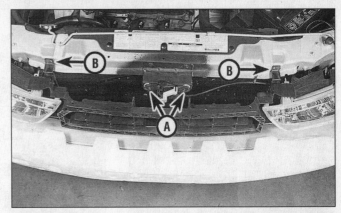

5.11 Remove the hood latch bolts (A) and the grille panel bolts (B)

Coolant reservoir

Refer to illustration 5.3

1   Disconnect the cable from the negative battery terminal.

2   Have a drain pan ready and disconnect the overflow hose at the radiator. Lower the hose into the drain pan to drain the coolant from the reservoir.

3   Remove the screws holding the reservoir to the right side of the radiator and remove the reservoir from the vehicle **(see illustration)**.

4   Prior to installation make sure the reservoir is clean and free of debris which could be drawn into the radiator (wash inside with soapy water and a long brush if necessary).

5   Installation is the reverse of removal. Fill the reservoir with the proper amount and mixture of antifreeze (see Chapter 1).

Radiator

Refer to illustrations 5.7a, 5.7b, 5.11, 5.12 and 5.13

6   Refer to Section 4 and remove the cooling fan and shroud assembly. Remove the coolant reservoir.

7   Using a special quick-disconnect coupling tool (available at most automotive parts stores), detach the transmission cooler lines from the radiator **(see illustrations)**. On 1999 and later models, also disconnect the auxiliary cooler lines from the radiator. Be careful not to damage the lines or fittings. Plug the ends of the disconnected lines to prevent leakage and stop dirt from entering the system. Have a drain pan ready to catch any spills.

8   Refer to Chapter 1 and drain the cooling system.

9   Disconnect the upper and lower radiator hoses.

10   Refer to Chapters 11 and 12 and remove the grille, headlights and turn signals, then remove the bolts from the hoodlatch support and pull the support away from the radiator.

11   Remove the bolts and the bracket at each side of the grille opening panel **(see illustration)**.

12   Remove the radiator mounting bolts **(see illustration)**.

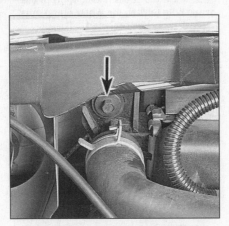

5.12 Remove the two upper radiator mounting bolts (arrow indicates left mounting bolt)

13   Remove the radiator support panel **(see illustration)**.

14   Remove the upper mounting bolts for the air conditioning condenser (see Section 17). **Warning:** *Do not disconnect the refrigerant hoses from the condenser.*

15   The condenser and radiator are held to each other by clips. Raise the condenser while pushing down on the radiator enough to disengage the clips, then remove the radiator.

16   Radiator installation is the reverse of removal. When installing the radiator, make sure it seats properly in the lower saddles and that the rubber mounts at the bottom are intact. **Note:** *Prior to installation of the radiator, replace any damaged hose clamps and radiator hoses.*

17   After installation, fill the system with the proper mixture of antifreeze, and also check the automatic transmission fluid level.

---

6   Coolant temperature sending unit - check and replacement

Refer to illustrations 6.1a and 6.1b
Warning: *Wait until the engine is completely cool before beginning this procedure.*

5.13 Remove the screws (arrows) on each side of the radiator support panel and remove the panel

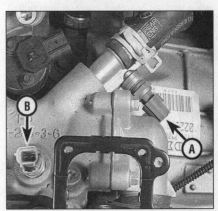

6.1a Coolant temperature sending unit (A) - (B) indicates the coolant temperature sensor for the Powertrain Control Module (PCM) (3.0L engine)

Check

1   The coolant temperature indicator system is composed of a temperature gauge mounted in the dash and a coolant temperature sending unit mounted on the engine **(see illustrations)**. Some vehicles have more than one sending unit, but only one is used for the indicator system and the other is used to send engine temperature information to the

3

6.1b On 3.8L models, the coolant
temperature sending unit (arrow) is at the
left end of the intake manifold

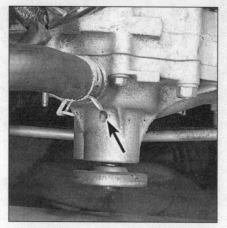

8.2 If there's coolant leaking from the
weep hole (arrow) the water pump must
be replaced

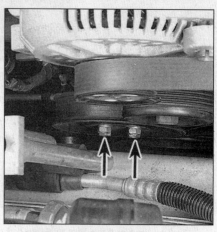

9.3 Remove the four bolts (arrows
indicate two) holding the water
pump pulley

Powertrain Control Module (PCM).

2   If an overheating indication occurs, check the coolant level in the system. Make sure the wiring between the gauge and the sending unit is secure and all fuses are intact.

3   To test the temperature sender, check that it reads in the cold range when the engine is cold. Disconnect the electrical connector from the sender and attach a jumper wire between the terminal corresponding to the red/white wire and ground. With the key ON, the gauge should now swing to full hot. If doesn't, the problem is in the circuit from the sender to the instrument panel. If the gauge responds when tested, measure the resistance of the sender, it should be between 10 ohms (hot) and 74 ohms (cold). If the resistance isn't as specified, replace the coolant temperature sending unit.

Replacement

4   If the sending unit must be replaced, allow the engine to cool, disconnect the electrical connector and unscrew the sensor from the engine. **Caution:** *The sending unit is made up of metal and plastic and is fragile. Use care not to damage the unit when removing it.* There will be some coolant loss as the unit is removed, so be prepared to catch it.

5   Use thread sealant on the threads and install the replacement. Check the coolant level after the replacement unit has been installed.

7   Engine oil cooler (3.8L engine only) - removal and installation

Removal

Warning: *The engine should be completely cool for this procedure.*

1   An engine oil cooler may be installed on some 3.8L models with a towing package. Coolant flows from the radiator through the cooler through two hoses attached to pipes on the cooler body.

2   To replace the oil cooler, refer to Chap-

ter 1 and remove the oil filter and drain the cooling system. **Note:** *If the oil cooler is not being completely removed from the vehicle, do not drain the cooling system or disconnect the coolant hoses at the oil cooler - detach the oil cooler from the oil filter adapter and position the unit aside.*

3   Disconnect the coolant hoses from the oil cooler.

4   Use a deep socket to remove the threaded adapter fitting retaining the oil cooler to the oil filter adapter and pull the oil cooler from the filter adapter.

Installation

5   Apply clean engine oil to the rubber seal and install the oil cooler, aligning its locator pins over the locating boss on the filter adapter. Tighten the threaded adapter fitting to the torque listed in this Chapter's Specifications.

9   The remainder of installation is the reverse of removal. Reconnect the coolant hoses and install a new oil filter. Refill and bleed the cooling system (see Chapter 1). Start the engine to check for oil and coolant leaks.

8   Water pump - check

Refer to illustration 8.2

1   Water pump failure can cause overheating and serious damage to the engine. There are three ways to check the operation of the water pump while it's installed on the engine. If any one of the following quick checks indicates water pump problems, it should be replaced immediately.

2   A seal protects the water pump impeller shaft bearing from contamination by engine coolant. If this seal fails, a weep hole in the water pump snout will leak coolant **(see illustration)** (an inspection mirror can be used to look at the underside of the pump if the hole isn't on top). If the weep hole is leaking, shaft bearing failure will follow. Replace the water pump immediately. **Note:** *A small amount of*

gray discoloration is normal. A wet area or heavy brown deposits indicate the pump seal has failed.

3   Besides contamination by coolant after a seal failure, the water pump impeller shaft bearing can also prematurely wear out. If a noise is coming from the water pump during engine operation, the shaft bearing has failed - replace the water pump immediately. **Note:** *Do not confuse drivebelt noise with bearing noise. Loose or glazed drivebelts may emit a high-pitched squealing noise.*

4   To identify excessive bearing wear before the bearing actually fails, grasp the water pump pulley (with drivebelt removed) and try to force it up-and-down or from side-to-side. If the pulley can be moved either horizontally or vertically, the bearing is nearing the end of its service life. Replace the water pump.

9   Water pump - removal and installation

Warning: *Wait until the engine is completely cool before beginning this procedure.*

Removal

1   Disconnect the cable from the negative battery terminal.

2   With the engine cold, drain the cooling system (see Chapter 1).

3.0L engine

Refer to illustrations 9.3 and 9.6

3   Remove the drivebelt(s) (see Chapter 1) and remove the water pump pulley **(see illustration)**. **Note:** *It's helpful to loosen the pulley bolts/nuts while the belt is still in place. It helps hold the pulley from turning.*

4   Disconnect the upper radiator hose from the water pump, and the bypass hose from the top of the pump.

5   Remove the bolt and two nuts and remove the drivebelt tensioner (see Chapter 1).

6   Remove the water pump retaining bolts

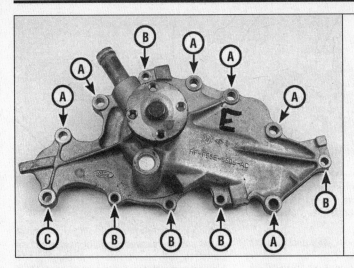

9.6 Water pump retaining bolts (3.0L engine) - bolts (A) and (C) are 8mm bolts of various lengths; bolts (B) are 6mm bolts; bolt C requires thread sealant upon pump installation

9.9 When disconnecting the coolant pipe from the pump, remove this bolt (lower arrow) and pull the pipe from the pump - upper arrow indicates second bolt of brace to be removed

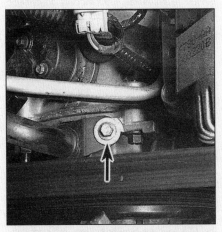

9.12 Remove the bolt (arrow) and detach the bypass tube from the water pump

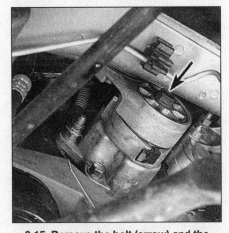

9.15 Remove the bolt (arrow) and the drivebelt tensioner

inches, until the water pump pulley can be removed.

15   Remove the drivebelt tensioner **(see illustration)**.

16   Remove the water pump nuts/bolts and detach the water pump from the engine **(see illustration)**. **Note:** *If the pump body seems to be stuck, strike it with a rubber hammer. Do not use a prybar to separate the pump from the front cover.*

Installation

17   Before installation, remove and clean all gasket or sealant material from the water pump and cylinder block.

18   Inspect the sealing surface of the water pump mounting area on the front cover for dirt and/or debris. Clean them thoroughly before reassembly.

19   Clean the back of the water pump of sealant and gasket material, then wipe with lacquer thinner. Coat a new gasket with a thin application of RTV sealant on both sides and adhere it to the pump.

20   Install the water pump and tighten the bolts/nuts to the torque listed in this Chap-

and remove the water pump **(see illustration)**. Take note of the installed positions of the various-length bolts and studs. **Note:** *If the pump body seems to be stuck, strike it with a rubber hammer. Do not use a prybar to separate the pump from the front cover.*

3.8L engine

Refer to illustrations 9.9, 9.12, 9.15 and 9.16

7   Loosen the water pump pulley bolts while the belt is still under tension, then refer to Chapter 1 and remove the drivebelt.

8   From below the vehicle, remove the nuts holding the engine mounts to the subframe (see Chapter 2A).

9   Disconnect the lower radiator hose from the water pump by first disconnecting it from the radiator, then removing the bolt from the metal tube and pulling it off the pump **(see illustration)**.

10   Refer to Chapter 5 and remove the alternator from the power steering mount.

11   Refer to Chapter 10 and disconnect the pressure hose from the power steering pump, draining the fluid into a drain pan, then remove the power steering pump bracket bolts and set power steering pump/bracket assembly aside.

12   Remove the bolt and pull the bypass hose tube from the water pump **(see illustration)**.

13   Remove the brace from the air-conditioning compressor mount **(see illustration 9.9)**.

14   Raise the engine from above with a 3-bar support fixture or engine hoist, about two

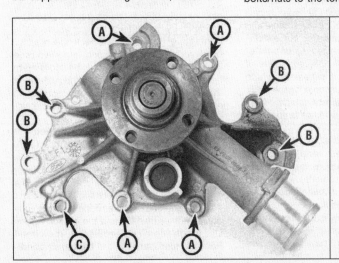

9.16 Water pump retaining nuts/bolts (3.8L engine) - bolts (A); nuts (B and C); if the stud (C) is removed, apply thread sealant upon installation

3

10.5b The thermal limiter (arrow) protects the components from excessive heat - check the thermal limiter for damage

10.6 Test the terminals of the blower resistor with an ohmmeter for continuity

10.5a Blower motor resistor (arrow) location on the evaporator housing - disconnect the wiring connector and remove the two screws

ter's Specifications. There are different torque values for the different size fasteners, and one fastener must be coated with Teflon thread sealant **(see illustrations 9.6 and 9.16)**.

21   The remainder of the installation is the reverse of the disassembly sequence. **Note:** *On 3.8L engines, install a new O-ring when connecting the lower radiator hose pipe to the pump.*

22   Fill the cooling system with the proper coolant mixture (see Chapter 1).

23   Start the engine and make sure there are no leaks. Check the level frequently during the first few weeks of operation to ensure there are no leaks and that the level in the system is stable.

10   Heater and air conditioning blower motor circuit - check

Refer to illustrations 10.5a, 10.5b, 10.6 and 10.9

Warning: *The models covered by this manual are equipped with Supplemental Restraint Systems (SRS), more commonly known as airbags. Always disconnect the negative battery cable, then the positive battery cable and wait two minutes before working in the vicinity of the impact sensors, steering column or instrument panel to avoid the possibility of accidental deployment of the airbag, which could cause personal injury (see Chapter 12). Do not use any electrical test equipment on any of the airbag system wires or tamper with them in any way.*

1   Check the fuse and all connections in the circuit for looseness and corrosion. Check the ground connections located under the right side of the instrument panel below the glove box and behind the left kick panel to the left of the brake pedal.

2   Make sure the battery is fully charged.

3   With the transmission in Park, the parking brake securely set, turn the ignition

switch to the On position. It isn't necessary to start the vehicle.

4   Switch the heater controls to VENT or FLOOR and the blower speed to HI. Feel for air output at the air ducts to verify the blower is operating. If it is, then switch the blower switch to each position and verify the motor operates at each speed. If the blower motor does not operate on any speed, the blower motor relay (1996 and later models), blower motor, blower switch or related wiring could be faulty. If the blower motor operates on high, but not on one or more lower speeds, the blower resistor or related wiring is probably faulty.

5   The blower motor resistor is located on the evaporator housing under the right side of the dash **(see illustrations)**. There are three resistor elements mounted on the resistor board to provide low and medium blower speeds (HI bypasses the resistor). The blower operates continuously, anytime the ignition switch is On and the mode switch is in any position other than Off. A thermal limiter resistor is integrated into the circuits to prevent heat damage to the components. If the thermal limiter circuit has been opened as a result of excessive heat, it should be replaced only with the identical replacement part. **Note:** *Do not replace your blower resistor with a resistor that does not incorporate the thermal limiter.*

6   With the resistor removed from the vehicle, visually check the limiter for damage, indicated by the material melting between the contacts of the limiter. Check the resistor block for continuity between terminals **(see illustration)**. There should be continuity between all terminals (with varying resistance at each set). If any of the resistor elements do not pass the tests, replace the blower resistor.

7   Disconnect the electrical connector at the blower motor **(see illustration 11.3)**. Using a voltmeter, connect the positive probe to the pink/white wire terminal of the blower motor connector and the negative probe to a good chassis ground; there should be at least 10 volts present with the mode switch in any position other than Off and the ignition switch On. If not, there is a problem in the circuit from the fuse panel to the heater/air condi-

tioning control panel (including the control panel and the blower relay), or from the fuse panel to the blower motor (including the blower relay). **Note:** *1995 models do not use a blower relay. Refer to the wiring diagrams at the end of Chapter 12.*

8   If there is voltage at the pink/white wire terminal, switch the meter to the ohms scale, connect the positive probe to the orange/black wire terminal of the blower motor connector and connect the negative probe to a good chassis ground; switch the blower speed switch to the high speed position, continuity should be indicated. If continuity is not indicated there is a problem in the ground circuit (including the blower motor switch). If battery power was present at the blower motor connector (Step 7) and the ground circuit is good but the blower motor does not operate when connected, replace the blower motor.

9   To test the blower switch, refer to Section 12 and remove the heater/air conditioning control panel. Disconnect the electrical connector from the back of the blower speed switch and test the terminals for continuity **(see illustration)**. In the Low (1) position, there should be NO continuity between terminal 2 and any other terminal; in Medium (2) position, there should be continuity between terminals 2 and 3; in Medium (3) position, there should be continuity between terminals 2 and 4; and in HI (4) position, there should be continuity between terminals 1 and 2. If the continuity is not as described, replace the

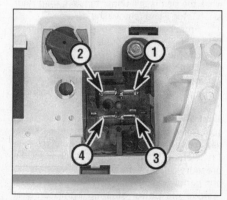

10.9 Blower speed switch terminal identification

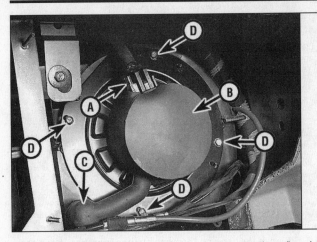

11.3 **Blower motor components**

A Electrical connector
B Motor cover
C Air tube
D Mounting screws

11.6 Pry off the retaining pushnut (arrow) retaining the blower fan to the blower motor shaft

blower speed switch.

10   Refer to Chapter 12 to test the blower relay (1996 and later models).

11 Heater and air conditioning blower motor - removal and installation

Warning: *The models covered by this manual are equipped with Supplemental Restraint Systems (SRS), more commonly known as airbags. Always disconnect the negative battery cable, then the positive battery cable and wait two minutes before working in the vicinity of the impact sensors, steering column or instrument panel to avoid the possibility of accidental deployment of the airbag, which could cause personal injury (see Chapter 12). Do not use any electrical test equipment on any of the airbag system wires or tamper with them in any way.*

Front blower motor

Refer to illustrations 11.3 and 11.6

1   Disconnect the negative battery cable.
2   Open the glove compartment and remove it (see Chapter 11).
3   Disconnect the blower motor electrical connector from the motor **(see illustration)**.
4   Remove the blower motor cover (if

equipped) and detach the air tube the blower motor.

5   Remove the four blower motor mounting screws **(see illustration 11.3)**, and withdraw the blower motor carefully from the housing.

6   If the blower motor is being replaced, the fan wheel should be transferred to the new motor at this time. It is attached to the blower motor shaft with a push nut. Grasp the nut with a pliers and pull it off or pry it off with a small screwdriver, being careful not to crack the push nut **(see illustration)**. To reinstall the nut, simply push it on to the shaft.

7   Installation is the reverse of removal.

Rear blower motor

Refer to illustration 11.9

8   Some models have an auxiliary heating air conditioning system located in the rear of the vehicle, behind the left rear interior trim panel. The rear blower motor and resistor can be checked in the same manner as the front system (see Section 10).

9   To access the auxiliary blower motor, remove the left rear interior trim panel **(see illustration)**.

10   Disconnect the electrical connector, detach the air tube, remove the mounting screws and withdraw the blower motor from the housing.

11   Installation is the reverse of removal.

12 Heater and air conditioning control assembly - removal and installation

Warning: *The models covered by this manual are equipped with Supplemental Restraint Systems (SRS), more commonly known as airbags. Always disconnect the negative battery cable, then the positive battery cable and wait two minutes before working in the vicinity of the impact sensors, steering column or instrument panel to avoid the possibility of accidental deployment of the airbag, which could cause personal injury (see Chapter 12). Do not use any electrical test equipment on any of the airbag system wires or tamper with them in any way.*

Removal

Refer to illustrations 12.2 and 12.3

1   Refer to Chapter 11 and remove of the center dash bezel.
2   Remove the four screws retaining the control assembly to the instrument panel **(see illustration)**.
3   Pull the control assembly out of the

3

11.9 Remove the cover in the left rear trim panel to access the optional auxiliary heating/air conditioning unit, blower motor (A) and blower resistor (B)

12.2 After the trim is removed, remove the four screws (arrows) and pull the control assembly out of the dash

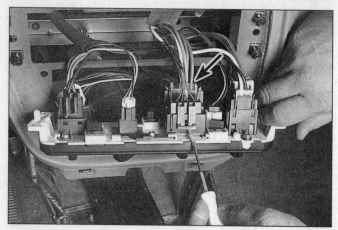

12.3 Disconnect the electrical connectors by gently prying up the clips and pulling the connectors off - arrow indicates the vacuum harness connection

13.2 Squeeze the tabs on the hose clamps together, move the clamps back and detach the heater hoses from the heater core inlet and outlet pipes (arrows)

instrument panel and disconnect the electrical connectors and vacuum harness **(see illustration)**. **Note:** *When disconnecting the vacuum harness be careful to avoid cracking the plastic connectors and causing a vacuum leak (possibly internal within the control head).*

4   Refer to Section 10 for electrical checks of the blower motor speed switch. The speed switch, function selector, and blend-control switch can all be removed from the control head by pulling the knob off from the front side, then removing the retaining screw on the back of the switch.

Installation

5   Installation is the reverse of the removal procedure.

13 Heater core - removal and installation

Warning 1: *The models covered by this manual are equipped with Supplemental Restraint Systems (SRS), more commonly known as*

airbags. Always disconnect the negative battery cable, then the positive battery cable and wait two minutes before working in the vicinity of the impact sensors, steering column or instrument panel to avoid the possibility of accidental deployment of the airbag, which could cause personal injury (see Chapter 12). *Do not use any electrical test equipment on any of the airbag system wires or tamper with them in any way.*

Warning 2: *The air conditioning system is under high pressure. DO NOT loosen any fittings or remove any components until after the system has been discharged. Air conditioning refrigerant should be properly discharged into an EPA-approved container at a dealer service department or an automotive air conditioning repair facility. Always wear eye protection when disconnecting air conditioning system fittings.*

Front heater core

Refer to illustrations 13.2, 13.5a, 13.5b, 13.6, 13.7a and 13.7b

1   Disconnect the cable from the negative battery terminal. Drain the cooling system (see Chapter 1).

2   Refer to Chapter 11 and remove the cowl vent panel, then disconnect the heater hoses from the heater core inlet and outlet tubes at the firewall **(see illustration)**.
3   On models so equipped, disconnect the engine ground strap attached to one of the heater core tubes.
4   Refer to Chapter 11 and remove the center dash support trim and the radio cassette box and ashtray.
5   Pull back the carpeting and pull the rear airflow duct away from the main floor duct **(see illustrations)**.
6   Remove the two instrument panel support brackets, and the heater main floor duct **(see illustration)**.
7   Remove the screws and the heater core cover, then remove the heater core **(see illustrations)**. Be careful not to tear the foam sealing material
8   When reinstalling the heater core in the housing, make sure the original foam sealing material is intact and in place.
9   The remainder of installation is the reverse of removal. **Note:** *Install new retainers and O-rings on the hoses that connect to the heater core at the firewall.*

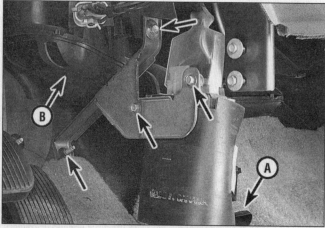

13.5a Pull the rear duct (A) away from the main floor duct (B) - remove the bolts (arrows) to remove the bracket

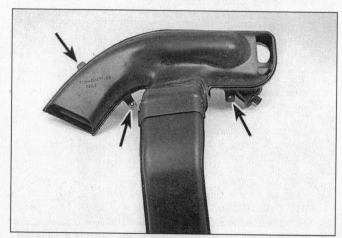

13.5b Rear duct, shown removed for clarity to indicate the screw locations (arrows)

13.6 Remove the instrument panel braces (arrows)

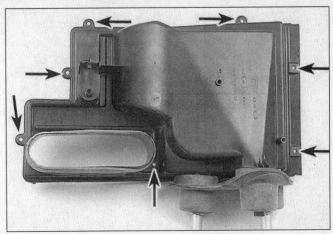

13.7a Remove the screws (arrows) and separate the cover from the heater core/evaporator core housing (cover removed for clarity)

13.7b Remove the heater core from the housing

13.11 The auxiliary rear heating/air conditioning unit is mounted behind the left rear trim panels

3

10   Fill the cooling system (see Chapter 1). Run the engine and check for coolant leaks.

Rear heater/air conditioning unit

Refer to illustrations 13.11, 13.14 and 13.16

11   Some models may be equipped with an optional rear heater/air conditioning system. the unit is located behind the left-rear interior trim panel (behind the left rear wheel housing). It includes a blower motor, heater core and evaporator core **(see illustration)**.

12   Have the air conditioning refrigerant discharged and recovered by a dealer or air conditioning shop. Drain the cooling system (see Chapter 1).

13   Refer to Chapter 11 for removal of the left rear trim panel and the rear seats.

14   Raise and suitably support the vehicle on jackstands, to access the hose connections under the mid-section of the vehicle chassis **(see illustration)**. Warning: *Some models covered by this manual are equipped with an air suspension system. Always disconnect the electrical power to the suspension system before lifting or towing (see*

Chapter 10). Failure to perform this procedure may result in unexpected shifting or movement of the vehicle, which could cause personal injury.

15   Remove the bolt retaining the hose bracket to the body. Using a spring-lock coupling tool (available at most automotive parts

stores), disconnect the refrigerant lines. Cap the open lines immediately to prevent contamination. Detach the heater hoses from the heater core tubes. Lower the vehicle.

16   Remove the screws retaining the ducts to the auxiliary heater/air conditioning unit, then remove the mounting screws and care-

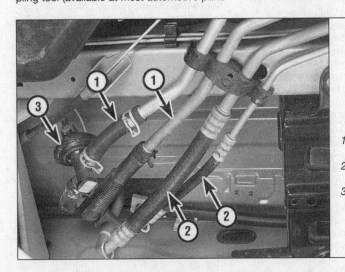

13.14 Auxiliary heater/air conditioning unit heater hoses and refrigerant lines

1 *Rear heater core coolant lines*
2 *Rear refrigerant lines*
3 *Heater control valve*

13.16 Remove the screws retaining the ducts (A) to the unit, then the screws retaining the unit to the body (B indicates the upper screw)

14.7 Place an accurate thermometer in the center dash vent, turn the air conditioning on and check the output temperature

fully remove the unit as an assembly **(see illustration)**.

17   Remove the screws retaining the heater core/evaporator core cover to the assembly, then remove the heater core or evaporator core.

18   Installation is the reverse of the removal procedure.

19   Refill the cooling system and have the system evacuated, recharged and leak tested by a dealer or air conditioning shop.

---

14   Air conditioning and heating system - check and maintenance

---

1   The following maintenance steps should be performed on a regular basis to ensure that the air conditioner continues to operate at peak efficiency.

  a) *Check the tension of the drivebelt and adjust if necessary (see Chapter 1).*
  b) *Check the condition of the hoses. Look for cracks, hardening and deterioration.* **Warning:** *Do not replace air conditioning hoses until the system has been discharged by a dealer or air conditioning shop.*
  c) *Check the fins of the condenser for leaves, bugs and other foreign material. A soft brush and compressed air can be used to remove them.*
  d) *Check the wire harness for correct routing, broken wires, damaged insulation, etc. Make sure the harness connections are clean and tight.*
  e) *Maintain the correct refrigerant charge.*

2   The system should be run for about 10 minutes at least once a month. This is particularly important during the winter months because long-term non-use can cause hardening of the internal seals.

3   Because of the complexity of the air conditioning system and the special equipment required to effectively work on it, accurate troubleshooting of the system should be left to a professional technician. One proba-

ble cause for poor cooling that can be determined by the home mechanic is low refrigerant charge. Should the system lose its cooling ability, the following procedure will help you pinpoint the cause.

Check

Refer to illustration 14.7

4   Warm the engine up to normal operating temperature.

5   Place the air conditioning temperature selector at the coldest setting and set the blower at the highest setting. Open the doors (to make sure the air conditioning system doesn't cycle off as soon as it cools the passenger compartment).

6   After the system reaches operating temperature, feel the two pipes connected to the evaporator at the firewall.

7   The inlet pipe, (thinner tubing) leading from the condenser outlet to the evaporator, should be cold and the evaporator outlet line (the thicker tubing that leads back to the compressor) should be slightly colder (3 to 10-degrees F). If the evaporator outlet is considerably warmer than the inlet, the system charge may be low. Insert a thermometer in the center air distribution duct **(see illustration)** while operating the air conditioning system - the temperature of the output air should be 35 to 40-degrees F below the ambient air temperature (down to approximately 40-degrees F). If the ambient (outside) air temperature is very high, say 110-degrees F, the duct air temperature may be as high as 60-degrees F, but generally the air conditioning is 35 to 40-degrees F cooler than the ambient air. If the air isn't as cold as it used to be, the system charge may be low. Further inspection or testing of the system is beyond the scope of the home mechanic and should be left to a professional.

Adding refrigerant

Refer to illustrations 14.9 and 14.12

8   **Note:** *All models covered by this manual use refrigerant R-134a. When recharging or replacing air conditioning components, use*

14.9 A basic charging kit for R-134a systems are available at most auto parts stores - it must say R-134a (not R-12) and so must the 12-ounce can of refrigerant

only refrigerant, refrigerant oil and seals compatible with this system. The seals and compressor oil used with older, conventional R-12 refrigerant are not compatible with the components in this system.

9   Buy an automotive charging kit at an auto parts store. A charging kit includes a 12-ounce can of R-134a refrigerant, a tap valve and a short section of hose that can be attached between the tap valve and the system low side service valve **(see illustration)**. Because one can of refrigerant may not be sufficient to bring the system charge up to the proper level, it's a good idea to buy a couple of additional cans. Try to find at least one can that contains red refrigerant dye. If the system is leaking, the red dye will leak out with the refrigerant and help you pinpoint the location of the leak.

Note: *New vehicles may be shipped with a leak-detect dye already in place. If your system has never been discharged, you can spot any refrigerant leaks with an ultraviolet light, which causes leaks to glow greenish-yellow.*

14.12 Add R-134a only to the low-side port (arrow) - the procedure is easier if you wrap the can with a warm, wet towel to prevent icing

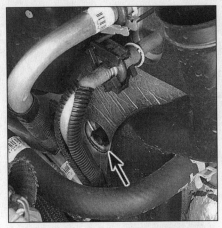

14.22 Check the evaporator drain tube (arrow) for blockage that could lead to mildew on the core - this view is of the right side of the firewall

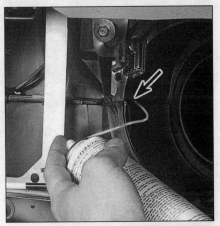

14.26 Remove the blower motor and spray disinfectant, aiming the long nozzle (arrow) toward the core to destroy mildew that causes air conditioning odors

The dye is said to be good for 500 hours of air conditioner operation, after which you can have more dye injected at your dealer.

10   Connect the charging kit by following the manufacturer's instructions.

11   Back off the valve handle on the charging kit and screw the kit onto the refrigerant can, making sure first that the O-ring or rubber seal inside the threaded portion of the kit is in place. **Warning:** *Wear protective eye wear when dealing with pressurized refrigerant cans.*

12   Remove the dust cap from the low-side charging port and attach the quick-connect fitting on the kit hose **(see illustration)**. **Warning:** *DO NOT hook the charging kit hose to the system high side! The fittings on the charging kit are designed to fit* **only** *on the low side of the system.*

13   Warm the engine to normal operating temperature and turn on the air conditioning. Keep the charging kit hose away from the fan and other moving parts.

14   Turn the valve handle on the kit until the stem pierces the can, then back the handle out to release the refrigerant. You should be able to hear the rush of gas. Add refrigerant to the low side of the system until both the outlet and the evaporator inlet pipe feel about the same temperature . Allow stabilization time between each addition. **Warning:** *Never add more than two cans of refrigerant to the system. The can may tend to frost up, slowing the procedure. Wet a shop towel with hot water and wrap it around the bottom of the can to keep it from frosting.*

15   Put your thermometer back in the center register and check that the output air is getting colder.

16   When the can is empty, turn the valve handle to the closed position and release the connection from the low-side port. Replace the dust cap.

17   Remove the charging kit from the can and store the kit for future use with the piercing valve in the UP position, to prevent inadvertently piercing the can on the next use.

Heating systems

Refer to illustration 14.22

18   If the air coming out of the heater vents isn't hot, the problem could stem from any of the following causes:

a) *The thermostat is stuck open, preventing the engine coolant from warming up enough to carry heat to the heater core. Replace the thermostat (see Section 3).*

b) *A heater hose is blocked, preventing the flow of coolant through the heater core. Feel both heater hoses at the firewall. They should be hot. If one of them is cold, there is an obstruction in one of the hoses or in the heater core, or the heater control valve is shut. Detach the hoses and back flush the heater core with a water hose. If the heater core is clear but circulation is impeded, remove the two hoses and flush them out with a water hose.*

c) *If flushing fails to remove the blockage from the heater core, the core must be replaced. (see Section 13).*

19   If the blower motor speed does not correspond to the setting selected on the blower switch, the problem could be a bad fuse, circuit, control panel or blower resistor (see Section 10).

20   If there isn't any air coming out of the vents:

a) *Turn the ignition ON and activate the fan control. Place your ear at the heating/air conditioning register (vent) and listen. Most motors are audible. Can you hear the motor running?*

b) *If you can't (and have already verified that the blower switch and the blower motor resistor are good), the blower motor itself is probably bad (see Section 11).*

21   If the carpet under the heater core is damp, or if antifreeze vapor or steam is coming through the vents, the heater core is leaking. Remove it (see Section 13) and install a

new unit (most radiator shops will not repair a leaking heater core).

22   Inspect the drain hose from the heater/evaporator assembly at the right center of the firewall, make sure it is not clogged **(see illustration)**. If there is a humid mist coming from the system ducts, this hose may be plugged with leaves or road debris.

Eliminating air conditioning odors

Refer to illustration 14.26

23   Unpleasant odors that often develop in air conditioning systems are caused by the growth of a fungus, usually on the surface of the evaporator core. The warm, humid environment there is a perfect breeding ground for mildew to develop.

24   The evaporator core on most vehicles is difficult to access, and dealerships have a lengthy, expensive process for eliminating the fungus by opening up the evaporator case and using a powerful disinfectant and rinse on the core until the fungus is gone. You can service your own system at home, but it takes something much stronger than basic household germ-killers or deodorizers.

25   Aerosol disinfectants for automotive air conditioning systems are available in most auto parts stores, but remember when shopping for them that the most effective treatments are also the most expensive. The basic procedure for using these sprays is to start by running the system in the RECIRC mode for ten minutes with the blower on its highest speed. Use the highest heat mode to dry out the system and keep the compressor from engaging by disconnecting the wiring connector at the compressor (see Section 16).

26   The disinfectant can usually comes with a long spray hose. Remove the blower motor (see Section 11), point the nozzle inside the hole and to the left towards the evaporator core, and spray according to the manufacturer's recommendations **(see illustration)**. Try to cover the whole surface of the evapo-

3

15.3 Disconnect the electrical connector (arrow) at the compressor clutch cycling switch

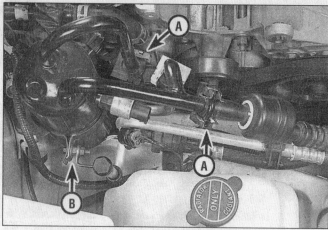

15.4a Disconnect the refrigerant lines (A) to the accumulator/drier and loosen the mounting clamp bolt (B)

rator core, by aiming the spray up, down and sideways. Follow the manufacturer's recommendations for the length of spray and waiting time between applications.

27   Once the evaporator has been cleaned, the best way to prevent the mildew from coming back again is to make sure your evaporator housing drain tube is clear **(see illustration 14.22)**.

15   Air conditioning accumulator/drier - removal and installation

Removal

Refer to illustrations 15.3, 15.4a, 15.4b and 15.4c

Warning: *The air conditioning system is under high pressure. DO NOT loosen any fittings or remove any components until after the system has been discharged. Air conditioning refrigerant should be properly discharged into an EPA-approved container at a dealer service department or an automotive air conditioning repair facility. Always wear eye protection when disconnecting air conditioning system fittings.*

1   The accumulator/drier stores refrigerant and removes moisture from the system. When any major air conditioning component (compressor, condenser, evaporator) is replaced, or the system has been apart and exposed to air for any length of time, the accumulator/drier must be replaced.

2   Take the vehicle to a dealer service department or automotive air conditioning shop and have the air conditioning system discharged. Disconnect the cable at the negative battery terminal.

3.   Disconnect the electrical connector at the compressor clutch cycling switch on top of the accumulator/drier **(see illustration)**. If the accumulator/drier is to be replaced with a new one, remove the cycling switch to transfer to the new drier.

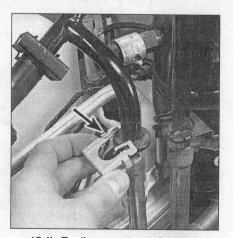

15.4b To disconnect a spring-lock coupling, remove the metal clip (arrow) at each connection . . .

15.4c . . . and use the spring-lock coupling tool to separate the line connections

4   Disconnect the refrigerant inlet and outlet lines **(see illustrations)**. Remove the metal clips first, then use spring-lock coupling tools to disconnect the two lines (one of which is connected to the accumulator/drier) from the evaporator core tubes at the firewall. To disconnect a fitting, close the two halves of the tool over the connection and push the tool towards the garter spring. This expands the spring to release its hold. While the spring is expanded and tool is still in place, pull in opposite directions on the two lines to separate the connection. Cap or plug the open lines immediately. **Note:** *Special spring-lock coupling tools are required to release the connectors used on the refrigerant lines throughout the air conditioning system, and are available at most auto parts stores in a set.*

5   Loosen the clamp-bolt on the mounting bracket and slide the accumulator/drier assembly up and out of the mounting bracket **(see illustration 15.4a)**.

Installation

6   If you are replacing the accumula-

tor/drier, drain the refrigerant oil from the old accumulator/drier. Add the same amount plus two ounces of clean refrigerant oil to the new accumulator. This will maintain the correct oil level in the system after the repairs are completed. **Note:** *The manufacturer recommends that to properly drain all of the oil from the old accumulator/drier for an accurate measurement, you should drill two 1/2-inch holes in the bottom of the accumulator/drier.*

7   Place the new accumulator/drier into position, tighten the mounting bracket bolt lightly, still allowing the accumulator drier to be turned to align the line connections.

8   Install the inlet and outlet lines. Lubricate the O-rings using clean refrigerant oil and reconnect the lines. Tighten the clamp bolt securely and reconnect the electrical connector.

9   Connect the cable to the negative terminal of the battery.

10   Have the system evacuated, recharged and leak tested by a dealer service department or an air conditioning repair facility.

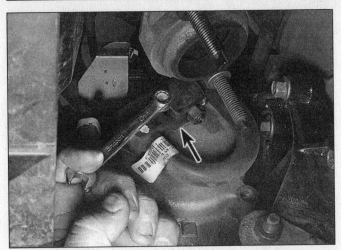

16.3 Unbolt the refrigerant line fitting block (arrow) from the back of the compressor

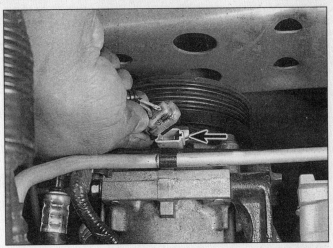

16.4 Disconnect the electrical connector (arrow) at the bottom front of the compressor

16 Air conditioning compressor - removal and installation

Warning: *The air conditioning system is under high pressure. DO NOT loosen any fittings or remove any components until after the system has been discharged. Air conditioning refrigerant should be properly discharged into an EPA-approved container at a dealer service department or an automotive air conditioning repair facility. Always wear eye protection when disconnecting air conditioning system fittings.*

Note: *Special spring-lock coupling tools are required to release the connectors used on the refrigerant lines throughout the air conditioning system. There are different special tools for each line size; these tools can usually be found at local auto parts stores, often in a set. See Section 15 for tool description and use.*

Removal

Refer to illustrations 16.3, 16.4 and 16.6

Caution: *Whenever a compressor is replaced, it will be necessary to replace the accumulator/drier and the expansion tube (see Sections 15 and 19).*

1   Take the vehicle to a dealer service department or automotive air conditioning shop and have the air conditioning system discharged. Disconnect the cable from the negative battery terminal.

2   Remove the accessory drivebelt(s) (see Chapter 1). Raise the vehicle and support it securely on jackstands. **Warning:** *Some models covered by this manual are equipped with an air suspension system. Always disconnect the electrical power to the suspension system before lifting or towing (see Chapter 10). Failure to perform this procedure may result in unexpected shifting or movement of the vehicle, which could cause personal injury.*

3   Unbolt the refrigerant line block from the compressor **(see illustration)**.

4   Disconnect the electrical connection at the compressor clutch **(see illustration)**.

5   On 1999 and later 3.0L engines, remove the power steering hose from the clip on the compressor.

6   Remove the four compressor mounting bolts (see illustration). Note: On 3.8L engines, the lower two compressor mounting bolts pass through the power steering hose support bracket. On some models, there may not be room to fully extract the bolts from the compressor. Just loosen them enough and remove the bolts with the compressor.

7   Remove the compressor from the mounting location. Drain and measure the refrigerant oil from the compressor.

Installation

8   If the compressor is being replaced, drain any shipping oil that may be in the new compressor.

9   If the amount of refrigerant oil drained from the old compressor was between 4 and 5 ounces, add that amount of new oil to the new compressor. If the amount drained was more than 5 ounces, add 5 ounces of new oil, and if the drained amount was less than 4 ounces, add 4 ounces of new oil to the new compressor.

10   Installation procedures are the reverse of those for removal. When installing the fitting block, use new O-rings and lubricate them with clean refrigerant oil.

11   After the compressor is installed, have the system evacuated, recharged and leak tested by a dealer service department or an air conditioning repair facility.

17 Air conditioning condenser - removal and installation

Warning 1: *The models covered by this manual are equipped with Supplemental Restraint Systems (SRS), more commonly known as airbags. Always disconnect the negative battery cable, then the positive battery cable and*

16.6 Remove the mounting bolts (arrows indicate two) and remove the air conditioning compressor from the engine compartment

wait two minutes before working in the vicinity of the impact sensors, steering column or instrument panel to avoid the possibility of accidental deployment of the airbag, which could cause personal injury (see Chapter 12). Do not use any electrical test equipment on any of the airbag system wires or tamper with them in any way.

Warning 2: *The air conditioning system is under high pressure. DO NOT loosen any fittings or remove any components until after the system has been discharged. Air conditioning refrigerant should be properly discharged into an EPA-approved container at a dealer service department or an automotive air conditioning repair facility. Always wear eye protection when disconnecting air conditioning system fittings.*

Note: *Special spring-lock coupling tools are required to release the connectors used on the refrigerant lines throughout the air conditioning system. There are different special tools for each line size; these tools can usually be found at local auto parts stores, often in a set. See Section 15 for tool description and use.*

3

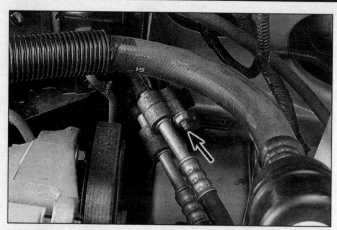

17.4 Disconnect the refrigerant lines from the condenser (arrow)

17.8 Remove the upper radiator support and the condenser mounting bolts (arrow indicates right side bolt)

Removal

Refer to illustrations 17.4 and 17.8
Caution: *Whenever a condenser is replaced, it will be necessary to replace the accumulator/drier (see Section 15).*
1   Take the vehicle to a dealer service department or automotive air conditioning shop and have the air conditioning system discharged.
2   Disconnect the negative battery cable (see Chapter 1).
3   Refer to Section 5 and follow Steps 10 through 14.
4   Disconnect the refrigerant lines at the right side of the condenser **(see illustration)**. **Note:** *Some models may have spring-lock couplings.*
5   On 1999 and later models, unclip the transmission lines.
6   On 1999 and later models, using a special quick-disconnect coupling tool (available at most automotive parts stores), disconnect the auxiliary cooler lines from the radiator. Be careful not to damage the lines or fittings. Plug the ends of the disconnected lines to prevent leakage and stop dirt from entering the system. Have a drain pan ready to catch any spills.
7   Remove the bolts and the upper radiator support **(see illustration 5.13)**.
8   Remove the two condenser mounting bolts at the radiator **(see illustration)**.
9   Carefully lift the condenser out of the bottom supports.

Installation

10   When replacing the condenser add 1.5 ounces of new refrigerant oil to the condenser before reassembly. This will maintain the correct oil level in the system after the repairs are completed.
11   The installation procedures are the reverse of those for removal. When installing the hose and fittings, use new O-rings and lubricate them with clean refrigerant oil.
12   After the condenser is installed have the system evacuated, recharged and leak tested by a dealer service department or an air conditioning repair facility.

18   Air conditioning evaporator core - removal and installation

Warning 1: *The models covered by this manual are equipped with Supplemental Restraint Systems (SRS), more commonly known as airbags. Always disconnect the negative battery cable, then the positive battery cable and wait two minutes before working in the vicinity of the impact sensors, steering column or instrument panel to avoid the possibility of accidental deployment of the airbag, which could cause personal injury (see Chapter 12). Do not use any electrical test equipment on any of the airbag system wires or tamper with them in any way.*
Warning 2: *The air conditioning system is under high pressure. DO NOT loosen any fittings or remove any components until after the system has been discharged. Air conditioning refrigerant should be properly discharged into an EPA-approved container at a dealer service department or an automotive air conditioning repair facility. Always eye wear protection when disconnecting air conditioning system fittings.*
Note: *Before replacing an evaporator core, determine for certain that the core is leaking by having a leak test performed with special equipment at dealer service department or automotive air conditioning repair facility.*

Removal

Refer to illustrations 18.4 and 18.9
1   The evaporator core is part of the large housing that contains the heater core, blower motor, evaporator and air control doors. The evaporator core is a "non-serviceable" component, it is part of the housing assembly and when replaced, the complete housing and evaporator core must be purchased as a assembly and all the necessary vacuum lines, vacuum control motors and air ducts must be transferred from the oil unit to the new unit.
Note: *Whenever the evaporator core is replaced with a new one, the accumulator/drier will also have to be replaced (see Section 15).*

18.4 Disconnect the refrigerant lines (arrows) at the firewall

2   On some models, there is an optional heater/air conditioning system located behind the left rear interior trim panel (behind the left rear wheel housing). The assembly includes a blower, heater core and evaporator core **(see illustration 11.9)**. Removal and installation procedures are similar to the procedures for the front evaporator core, although the housing does not have to be replaced. The rear evaporator core can be removed when the housing is disassembled by taking out the screws holding the two halves together.
3   Have the refrigerant discharged and recovered at a dealer or air conditioning service shop.
4   Disconnect the refrigerant lines at the firewall **(see illustration)**.
5   Drain the cooling system and disconnect the heater hoses at the firewall (see Section 13). Refer to Section 13 and remove the heater core. Remove the blower motor (see Section 11).
6   Tag and disconnect all vacuum hoses, electrical connectors, and ground wires from the heater/air conditioning housing.
7   Refer to Section 13 and pull back the

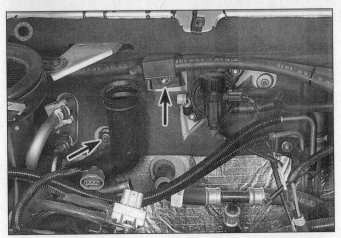

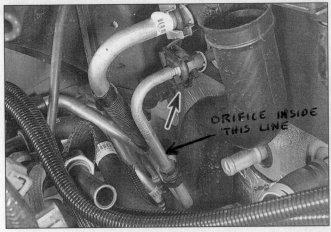

18.9 Remove the nuts on the engine side of the firewall (arrows indicate two, one is behind the accumulator/drier and another above the fresh air duct opening in the firewall)

19.2 Disconnect the high-pressure line at this point (arrow) to access the expansion tube (orifice)

rear floor duct and remove the main floor duct.

8   Remove the nut/bolt and bracket at the left side of the housing, near the throttle pedal **(see illustration 13.5a)**.

9   On the engine compartment side of the firewall, remove the four nuts from the housing studs **(see illustration)**. Carefully remove the evaporator housing assembly from the vehicle.

Installation

10   Installation is the reverse of the removal process. Transfer all of the vacuum motors, hoses, heater core, drain tube, blower motor and blower resistor to the new housing. Add 4 ounces of new refrigerant oil to the (new) accumulator/drier inlet tube when a new evaporator core is installed. Also, before the lines are reconnected, it 's a good idea to replace the expansion tube (orifice) (see Section 19).

11   Have the system evacuated, recharged and leak tested by the dealer service department or an air conditioning repair facility.

19   Air conditioning expansion tube (orifice) - removal and installation

Warning: *The air conditioning system is under high pressure. DO NOT loosen any fittings or remove any components until after the system has been discharged. Air conditioning refrigerant should be properly discharged into an EPA-approved container at a dealership service department or an automotive air conditioning repair facility. Always wear eye protection when disconnecting air conditioning system fittings.*

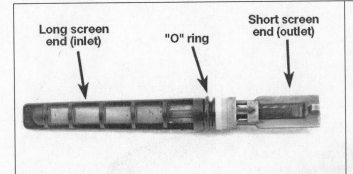

Long screen end (inlet)   "O" ring   Short screen end (outlet)

19.3 The expansion tube (orifice) contains a precise orifice and several screen filters - it should be replaced whenever a compressor is replaced

Note: *Whenever the expansion tube is replaced, the accumulator/drier should also be replaced (see Section 15).*

Removal

Refer to illustrations 19.2 and 19.3

1   Have the refrigerant discharged and recovered at a dealer or air conditioning service shop. Disconnect the cable from the negative terminal of the battery.

2   Using a spring-lock coupling tool (see Section 15), disconnect the condenser-to-evaporator refrigerant line at the firewall **(see illustration)**.

3   The expansion tube (orifice) is a tube with a fixed-diameter orifice and a mesh filter at each end **(see illustration)**. When you separate the pipe at the fitting you will see one end of the orifice inside the pipe leading to the evaporator. Use a special orifice removal tool (available at most automotive parts stores) to remove the orifice. It may be possible to remove an orifice with needle-nose pliers, but if the orifice breaks on removal the broken end must be removed with a special broken orifice removal tool.

Caution: *Pull the orifice straight out, do not twist it.*

4   The orifice acts to meter the refrigerant, changing it from high-pressure liquid to low-pressure liquid. It is possible to reuse the orifice if:

A   *The screens aren't plugged with grit or foreign material*
B   *Neither screen is torn*
C   *The plastic housing over the screens is intact*
D   *The brass orifice inside the plastic housing is unrestricted*

Installation

5   Installation is the reverse of removal. Be sure to insert the expansion tube (orifice) with the shorter end in first, toward the evaporator, and lubricate the refrigerant line and the orifice with clean refrigerant oil to aid assembly. **Caution:** *Always use a new O-ring when installing the expansion tube (orifice).*

6   Connect the refrigerant line, then have the system evacuated, recharged and leak-tested by the shop that discharged it.

3

Notes

Chapter 4
Fuel and exhaust systems

Contents

Specifications

4

Fuel pressure

Key on, engine off	35 to 45 psi
Fuel system pressure (at idle)	
Vacuum hose attached	28 to 45 psi
Vacuum hose detached	38 to 50 psi
Fuel system hold pressure (after five minutes)	30 to 40 psi
Fuel pump pressure (maximum)	65 psi
Injector resistance (approximate)	13.5 to 19 ohms

Torque specifications

Ft-lbs (unless otherwise indicated)

Air intake plenum mounting bolts	
3.0L engine	
1995 through 1998	15 to 22
1999 on	
Step 1	14
Step 2	18
3.8L engine	
1995	
Step 1	96 in-lbs
Step 2	15
Step 2	24
1996 through 1998	71 to 106 in-lbs
1999 on	
Upper intake manifold spacer bolts	89 in-lbs
Upper intake manifold bolts	89 in-lbs
Throttle body mounting nuts (3.8L engine)	
1995	15 to 22
1996 and later	71 to 97 in-lbs
Fuel rail mounting bolts	71 to 97 in-lbs

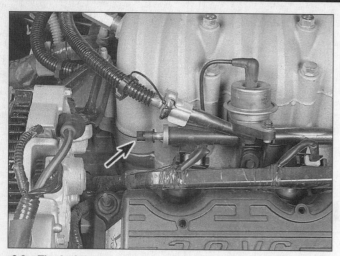

2.2a The fuel pressure test port (arrow) is located on the fuel rail facing the front of the engine compartment (3.0L engine)

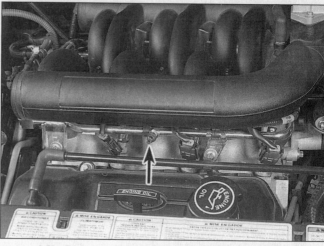

2.2b Fuel pressure test port location (1995 through 1998 3.8L engines)

1   General information

The fuel system consists of a fuel tank, an electric fuel pump (located in the fuel tank), a fuel pump relay, the fuel rail and fuel injectors, an air cleaner assembly and a throttle body unit. All models are equipped with a Sequential Electronic Fuel Injection (SEFI) system.

Sequential Electronic Fuel Injection (SEFI) system

Sequential Electronic Fuel Injection uses timed impulses to inject the fuel directly into the intake port of each cylinder according to its firing order. The injectors are controlled by the Powertrain Control Module (PCM). The PCM monitors various engine parameters and delivers the exact amount of fuel required into the intake ports. The throttle body serves only to control the amount of air passing into the system. Because each cylinder is equipped with its own injector, much better control of the fuel/air mixture ratio is possible.

Fuel pump and lines

Fuel is circulated from the fuel tank to the fuel injection system, and back to the fuel tank, through a pair of metal lines running along the underside of the vehicle. An electric fuel pump and fuel level sending unit is located inside the fuel tank. A vapor return system routes all vapors back to the fuel tank through a separate return line.

The fuel pump relay is equipped with a primary and secondary voltage circuit. The primary circuit is controlled by the PCM and the secondary circuit is linked directly to battery voltage from the ignition switch. With the ignition switch ON (engine not running), the PCM will ground the relay for one second. During cranking, the PCM grounds the fuel pump relay as long as the camshaft position sensor (CMP) sends its position signal (see Chapter 6). If there are no reference pulses,

the fuel pump will shut off after two or three seconds.

The inertia switch will disable the fuel pump circuit in the event of a collision. The inertia switch is a cylindrical magnet with a steel ball that will release (breakaway) and trip a shutdown lever when the vehicle inertia reaches a certain peak value.

Exhaust system

The exhaust system includes a pair of exhaust manifolds, a diverter (Y) pipe fitted with two upstream (before catalytic converter) and two downstream (after catalytic converter) oxygen sensors, dual catalytic converters, a muffler and a tail pipe.

The catalytic converters are an emission control device added to the exhaust system to reduce pollutants. A single-bed converter is used in combination with a three-way (reduction) catalyst. Refer to Chapter 6 for more information regarding the catalytic converters.

2   Fuel pressure relief procedure

Warning: *Gasoline is extremely flammable, so take extra precautions when you work on any part of the fuel system. Don't smoke or allow open flames or bare light bulbs near the work area, and don't work in a garage where a natural gas-type appliance (such as a water heater or a clothes dryer) with a pilot light is present. Since gasoline is carcinogenic, wear latex gloves when there's a possibility of being exposed to fuel, and, if you spill any fuel on your skin, rinse it off immediately with soap and water. Mop up any spills immediately and do not store fuel-soaked rags where they could ignite. The fuel system is under constant pressure, so, if any fuel lines are to be disconnected, the fuel pressure in the system must be relieved first. When you perform any kind of work on the fuel system, wear safety glasses and have a Class B type fire extinguisher on hand.*

2.2c Fuel pressure test port location (1999 and later 3.8L engines)

Note: *After the fuel pressure has been relieved. it's a good idea to lay a shop towel over any fuel connection to be disassembled, to absorb the residual fuel that may leak out when servicing the fuel system.*

1   There are two methods for relieving the fuel system pressure; the easiest and most accessible is using a fuel pressure gauge

2.3 Install a special fuel pressure gauge onto the test port connector and bleed the fuel into a suitable container

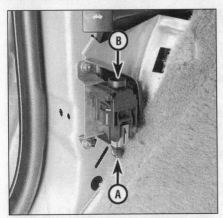

2.7a Typical inertia shut off switch - disconnect the electrical connector (A) to disable the fuel pump - to energize the fuel pump, simply plug the connector back into the switch and push the reset button (B) if necessary (1998 model shown)

2.7b On 1999 and later models, the inertia switch is located on the right side of the luggage compartment, behind this access door

2.7c After opening the door, you'll find a plastic liner

equipped with a bleed-off valve. This tool can be purchased at most auto parts stores. In the event the tool is not available, locate the inertia switch and disable the fuel pump.

Fuel pressure gauge bleeding method

Refer to illustrations 2.2a, 2.2b, 2.2c and 2.3

2   Locate the fuel pressure test port on the fuel rail and connect the fuel pressure gauge to the Schrader valve **(see illustrations)**.

3   Direct the bleed-off hose into a metal cup or suitable container for gasoline storage **(see illustration)**.

4   Turn the valve and allow the excess fuel to bleed into the container.

5   Close the valve, remove the fuel pressure gauge and cap the test port.

Inertia switch method

Refer to illustrations 2.7a, 2.7b, 2.7c, 2.7d and 2.7e

6   The fuel pump switch - sometimes called the "inertia switch" - which shuts off fuel to the engine in the event of a collision,

affords a simple and convenient means by which fuel pressure can be relieved before servicing fuel injection components.

7   On 1995 through 1997 models the inertia switch is located at the rear of the vehicle in the spare tire/jack compartment. On 1998 models the inertia switch is located at the front of the vehicle behind the driver's side kick panel **(see illustration)**. On 1999 and later models, the inertia switch is located on the right side of the luggage compartment, behind the jack access panel **(see illustrations)**.

8   Start the engine and allow it to run until it stops. This should take only a few seconds.

9   The fuel system pressure is now relieved. When you're finished working on the fuel system, simply plug the electrical connector back into the switch. If the inertia switch was "popped" (activated) during this procedure, push the reset button on the top of the switch **(see illustration 2.7)**.

---

3   Fuel pump/fuel pressure - check

Warning: *Gasoline is extremely flammable, so take extra precautions when you work on*

any part of the fuel system. See the **Warning** in Section 2.

Note 1: *To perform the fuel pressure test, you will need to obtain a fuel pressure gauge and adapter set (fuel line fittings).*

Note 2: *The fuel pump will operate as long as the engine is cranking or running and the PCM is receiving ignition reference pulses from the electronic ignition system. If there are no reference pulses, the fuel pump will shut off after two or three seconds.*

Note 3: *After the fuel pressure has been relieved, it's a good idea to lay a shop towel over any fuel connection to be disassembled, to absorb the residual fuel that may leak out when servicing the fuel system.*

Preliminary inspection

Refer to illustrations 3.2, 3.3a and 3.3b

1   Should the fuel system fail to deliver the proper amount of fuel, or any fuel at all, inspect it as follows. Remove the fuel filler cap. Have an assistant turn the ignition key to the On position (engine not running) while you listen at the fuel filler opening. You should hear a whirring sound that lasts for a couple of seconds.

2   If you don't hear anything, check the fuel pump fuse **(see illustration)**. If the fuse is blown, replace it and see if it blows again.

4

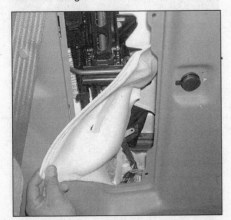

2.7d Peel back the plastic liner . . .

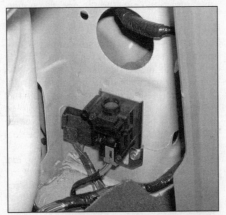

2.7e . . . and then push the button on top of the inertia switch

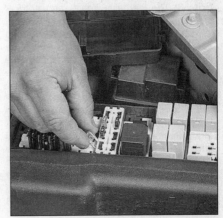

3.2 Remove the fuel pump fuse from the engine compartment fuse box - make sure it is not blown

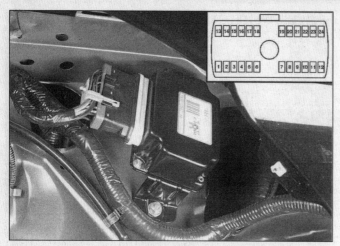

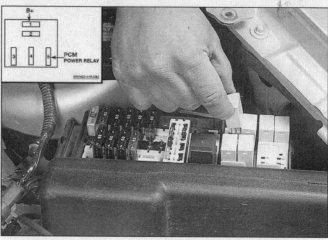

3.3a On 1995 through 1997 models, disconnect the CCRM harness connector and check for battery voltage at terminal 11 with the ignition key OFF, then check for battery voltage at terminal 13 with the ignition key ON (engine not running). There should also be power at terminal 5 with the CCRM plugged in and the key ON

3.3b On 1998 models, remove the fuel pump relay from the engine compartment fuse box and check for battery voltage at terminal 1 with the ignition key in the OFF position. Then check for battery voltage at terminal 5 with the ignition key in the ON position (engine not running)

If it does, trace the fuel pump circuit for a short. Refer to the wiring diagrams at the end of Chapter 12.

3   Check for battery voltage to the CCRM/fuel pump relay connector with the ignition key in the OFF position. Then turn the ignition key to the ON position and check for battery voltage from the PCM power relay **(see illustrations). Note 1:** *Early models are equipped with a Constant Control Relay Module (CCRM) which incorporates five different relays (the fuel pump relay, the EDF relay(s), the EDF relay control and the A/C clutch control relay) into a single module. The CCRM is located in the left front corner of the engine compartment, next to the battery. Disconnect the connector from the CCRM and perform the tests described.* **Note 2:** *On 1998 models the fuel pump relay is located in the engine compartment fuse box; simply remove the relay and perform the tests described. If there is battery voltage present at the relay, have the relay or the CCRM tested at a dealer service department or other qualified automotive repair shop.*

Note 3: *The inertia switch is an electrical device wired into the fuel pump circuit that will shut down power to the fuel pump in an accident. Be sure to check that the inertia switch is activated and in working order if the fuel pump is not receiving the proper voltage (see Section 2).* **Note 4:** *The fuel pump relay is equipped with a primary and secondary circuit. The primary circuit is controlled by the PCM and the secondary circuit provides battery voltage to the fuel pump as the relay is energized. With the ignition switch ON (engine not running), the PCM will ground the relay for one second. During cranking, the PCM grounds the fuel pump relay as long as the camshaft position sensor (CMP) sends its position signal (see Chapter 6). If there are no reference pulses, the fuel pump will shut off after two or three seconds. Refer to the wiring schematics at the end of Chapter 12 for addi-*

tional information on the wiring color designations for the fuel pump relay.

4   If there is no voltage present, check the fuse(s) and the wiring circuit for the fuel pump relay and/or PCM power relay (see Chapter 12). If voltage is present, check for battery voltage at the fuel pump harness connector located near the fuel tank. If voltage is reaching the fuel pump, remove the fuel pump and have it checked by a dealer service department or other qualified automotive repair facility.

Operating pressure check

Refer to illustrations 3.7, 3.13 and 3.15

5   Relieve the fuel system pressure (see Section 2).

6   Detach the cable from the negative battery terminal.

7   Remove the cap from the fuel pressure test port and attach a fuel pressure gauge **(see illustration).** If you don't have the correct adapter for the test port, remove the Schrader valve and connect the gauge hose to the fitting, using a hose clamp.

8   Attach the cable to the negative battery terminal.

9   Start the engine.

10   Check the fuel pressure at idle. Compare your readings with the values listed in this Chapter's Specifications. Disconnect the vacuum hose from the fuel pressure regulator and watch the fuel pressure gauge - the fuel pressure should increase considerably as soon as the hose is disconnected. If it doesn't, check for a vacuum signal to the fuel pressure regulator (see Step 15).

11   If the fuel pressure is low, pinch the fuel return line shut and watch the gauge. If the pressure doesn't rise, the fuel pump is defective or there is a restriction in the fuel feed line. If the pressure rises sharply, replace the pressure regulator. **Note:** *If the vehicle is equipped with a nylon fuel return line (or fuel lines made up of steel or other rigid material),*

it will be necessary to install a special fuel testing harness between the fuel rail and the return line. This can be made up from compatible fuel line connectors (available at the dealer parts department and some auto parts stores), fuel hose and hose clamps.

12   If the fuel pressure is too high, turn the engine off. Disconnect the fuel return line and blow through it to check for a blockage. If there is no blockage, replace the fuel pressure regulator.

13   Hook up a hand-held vacuum pump to the port on the fuel pressure regulator **(see illustration).**

14   Read the fuel pressure gauge with vacuum applied to the fuel pressure regulator and also with no vacuum applied. The fuel pressure should decrease as vacuum increases (and increase as vacuum decreases).

15   Connect a vacuum gauge to the pressure regulator vacuum hose. Start the engine and check for vacuum **(see illustration).** If

3.7 If you don't have the correct adapter, it is possible to remove the Schrader valve from the fitting and install a standard fuel pressure gauge, using a hose clamp

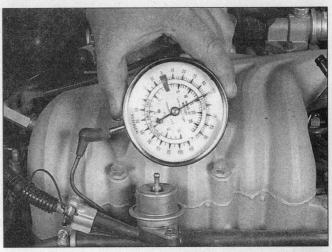

3.13 Connect a hand-held vacuum pump to the fuel pressure regulator and read fuel pressure with vacuum applied. Pressure should decrease as vacuum is increased

3.15 Detach the vacuum line from the fuel pressure regulator and verify vacuum is present when the engine is running

there isn't vacuum present, check for a clogged hose or vacuum port. If the amount of vacuum is adequate, replace the fuel pressure regulator.

16 Turn the ignition switch to OFF, wait five minutes and recheck the pressure on the gauge. Compare the reading with the hold pressure listed in this Chapter's Specifications. If the hold pressure is less than specified:

a) *The fuel lines may be leaking.*

b) *The fuel pressure regulator may be allowing the fuel pressure to bleed through to the return line*

c) *A fuel injector (or injectors) may be leaking.*

d) *The fuel pump may be defective.*

---

4 Fuel lines and fittings - general information

Warning: *Gasoline is extremely flammable, so take extra precautions when you work on any part of the fuel system. See the* **Warning** *in Section 2.*

Note: *The majority of fuel line fittings on these vehicles require special disconnect tools to allow removal of the fuel lines. These tools are available at most auto part stores.*

Push-connect fittings - disassembly and reassembly

1 Ford uses two different push-connect fitting designs. Fittings used with 3/8 and 5/16-inch diameter lines have a "hairpin" type clip; fittings used with 1/4-inch diameter lines have a "duck bill" type clip. The procedure used for releasing each type of fitting is different. The clips should be replaced whenever a connector is disassembled.

2 Disconnect all push-connect fittings from fuel system components such as the fuel filter, the fuel charging assembly, the fuel tank, etc. before removing the assembly.

3/8 and 5/16-inch fittings (hairpin clip)

Refer to illustration 4.5

3 Inspect the internal portion of the fitting for accumulations of dirt. If more than a light coating of dust is present, clean the fitting before disassembly.

4 Some adhesion between the seals in the fitting and the line will occur over a period of time. Twist the fitting on the line, then push and pull the fitting until it moves freely.

5 Remove the hairpin clip from the fitting by bending the shipping tab down until it clears the body. Then spread each leg about 1/8-inch to disengage the body and push the legs through the fitting. If possible, don't use any tools to do this, to prevent damage to the plastic fitting on the fuel line. Finally, pull lightly on the triangular end of the clip and work it clear of the line and fitting (**see illustration**).

6 Grasp the fitting and hose and pull it straight off the line.

7 Do not reuse the original clip in the fitting. A new clip must be used.

8 Before reinstalling the fitting on the line, wipe the line end with a clean cloth. Inspect

the inside of the fitting to ensure that it's free of dirt and/or obstructions.

9 To reinstall the fitting on the line, align them and push the fitting into place. When the fitting is engaged, a definite click will be heard. Pull on the fitting to ensure that it's completely engaged. To install the new clip, insert it into any two adjacent openings in the fitting with the triangular portion of the clip pointing away from the fitting opening. Using your index finger, push the clip in until the legs are locked on the outside of the fitting.

1/4-inch fittings (duck bill clip)

Refer to illustrations 4.10, 4.13 and 4.14

10 The duck bill clip type fitting consists of a body, spacers, O-rings and the retaining clip (**see illustration**). The clip holds the fitting securely in place on the line. One of the two following methods must be used to disconnect this type of fitting.

11 Before attempting to disconnect the fitting, check the visible internal portion of the fitting for accumulations of dirt. If more than a light coating of dust is evident, clean the fitting before disassembly.

12 Some adhesion between the seals in the

4

4.5 A hairpin clip type push-connect fitting

4.10 A push-connect fitting with a duck bill clip

4.13 Remove the safety clip

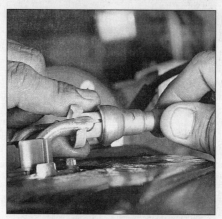

4.14 Duck bill clip fitting disassembly using the special tool

4.26a If the spring lock couplings are equipped with safety clips, pry them off with a small screwdriver

fitting and line will occur over a period of time. Twist the fitting on the line, then push and pull the fitting until it moves freely.

13   Remove the safety clip from the fuel line **(see illustration)**.

14   The preferred method used to disconnect the fitting requires a special tool available at most auto part stores. To disengage the line from the fitting, align the slot in the push-connect disassembly tool with either tab on the clip (90-degrees from the slots on the side of the fitting) and insert the tool **(see illustration)**. This disengages the duck bill from the line. **Note:** *Some fuel lines have a secondary bead which aligns with the outer surface of the clip. The bead can make tool insertion difficult. If necessary, use the alternative disassembly method described in Step 16. Holding the tool and the line with one hand, pull the fitting off. **Note:** *Only moderate effort is necessary if the clip is properly disengaged. The use of anything other than your hands should not be required.*

15   After disassembly, inspect and clean the line sealing surface. Also inspect the inside of the fitting and the line for any internal parts that may have been dislodged from the fitting. Any loose internal parts should be immediately reinstalled (use the line to insert the parts).

16   The alternative disassembly procedure requires a pair of small adjustable pliers. The pliers must have a jaw width of 3/16-inch or less.

17   Align the jaws of the pliers with the openings in the side of the fitting and compress the portion of the retaining clip that engages the body. This disengages the retaining clip from the body (often one side of the clip will disengage before the other - both sides must be disengaged).

18   Pull the fitting off the line. **Note:** *Only moderate effort is required if the retaining clip has been properly disengaged. Do not use any tools for this procedure.*

19   Once the fitting is removed from the line end, check the fitting and line for any internal parts that may have been dislodged from the fitting. Any loose internal parts should be immediately reinstalled (use the line to insert the parts).

20   The retaining clip will remain on the line. Disengage the clip from the line bead to remove it. Do not reuse the retaining clip - install a new one!

21   Before reinstalling the fitting, wipe the line end with a clean cloth. Check the inside of the fitting to make sure that it's free of dirt and/or obstructions.

22   To reinstall the fitting, align it with the line and push it into place. When the fitting is engaged, a definite click will be heard. Pull on the fitting to ensure that it's fully engaged.

23   Install the new replacement clip by inserting one of the serrated edges on the duck bill portion into one of the openings. Push on the other side until the clip snaps into place.

Spring lock couplings - disassembly and reassembly

Refer to illustrations 4.26a, 4.26b and 4.26c

24   The fuel supply and return lines used on SEFI engines utilize spring lock couplings at the engine fuel rail end instead of plastic push-connect fittings. The male end of the spring lock coupling, which is girded by two O-rings, is inserted into a female flared end

engine fitting. The coupling is secured by a garter spring which prevents disengagement by gripping the flared end of the female fitting. A cup-tether assembly provides additional security.

25   To disconnect the 1/2-inch spring lock coupling supply fitting or the 3/8-inch return fitting, you will need to obtain the appropriate spring lock coupling tool from your local auto parts store. The tools come in two different sizes - 3/8 and 1/2 inch - which correspond with the diameter of the fuel supply and return lines.

26   Study the accompanying illustrations carefully before detaching either spring lock coupling fitting **(see illustrations)**.

---

5   Fuel tank - removal and installation

Refer to illustrations 5.5, 5.6, 5.7 and 5.9
Warning: *Gasoline is extremely flammable, so take extra precautions when you work on any part of the fuel system. See the* **Warning** *in Section 2.*

4.26b Open the spring-loaded halves of the spring lock coupling tool and place it in position around the coupling, then close it

4.26c To disconnect the coupling, push the tool into the cage opening to expand the garter spring and release the female fitting, then pull the male and female fittings apart

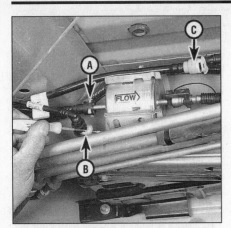

5.5 Disconnect the fuel feed line (A) from the fuel filter, and also disconnect the fuel return line (B) and the vapor return line (C)

5.6 Disconnect the fuel line retaining clip and the fuel pump/sending unit electrical connector (arrow) from the rear crossmember

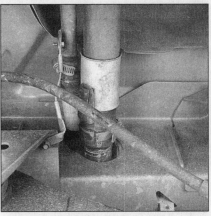

5.7 Remove the fuel filler hose, vent hose and the ground strap from the filler neck assembly

Note: *Don't begin this procedure until the gauge indicates that the tank is empty or nearly empty. If the tank must be removed when it's full (for example, if the fuel pump malfunctions), siphon any remaining fuel from the tank prior to removal.*

Note: *This procedure requires a special fuel line disconnect tool which is available at most auto part stores. Some models may be equipped with a fuel line disconnect tool which is installed on the vehicle about ten inches behind the fuel filter.*

1   Unless the vehicle has been driven far enough to completely empty the tank, it's a good idea to siphon the residual fuel out before removing the tank from the vehicle. **Warning:** *DO NOT start the siphoning action by mouth! Use a siphoning kit (available at most auto parts stores).*

2   Relieve the fuel pressure (see Section 2).

3   Disconnect the cable from the negative terminal of the battery.

4   Raise the vehicle and support it securely on jackstands. **Warning:** *Some models covered by this manual are equipped with air suspension systems. Always disconnect electrical power to the suspension system before lifting or towing the vehicle (see Chapter 10). Failure to perform this procedure may result in unexpected shifting or movement of the vehicle which could cause personal injury.*

5   Disconnect the fuel feed line from the rear of the fuel filter, the return line fitting just below the fuel filter, and the fuel vapor return line just above the filter (see Section 4 if necessary) **(see illustration)**. Then detach any remaining clips securing the fuel lines.

6   Disconnect the electrical connector attached on the rear crossmember **(see illustration)**.

7   Loosen the hose clamps and detach the fuel tank filler hose and vapor hose from the fuel filler neck and the fuel tank **(see illustration)**. Detach the fuel filler neck ground strap if equipped.

8   Place a floor jack under the tank and position a wood plank between the jack pad and the tank. Raise the jack until it's support-

ing the tank.

9   Remove the bolts that retain the fuel tank mounting straps **(see illustration)**. The straps are hinged at the other end so you can swing them out of the way. **Note:** *20 gallon fuel tanks are equipped with two retaining straps, while 25 gallon tanks are equipped with three retaining straps.*

10   Slowly lower the jack while guiding the fuel tank past the muffler and tail pipe assembly. Remove the tank from the vehicle.

11   If you're replacing the tank, or having it cleaned or repaired, refer to Section 6.

12   Refer to Section 7 to remove and install the fuel pump or sending unit, if necessary.

13   Installation is the reverse of removal. Clean engine oil can be used as an assembly aid when pushing the fuel filler hose back onto the fuel tank. **Warning:** *Fuel tank heat shields must be reinstalled in their original positions to prevent heat damage to the fuel tank.*

---

6   Fuel tank - cleaning and repair

1   The fuel tanks installed in the vehicles covered by this manual are made of plastic and are not repairable.

2   If the fuel tank is removed from the vehicle, it should not be placed in an area where sparks or open flames could ignite the fumes coming out of the tank. Be especially careful inside a garage where a natural gas-type appliance is located, because the pilot light could cause an explosion.

---

7   Fuel pump - removal and installation

Refer to illustrations 7.6 and 7.7

Warning: *Gasoline is extremely flammable, so take extra precautions when you work on any part of the fuel system. See the **Warning** in Section 2.*

Note 1: *This procedure requires a special fuel line disconnect tool which is available at most*

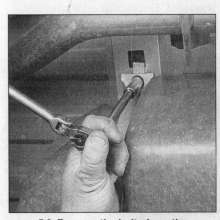

5.9 Remove the bolts from the fuel tank straps

auto part stores. Some models may be equipped with a fuel line disconnect tool which is installed on the vehicle about ten inches behind the fuel filter.

Note 2: *The following removal and installation procedure applies to all fuel pumps, but on 1999 and later models, the fuel pump module is a single, non-serviceable module, i.e. it cannot be disassembled like earlier pumps.*

1   Unless the vehicle has been driven far enough to completely empty the tank, it's a good idea to siphon out the residual fuel before removing the fuel pump from the vehicle. **Warning:** *DO NOT start the siphoning action by mouth! Use a siphoning kit (available at most auto parts stores).*

2   Relieve the fuel pressure (refer to Section 2).

3   Disconnect the cable from the negative terminal of the battery.

4   Raise the vehicle and support it securely on jackstands. **Warning:** *Some models covered by this manual are equipped with air suspension systems. Always disconnect electrical power to the suspension system before lifting or towing the vehicle (see Chapter 10). Failure to perform this procedure may result in unexpected shifting or movement of the vehicle which could cause personal injury.*

4

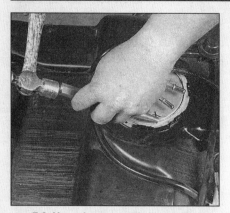

7.6 Use a brass punch to turn the locking ring

5   Remove the fuel tank from the vehicle (see Section 5). Detach the fuel feed line and return lines from the fuel pump module (see Section 4).

6   Using a brass punch or wood dowel only, tap the fuel pump module lock-ring counterclockwise until it's loose **(see illustration)**.

7   Lift the fuel pump/sending unit assembly out of the tank **(see illustration). Caution:** *The fuel level float and sending unit are delicate. Don't bump or bend them during removal or the accuracy of the sending unit may be affected. On 1998 models, DO NOT lift the fuel pump module out of the tank by the hoses.*

8   Inspect the condition of the O-ring around the opening of the tank. If it is dried, cracked or deteriorated, replace it.

1995 through 1997 models

Refer to illustrations 7.9, 7.10 and 7.11

9   Remove the fuel pump mounting clamp bolt **(see illustration)**.

10   To separate the fuel pump from the assembly, remove the fuel hose lower clamp and disconnect the electrical connector from the fuel pump **(see illustration)**.

11   Remove the strainer from the lower end of the fuel pump **(see illustration)**. If it's dirty, remove it and clean it with carburetor cleaner spray. If it's too dirty to be cleaned, replace it. **Note:** *Most new fuel pump assemblies come equipped with a new strainer.*

12   Installation of the fuel pump to the sending unit is the reverse of removal.

13   Clean the fuel pump mounting flange and the tank mounting surface and seal ring groove. Apply a thin coat of heavy grease to the new seal ring to hold it in place during assembly.

14   Position the O-ring around the opening in the fuel tank and guide the fuel pump/sending unit assembly into the tank.

15   Turn the lock-ring clockwise until the locking cams are fully engaged by the retaining tangs. **Note:** *If you've installed a new O-ring, it may be necessary to push down on the lock-ring until the locking cams slide under the retaining tangs.*

16   Install the fuel tank (see Section 5).

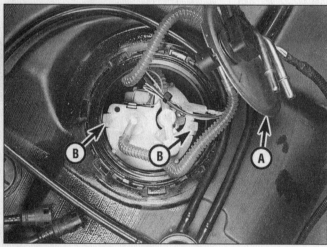

7.7 On 1998 models, lift off the fuel line retainer (A), then release the locking tabs (B) and carefully angle the fuel pump module out of the fuel tank without damaging the fuel level sending arm and float

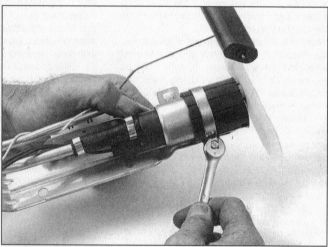

7.9 Loosen the fuel pump mounting clamp, then slide it down past the end of the mounting bracket

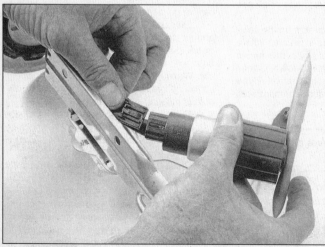

7.10 Disconnect the fuel pump electrical connector and the fuel line lower hose clamp from the fuel pump

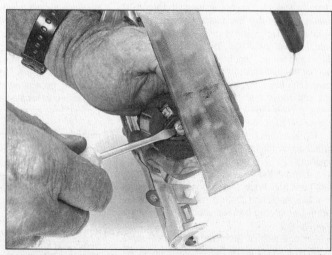

7.11 Remove the C-clip from the base of the fuel pump then separate the strainer from the fuel pump

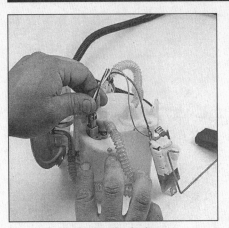

7.17 Unplug the fuel pump electrical connector

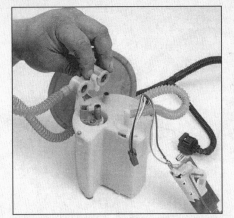

7.18 Remove the retaining screws and detach the fuel feed line from the top of the fuel pump and strainer assembly

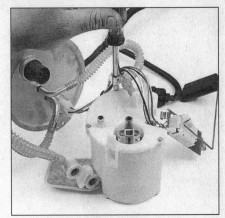

7.19 Remove the retaining screws and detach the top half of the strainer assembly

1998 models

Refer to illustrations 7.17, 7.18, 7.19, 7.20 and 7.24

17   Detach the sending unit from the strainer assembly. Unplug the fuel pump electrical connector **(see illustration)**.

18   Remove the fuel feed line from the top of the fuel pump and strainer assembly **(see illustration)**.

19   Remove the screws and detach the top of the strainer assembly **(see illustration)**.

20   Detach the fuel pump from the lower half of the strainer assembly **(see illustration)**.

21   Installation of the fuel pump-to-strainer assembly is the reverse of removal.

22   Clean the fuel line retainer mounting flange and the tank mounting surface and seal ring groove. Apply a thin coat of heavy grease to the new seal ring to hold it in place during assembly.

23   Position the O-ring around the opening in the fuel tank, then install the fuel pump/sending unit module into the tank making sure the locking tabs on the strainer are aligned with the tabs on the fuel tank. Snap the fuel pump module into place.

24   Push the fuel feed and return lines from

the module into the tank and install the fuel line retainer into place on the fuel tank **(see illustration)**. Turn the lock-ring clockwise until the locking cams are fully engaged by the retaining tangs. **Note:** *If you've installed a new O-ring, it may be necessary to push down on the lock-ring until the locking cams slide under the retaining tangs.*

25   Install the fuel tank (see Section 5).

8   Fuel level sending unit - check and replacement

Check

Refer to illustrations 8.3 and 8.5

1   Raise the vehicle and support it securely on jackstands. **Warning:** *Some models covered by this manual are equipped with self leveling suspension systems. Always disconnect electrical power to the suspension system before lifting or towing the vehicle (see Chapter 10). Failure to perform this procedure may result in unexpected shifting or movement of the vehicle which could cause personal injury.*

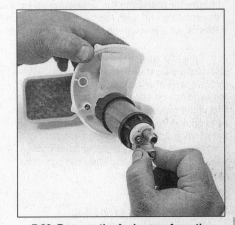

7.20 Remove the fuel pump from the lower half of the strainer assembly

2   Disconnect the electrical connector from the rear crossmember for the fuel level sending unit **(see illustration 5.6)**.

3   Position the ohmmeter probes into the electrical connector and check the resistance **(see illustration)**. Use the 200-ohm scale on the ohmmeter.

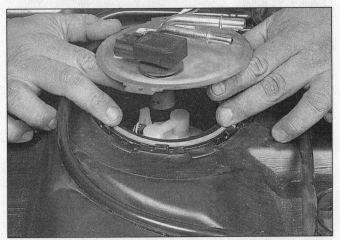

7.24 Push the fuel feed and return lines from the module into the tank making sure the lines are not twisted or kinked, then install the fuel line retainer into place on the fuel tank

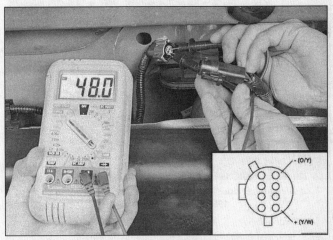

8.3 Using an ohmmeter, probe the indicated terminals of the fuel sending unit connector to check the resistance

4

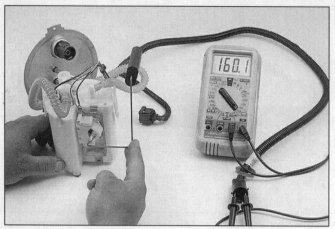

8.5 A more accurate check of the fuel level sending unit can be performed with the assembly on the bench. Connect the ohmmeter probes to the connector and check the resistance of the sending unit with the float positioned on "empty" and "full". Check for a smooth change in resistance as the float is moved

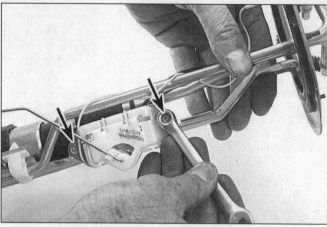

8.8a On 1995 through 1997 models, remove the fuel level sending unit bolts (arrows) from the fuel pump assembly frame, then detach the harness connectors

4   With the fuel tank completely full, the resistance should be about 160.0 ohms. With the fuel tank nearly empty, the resistance of the sending unit should be about 15.0 ohms.

5   If the readings are incorrect, replace the sending unit. **Note:** *A more accurate check of the sending unit can be made by removing it from the fuel tank and checking its resistance while manually operating the float arm* **(see illustration)**.

Replacement

Refer to illustrations 8.8a and 8.8b

6   Remove the fuel tank (see Section 5).

7   Remove the fuel pump module from the fuel tank (see Section 7).

8   Remove the sending unit mounting bolt(s) **(see illustrations)** and harness connectors from the fuel pump module.

9   Installation is the reverse of removal.

9   Air cleaner housing - removal and installation

Refer to illustrations 9.2a and 9.2b

1   Detach the cable from the negative terminal of the battery.

2   Disconnect the IAT sensor and the MAF sensor wiring **(see illustrations)**,

3   On 3.0L and 1995 3.8L engines, remove the crankcase ventilation hose and the retaining clamps from the air cleaner outlet tube. Detach the air outlet tube from the throttle body and the MAF sensor. Remove the air cleaner housing retaining bolt and the intake air duct retaining nut, then lift the air cleaner assembly from the vehicle.

4   On 1996 and later 3.8L engines, loosen the air cleaner outlet tube retaining clamp and detach it from the throttle body. Carefully

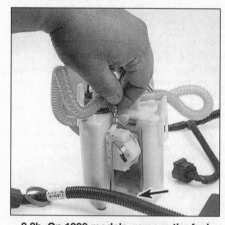

8.8b On 1998 models, remove the fuel level sending unit retaining screw (arrow) from the strainer assembly, then detach the harness connectors

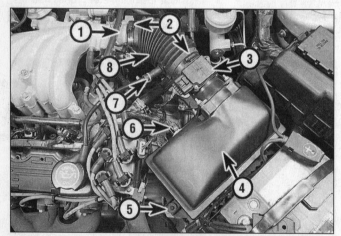

9.2a Air cleaner housing mounting details (3.0L engine shown, 1995 3.8L engine similar)

1	Throttle body	6	Intake Air Temperature
2	Retaining clamp		(IAT) sensor
3	Mass airflow (MAF) sensor	7	Crankcase ventilation tube
4	Air cleaner housing	8	Air cleaner outlet tube
5	Retaining bolt		

9.2b Air cleaner housing mounting details (1996 and later 3.8L engines)

1	Retaining clamp	4	Intake air duct
2	Intake Air Temperature	5	Mass airflow (MAF) sensor
	(IAT) sensor	6	Air cleaner outlet tube
3	Air cleaner housing	7	Throttle body

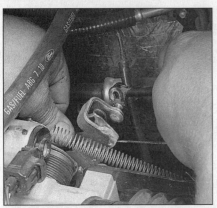

10.2 Detach the cruise control cable from the throttle lever if equipped

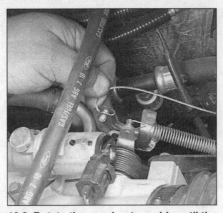

10.3 Rotate the accelerator cable until the slot in the throttle lever aligns with the cable, then pass the cable casing through the slot

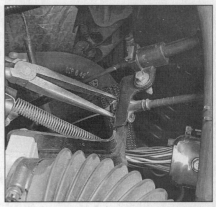

10.4 Depress the clips and disengage the cable casing from the bracket

lift upward on the air cleaner housing to release the rubber grommets securing the housing to the battery tray. Detach the air cleaner housing from intake air duct and remove the housing from the vehicle.

5 Installation is the reverse of removal.

10 Accelerator cable - removal, installation and adjustment

Removal

Refer to illustrations 10.2, 10.3, 10.4 and 10.5

1 Remove the throttle lever protection shield (if equipped).

2 Detach the accelerator cable return spring (if necessary) and the cruise control cable end from the throttle lever **(see illustration)**.

3 Detach the accelerator cable from the throttle lever **(see illustration)**.

4 Separate the accelerator cable casing from the cable bracket **(see illustration)**.

5 Pull the cable end out from the accelerator pedal arm, then pass the cable through the slot in the arm **(see illustration)**.

6 Disconnect any cable clips or brackets

securing the accelerator cable.

7 Remove the cable through the firewall from the engine compartment.

Installation

8 Installation is the reverse of removal. Be sure the cable is routed correctly and the grommet seats completely in the firewall. **Note:** *On 3.0L engines the throttle return spring must be installed with the open ends of the spring facing the firewall. On 3.8L engines the throttle return spring must be installed with the open ends of the spring facing the throttle body.*

9 If necessary, at the engine compartment side of the firewall, apply sealant around the accelerator cable grommet to prevent water from entering the passenger compartment.

Adjustment

10 Measure the freeplay by firmly gripping the cable and pressing it down from the cable housing. There should be a slight amount of cable freeplay.

11 If there is no cable freeplay or the cable is binding and not allowing the throttle lever to completely close, replace the cable.

11 Fuel injection system - general information

Sequential Electronic Fuel Injection (SEFI)

Refer to illustrations 11.1a, 11.1b and 11.1c

The Sequential Electronic Fuel Injection (SEFI) system is a multi-point fuel injection system. On the SEFI system, fuel is metered into each intake port in sequence with the engine firing order in accordance with engine demand through one injector per cylinder mounted on a tuned intake manifold. The intake manifold incorporates an air intake plenum to aid in air flow and distribution. Each engine uses a slightly different plenum design and fuel rail arrangement. The 1995 through 1998 3.0L engine incorporates a one-piece air intake plenum and throttle body assembly, while 1999 and later 3.0L engines and all 3.8L engines are equipped with a separate air intake plenum and throttle body **(see illustrations)**. All air intake plenums bolt to the intake manifold which sits directly in the middle of the engine block. The air intake plenum on 1999 and later 3.8L engines is a

4

10.5 Working under the dash, pull the cable end from the accelerator pedal recess and lift it through the slot (arrow)

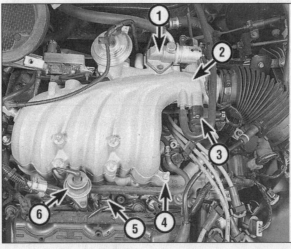

11.1a Fuel injection components (1995 through 1998 3.0L engine shown; 1999 and later models similar, except that throttle body is a separate component bolted to the plenum)

1 *Idle Air Control (IAC) valve*
2 *Throttle body/air intake plenum*
3 *Throttle Position Sensor (TPS)*
4 *Fuel rail*
5 *Fuel injector*
6 *Fuel pressure regulator*

11.1b Fuel injection components (1996 through 1998 3.8L models)

1	Idle Air Control (IAC) valve	4	Air intake plenum
2	Throttle body	5	Fuel rail
3	Throttle Position Sensor (TPS)	6	Fuel injector
		7	Fuel pressure regulator

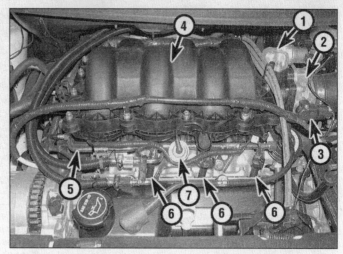

11.1c Fuel injection components (1999 and later 3.8L models)

1	Idle Air Control (IAC) valve	4	Air intake plenum (upper intake manifold)
2	Throttle body	5	Fuel rail
3	Throttle Position Sensor (TPS)	6	Fuel injector
		7	Fuel pressure regulator

two-piece design consisting of the upper intake manifold and the upper intake manifold spacer. On these engines, the throttle body is bolted to the upper intake manifold *spacer*, not to the upper intake manifold.

1996 and later 3.8L engines are equipped with the Intake Manifold Runner Control (IMRC) system. The IMRC system controls the air intake charge by opening or closing the butterfly valve on the secondary intake valve directly at the intake manifold. By closing the butterfly to the secondary intake valves under 3,000 rpm, low end driveability is improved. Above 3,000 rpm the butterfly valves open to increase high-end performance. The butterfly valves are controlled by the IMRC actuator and cable assembly.

The Sequential Electronic Fuel Injection system incorporates an on-board Electronic Engine Control (EEC-V) computer that accepts inputs from various engine sensors to compute the required fuel flow rate necessary to maintain a prescribed air/fuel ratio throughout the entire engine operational range. The computer then outputs a command to the fuel injectors to meter the approximate quantity of fuel. The system automatically senses and compensates for changes in altitude, load and speed. **Note:** *The computer terminology has changed from Electronic Control Module (ECM) to the Powertrain Control Module (PCM) due to standardization of the Self Diagnosis system within the automotive industry.*

The fuel delivery systems include an electric in-tank fuel pump which forces pressurized fuel through a series of metal and plastic lines and an inline fuel filter to the fuel rail assembly. The SEFI system uses a single high-pressure pump mounted inside the tank.

The fuel rail assembly incorporates an electrically actuated fuel injector directly above each intake port. When energized, the injectors spray a metered quantity of fuel into the intake air stream.

A constant fuel pressure is supplied to the injectors by a pressure regulator. The regulator is positioned downstream from the fuel injectors. Excess fuel passes through the regulator and returns to the fuel tank through a fuel return line.

On the SEFI system, each injector is energized once every other crankshaft revolution in sequence with engine firing order. The period of time that the injectors are energized (known as "on time" or "pulse width") is controlled by the PCM. Air entering the engine is sensed by speed, pressure and temperature sensors. The outputs of these sensors are processed by the PCM. The computer determines the needed injector pulse width and outputs a command to the injector to meter the exact quantity of fuel.

12 Fuel injection system - check

Warning: *Gasoline is extremely flammable, so take extra precautions when you work on any part of the fuel system. See the* **Warning** *in Section 2.*
Note: *The following procedure is based on the assumption that the fuel pump is working and the fuel pressure is adequate (see Section 3).*

Preliminary checks

1   Check all electrical connectors that are related to the system. Loose electrical connectors and poor grounds can cause many problems that resemble more serious malfunctions.

2   Check to see that the battery is fully charged, as the control unit and sensors depend on an accurate supply voltage in order to properly meter the fuel.
3   Check the air filter element - a dirty or partially blocked filter will severely impede performance and economy (see Chapter 1).
4   If a blown fuse is found, replace it and see if it blows again. If it does, search for a grounded wire in the harness to the fuel pump (see Chapter 12).

System checks

Refer to illustrations 12.7, 12.8 and 12.9
5   Check the condition of the vacuum hoses connected to the intake manifold.
6   Remove the air intake duct from the throttle body and check for dirt, carbon or other residue build-up in the throttle body, particularly around the throttle plate. **Caution:** *The throttle body on these models is coated with a sludge-resistant material designed to protect the bore and throttle plate. Do not attempt to clean the interior of the throttle body with carburetor or other spray cleaners. This throttle body is designed to resist sludge accumulation and cleaning may impair the performance of the engine.*
7   With the engine running, place an automotive stethoscope against each injector, one at a time, and listen for a clicking sound, indicating operation **(see illustration)**. If you don't have a stethoscope, you can place the tip of a long screwdriver against the injector and listen through the handle.
8   If an injector isn't functioning (not clicking), purchase a special injector test light (sometimes called a "noid" light) and install it into the injector electrical connector **(see illustration)**. Start the engine and check to see if the noid light flashes. If it does, the injector is receiving proper voltage. If it does-

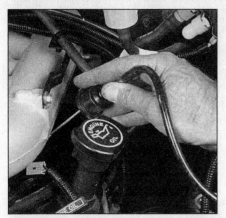

12.7 Use a stethoscope or screwdriver to determine if the injectors are working properly - they should make a steady clicking sound that rises and falls as engine speed changes

12.8 Install the fuel injector test light or "noid light" into the fuel injector electrical connector and confirm that it blinks when the engine is cranked or running

12.9 Measure the resistance of each injector. It should be within Specifications

n't flash, further diagnosis should be performed by a dealer service department or other properly equipped repair facility.

9   With the engine OFF and the fuel injector electrical connectors disconnected, measure the resistance of each injector **(see illustration)**. Check the Specifications listed in this Chapter for the correct injector resistance.

10   The remainder of the system checks can be found in Section 13 and Chapter 6.

---

13   Sequential Electronic Fuel Injection (SEFI) system - component check and replacement

Throttle body/air intake plenum

Caution: *The throttle body is coated with a sludge-resistant material designed to protect the bore and throttle plate. Do not attempt to clean the interior of the throttle body. The throttle body is designed to resist sludge*

accumulation and cleaning may impair the performance of the engine.

Note: *The throttle body and the air intake plenum on 1995 through 1998 3.0L engines is a one piece design. All 1998 and later 3.0L and 3.8L engines are equipped with a separate throttle body which can be unbolted from the air intake plenum.*

3.0L engine

Refer to illustrations 13.5, 13.8, 13.9 and 13.12

1   Detach the cable from the negative terminal of the battery.

2   Unplug the electrical connectors at the IAC valve, Throttle Position Sensor (TPS), IAT sensor and the EGR backpressure transducer **(see illustration 11.1a)**.

3   Remove the air cleaner outlet tube (see Section 9).

4   Detach the accelerator cable (see Section 10) and cruise control cable (if equipped) from the throttle body assembly.

5   Remove the accelerator cable bracket from the air intake plenum **(see illustration)** and position the bracket and cables out of the way.

13.5 Remove the accelerator cable bracket mounting nuts and lift the bracket from the plenum

4

6   Clearly label, then detach, the vacuum lines from the air intake plenum and the EGR valve.

7   Remove the PCV vent hose from the throttle body.

8   Remove the EGR tube **(see illustration)**.

9   Remove the alternator support bracket **(see illustration)**.

10   Remove the plenum mounting nuts and

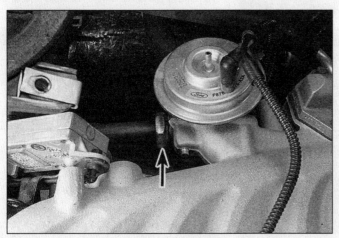

13.8 Detach the EGR tube retaining nut (arrow)

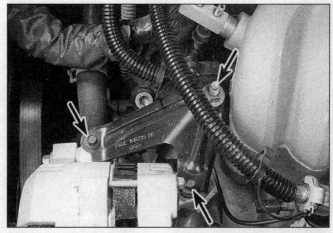

13.9 Remove the alternator mounting brace bolts (arrows) and separate the brace from the air intake plenum

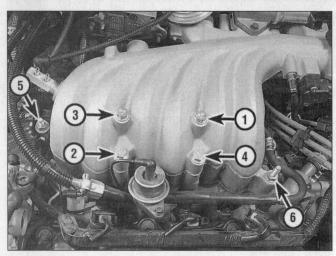

13.12 Tighten the plenum mounting bolts in the order shown - 3.0L engine

13.17 Throttle body and accelerator cable bracket mounting bolts (1996 through 1998 3.8L model shown; 1999 and later 3.8L models similar)

bolts, working in an order opposite that of the tightening sequence **(see illustration 13.12)**.

11   Lift the throttle body/air intake plenum from the lower in-take manifold.

12   Installation is the reverse of removal. Be sure to clean the gasket mating surfaces. If scraping is necessary, be careful not to damage the gasket surfaces or allow material to drop into the manifold. Tighten the air intake plenum mounting bolts in the proper sequence and to the torque listed in this Chapter's Specifications **(see illustration)**.

3.8L engine

13   Detach the cable from the negative terminal of the battery.

Throttle body

Refer to illustrations 13.17 and 13.18

14   Detach the throttle position sensor (TPS) electrical connector, the Idle Air Control (IAC) valve electrical connector **(see illustration 11.1b)**.

15   Remove the air cleaner outlet tube (see Section 9).

16   Detach the accelerator cable (see Section 10) and cruise control cable (if equipped)

from the throttle body assembly.

17   On 1996 and later models, remove the accelerator cable bracket from the throttle body **(see illustration)** and position the bracket and cables out of the way.

18   Remove the throttle body mounting nuts **(see illustration)**.

19   Remove and discard the throttle body gasket.

20   Installation is the reverse of removal. Be sure to clean the gasket mating surfaces. If scraping is necessary, be careful not to damage the gasket surfaces or allow material to drop into the manifold. Tighten the throttle body mounting bolts to the torque listed in this Chapter's Specifications.

Air intake plenum

Refer to illustrations 13.26, 13.28, 13.29a, 13.29b and 13.29c

21   On 1995 models, remove the cowl vent panel (see Chapter 11).

22   Remove the throttle body (see Step 13).

23   Clearly label, then detach, the vacuum lines from the air intake plenum, the EGR valve and the fuel pressure regulator.

24   On 1995 models, detach the EGR tube

from the EGR valve and the exhaust manifold. On 1996 and later models, disconnect the EGR tube from the air intake plenum.

25   Remove the PCV vent hose from the air intake plenum and the PCV valve from the valve cover.

26   On 1995 models, remove the engine drivebelt and the alternator support brace. Then remove the front and rear support braces that secure the air intake plenum to the lower intake manifold **(see illustration)**.

27   Remove the air intake plenum mounting bolts in the reverse order of the tightening sequence **(see illustration 13.29a, 13.29b and 13.29c)**. Separate the air intake plenum from the lower intake manifold (or, on 1999 and later 3.8L engines, from the upper intake manifold spacer).

28   On 1999 and later 3.8L engines, the air intake plenum is a two-piece assembly consisting of the "upper intake manifold" and the "upper intake manifold spacer." On these models, the upper intake manifold is bolted to the upper intake manifold spacer and the spacer is bolted to the lower intake manifold. To remove the upper intake manifold spacer, remove the spacer bolts in the reverse order of the tightening sequence **(see illustration)**.

29   Installation is the reverse of removal. Be sure to clean and inspect the mounting faces of the lower intake manifold (see Chapter 2A) and the air intake plenum before positioning the new gasket(s) onto the lower intake mounting face. The use of alignment studs may be helpful. Install the air intake plenum and throttle body assembly onto the lower intake manifold. Ensure the gasket remains in place (if alignment studs are not used). Install the air intake plenum retaining bolts and tighten them to the torque listed in this Chapter's Specifications, in the proper sequence **(see illustrations)**.

Throttle Position (TP) sensor

30   Refer to Chapter 6 for the check and replacement procedures for the TP sensor.

13.18 Throttle body mounting fasteners (arrows) (1995 3.8L engine)

13.26 On 1995 models, remove the air intake plenum support brace bolts

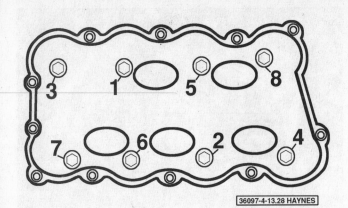

13.28 Upper intake manifold spacer bolt tightening sequence (1999 and later 3.8L engines)

13.29a Air intake plenum tightening sequence (1995 3.8L engine)

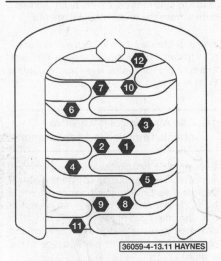

13.29b Air intake plenum tightening sequence (1996 through 1998 3.8L engines)

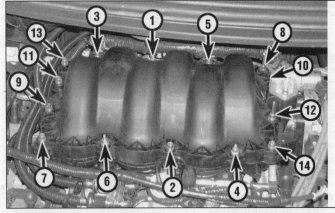

13.29c Air intake plenum (upper intake manifold) bolt tightening sequence (1999 and later 3.8L engines)

Idle Air Control (IAC) valve

31   Refer to Section 14 for the check and replacement procedures for the IAC valve.

Fuel rail and fuel injector assembly

Refer to illustrations 13.35, 13.36a, 13.36b, 13.37, 13.39a and 13.39b

Removal

32   Relieve the fuel pressure (see Section 2).
33   Detach the cable from the negative terminal of the battery.
34   Remove the air intake plenum from the lower intake manifold (see Step 1 for 3.0L engines or Step 21 for 3.8L engines).
35   Using the special spring-lock coupling tool (available at most auto parts stores), disconnect the fuel feed and return lines from the fuel rail assembly **(see illustration)**. **Note:** *Refer to Section 4 for additional information on disconnecting fuel lines.*
36   Disconnect the fuel injector connectors and remove the fuel rail retaining bolts (two on each side) **(see illustrations)**.

4

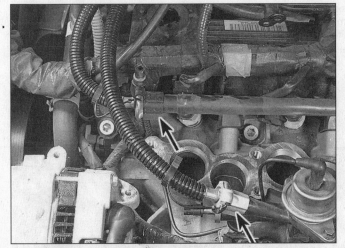

13.35 Disconnect the fuel feed and return line spring-lock couplings (arrows) using the method described in Section 4 (3.0L engine shown, 3.8L engine similar)

13.36a Fuel rail mounting bolt locations (arrows) (two on each side) (3.0L engine)

13.36b Fuel rail mounting bolt locations (arrows) (1996 and later 3.8L engines)

13.37 Using a rocking, side to side motion, carefully lift the fuel rail and the fuel injectors from the lower intake manifold

37 Using a rocking, side to side motion, carefully lift the fuel rail and the fuel injectors as an assembly from the lower intake manifold **(see illustration)**.

38 To remove a fuel injector from the fuel rail, simply rock the injector body from side-to-side and detach it from the fuel rail.

39 Inspect the injector O-rings (two per injector) for signs of deterioration **(see illustrations)**. Replace as required. **Note:** *As long as you have the fuel rail off, it's a good idea to replace all of the O-rings.*

40 Inspect the injector plastic "hat" (covering the injector pintle) and washer for signs of deterioration. Replace as required. If the hat is missing, look for it in the intake manifold.

Installation

41 Ensure that the injector caps are clean and free of contamination.

42 Lubricate the new O-rings with light grade oil and install two on each injector. **Caution:** *Do not use silicone grease. It will clog the injectors.*

43 Using a light twisting motion, install the injector(s) into the lower intake manifold.

44 Place the fuel rail over each of the injectors and seat the injectors into the fuel rail.

Ensure that the injectors are well seated in the fuel rail. **Note:** *It may be easier to seat the injectors in the fuel rail and then seat the entire assembly in the lower intake manifold.*

45 Secure the fuel rail assembly with the four retaining bolts and tighten them to the torque listed in this Chapter's Specifications.

46 The remainder of installation is the reverse of removal.

Fuel pressure regulator

Check

47 Refer to the fuel system pressure checks in Section 3.

Replacement

Refer to illustration 13.51

48 Relieve the fuel pressure from the system (see Section 2). Disconnect the cable from the negative terminal of the battery.

49 Clean any dirt from around the fuel pressure regulator.

50 Detach the vacuum hose from the fuel pressure regulator.

51 Remove the two bolts retaining the fuel pressure regulator **(see illustration)** and detach the regulator from the fuel rail.

52 Install new O-rings on the pressure regu-

lator and lubricate them with a light coat of oil.

53 Installation is the reverse of removal. Tighten the pressure regulator mounting bolts securely.

---

14   Idle Air Control (IAC) valve - check, removal and adjustment

Caution: *The throttle body on these models is coated with a sludge-resistant material designed to protect the bore and throttle plate. Do not attempt to clean the interior of the throttle body. The throttle body is designed to resist sludge accumulation and cleaning may impair the performance of the engine.*

Note: *The minimum idle speed is preset at the factory and is not adjustable under normal circumstances. If the idle fluctuates, stalls, idles high or speeds out of control, follow these quick checks to determine if the IAC valve is damaged. Because idle problems involve possible air leaks, fuel injector problems, malfunctioning TPS, PCM problems, etc. have the IAC valve and system diagnosed by a dealer service department or other qualified repair shop.*

13.39a Remove the O-ring from the top of the fuel injector . . .

13.51 Remove the fuel pressure regulator bolts (3.8L engine shown, 3.0L similar)

13.39b . . . then remove the lower O-ring from the injector

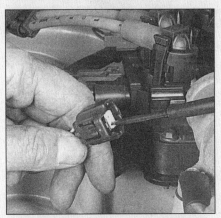

14.2 Disconnect the electrical connector from the IAC valve and check for voltage with the ignition key ON (engine not running)

Check

Refer to illustration 14.2

1   The Idle Air Control (IAC) valve controls the amount of air that bypasses the throttle valve, which controls the engine idle speed. This output actuator is mounted on the throttle body and is controlled by voltage pulses sent from the PCM (computer). The IAC valve within the body moves in or out, allowing more or less intake air into the system. To increase idle speed, the PCM extends the IAC valve from the seat and allows more air to bypass the throttle bore. To decrease idle speed, the PCM retracts the IAC valve towards the seat, reducing the air flow.

2   To check the system, first check for the voltage signal from the PCM. Turn the ignition key On (engine not running) and with a voltmeter, probe the wires of the IAC valve electrical connector (harness side). It should be approximately 10.5 to 12.5 volts **(see illustration)**. This indicates that the IAC valve is receiving the proper signal from the PCM.

3   If the IAC valve is receiving proper voltage, check the condition of the valve itself. Measure the resistance across the terminals on the IAC valve. There should be 6.0 to 13.0 ohms. If the resistance is incorrect, replace the IAC valve.

4   Check the IAC valve for an internal short circuit. Measure resistance from either terminal to the IAC body. There should be 10,000 ohms or greater. If less, the internal circuitry is grounding against the case; replace the IAC valve.

5   Next, remove the valve (proceed to Step 7) and check the pintle for excessive carbon deposits. If necessary, clean it with a soft rag. Also clean the valve housing to remove any deposits.

Adjustment

6   The idle speed is not adjustable. This procedure requires a special SCAN tool to extract working parameters from the EEC-V system while it is running. Have the procedure performed by a dealer service department or other qualified repair shop.

Removal

Refer to illustration 14.8

7   Unplug the electrical connector from the IAC valve **(see illustrations 11.1a and 11.1b)**.

8   Remove the two valve attaching screws and withdraw the valve from the throttle body **(see illustration)**.

9   Check the condition of the O-ring. If it's hardened or deteriorated, replace it.

10   Clean the sealing surface and the bore of the throttle body with a shop rag or soft cloth to ensure a good seal. **Caution:** *The IAC valve itself is an electrical component and must not be soaked in any liquid cleaner, as damage may result.*

Installation

11   Position the new O-ring on the IAC valve. Lubricate the O-ring with a light film of engine oil.

12   Install the IAC valve and tighten the screws securely.

13   Plug in the electrical connector at the IAC valve.

15 Intake Manifold Runner Control (IMRC) system (1996 and later 3.8L engines) - general information, removal and installation

General information

1   The IMRC system controls the air intake charge by opening or closing the butterfly valve on the secondary intake valve directly at the intake manifold. By closing the butterfly to the secondary intake valves under 3,000 rpm, low end driveability is improved. Above 3,000 rpm the butterfly valves open to increase high-end performance. The butterfly valves are controlled by the IMRC actuators.

2   The IMRC system is difficult to check and requires a special SCAN tool to access the PCM for information and operating conditions. Have the system diagnosed by a dealer service department or other qualified repair shop.

Actuator removal and installation

Refer to illustration 15.4

3   Remove the air intake plenum (see Section 13).

4   Remove the bolts that retain the IMRC actuator assembly to the cylinder head/intake manifold area **(see illustration)**.

5   Disconnect the actuator rod from the lever on the intake manifold runner.

6   Installation is the reverse of removal.

16 Exhaust system servicing - general information

Refer to illustrations 16.4a, 16.4b, 16.4c and 16.4d

Warning 1: *Inspection and repair of exhaust system components should be done only after enough time has elapsed after driving the vehicle to allow the system components to cool completely. Also, when working under the vehicle, make sure it is securely supported on jackstands.*

 4

14.8 Remove the IAC mounting bolts (3.0L engine shown, 3.8L similar)

15.4 Remove the IMRC actuator mounting bolts (arrows)

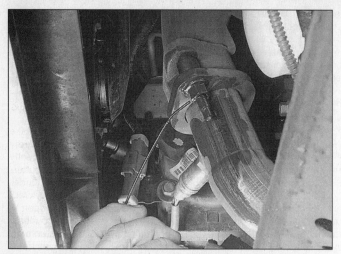

16.4a Be sure to apply penetrating lubricant to the exhaust system fasteners before attempting to remove them

16.4b Remove the exhaust clamp from the center section of the exhaust system - when installing the flex tube, always make sure it is positioned one inch from the end of the muffler

Warning 2: *All models covered by this manual are equipped with an exhaust system flex tube which is extremely sensitive to sharp bends. Do not allow the flex tube to hang downward during servicing or damage will occur.*

1   The exhaust system consists of the exhaust manifold(s), the catalytic converter, the muffler, the tailpipe and all connecting pipes, brackets, hangers and clamps. The exhaust system is attached to the body with mounting brackets and rubber hangers. If any of the parts are improperly installed, excessive noise and vibration will be transmitted to the body.

2   Conduct regular inspections of the exhaust system to keep it safe and quiet. Look for any damaged or bent parts, open seams, holes, loose connections, excessive corrosion or other defects which could allow exhaust fumes to enter the vehicle. Deterio-

rated exhaust system components should not be repaired; they should be replaced with new parts.

3   If the exhaust system components are extremely corroded or rusted together, welding equipment will probably be required to remove them. The convenient way to accomplish this is to have a muffler repair shop remove the corroded sections with a cutting torch. If, however, you want to save money by doing it yourself (and you don't have a welding outfit with a cutting torch), simply cut off the old components with a hacksaw. If you have compressed air, special pneumatic cutting chisels can also be used. If you do decide to tackle the job at home, be sure to wear safety goggles to protect your eyes from metal chips and work gloves to protect your hands.

4   Here are some simple guidelines to follow when repairing the exhaust system:

a) *Work from the back to the front when removing exhaust system components* **(see illustrations)**.

b) *Apply penetrating oil to the exhaust system component fasteners to make them easier to remove.*

c) *Use new gaskets, hangers and clamps when installing exhaust systems components.*

d) *Apply anti-seize compound to the threads of all exhaust system fasteners during reassembly.*

e) *Be sure to allow sufficient clearance between newly installed parts and all points on the underbody to avoid overheating the floor pan and possibly damaging the interior carpet and insulation. Pay particularly close attention to the catalytic converter and heat shield.*

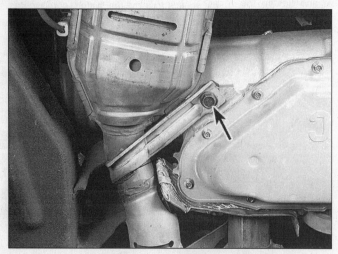

16.4c Remove the exhaust system bracket bolt (arrow) from the transmission housing

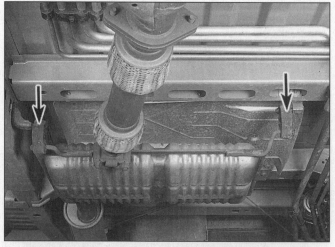

16.4d Check the condition of the rubber hangers (arrows) that support the exhaust system

Chapter 5
Engine electrical systems

Contents

Specifications

Battery voltage
Engine off .... 12 volts
Engine running .... 14-to-15 volts

Firing order (all models) 1-4-2-5-3-6

Spark plug wire resistance (approximate) .... 5,000 ohms per foot

Ignition coil resistance
Primary resistance .... 0.3 to 1.0 ohms
Secondary resistance .... 6.5 to 11.5 k-ohms

Ignition timing (base setting)
All engines .... 10-degrees BTDC with SPOUT disconnected

Alternator brush length
New .... 1/2 inch
Minimum .... 1/4 inch

1 General information

The engine electrical systems include all ignition, charging and starting components. Because of their engine-related functions, these components are considered separately from chassis electrical devices like the lights, instruments, etc.

Be very careful when working on the engine electrical components. They are easily damaged if checked, connected or handled improperly. The alternator is driven by an engine drivebelt which could cause serious injury if your hands, hair or clothes become entangled in it with the engine running. Both the starter and alternator are connected directly to the battery and could arc or even cause a fire if mishandled, overloaded or shorted out.

Never leave the ignition switch on for long periods of time with the engine off. Don't disconnect the battery cables while the engine is running. Correct polarity must be maintained when connecting battery cables from another source, such as another vehicle, during jump starting. Always disconnect the negative cable first and hook it up last or the battery may be shorted by the tool being used to loosen the cable clamps.

Additional safety related information on the engine electrical systems can be found in *Safety first* near the front of this manual. It should be referred to before beginning any operation included in this Chapter.

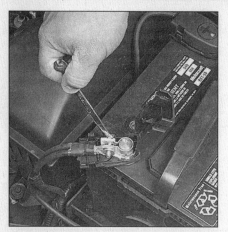

2.1 Removing the cable from a battery post with a wrench - sometimes special battery pliers are required for this procedure if corrosion has caused deterioration of the nut hex (always remove the ground cable first and hook it up last!)

2   Battery - removal and installation

Refer to illustrations 2.1, 2.2 and 2.4

1   Disconnect both cables from the battery terminals **(see illustration)**. **Warning:** *Always disconnect the negative cable first and hook it up last or the battery may be shorted by the tool being used to loosen the cable clamps.* **Note 1:** *When the battery is disconnected and reconnected, the vehicle may experience abnormal driving symptoms while the computer (PCM) relearns its adaptive strategy. The vehicle may need to be driven 10 miles or more to regain smooth operation.* **Note 2:** *Devices known as "memory-savers" can be used to avoid some of the above problems. Precise details vary according to the device used. Typically, it is plugged into the cigarette lighter, and is connected by its own wires to a spare battery; the vehicle's own battery is then disconnected from the electrical system, leaving the "memory-saver" to pass sufficient current to maintain audio unit security codes and PCM memory values, and also to run perma-*

2.2 Remove the bolt (arrow) and the wedge that holds the base of the battery to the battery tray

nently live circuits such as the clock, all the while isolating the battery in the event of a short-circuit occurring while work is carried out. **Warning:** *Some of these devices allow a considerable amount of current to pass, which can mean that many of the vehicle's systems are still operational when the main battery is disconnected. If a "memory-saver" is used, ensure that the circuit concerned is actually "dead" before carrying out any work on it!*

2   Remove the bolt and wedge from the battery tray **(see illustration)**.
3   Lift out the battery. Be careful - it's heavy. Special lifting straps that attach to the battery posts are available at auto parts stores - lifting and moving the battery is much easier if you use one.
4   If necessary for access to other components, remove the bolts that secure the battery tray **(see illustration)**.
5   Installation is the reverse of removal.

3   Battery - emergency jump starting

Refer to the *Booster battery (jump) starting* procedure at the front of this manual.

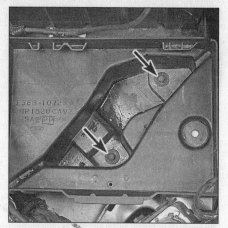

2.4 Remove the mounting bolts (arrows) for the battery tray - the lower bolts that secure the battery tray to the side of the left frame rail are not visible in this photo

4   Battery cables - check and replacement

Refer to illustrations 4.4a, 4.4b, 4.4c, 4.4d and 4.4e

1   Periodically inspect the entire length of each battery cable for damage, cracked or burned insulation and corrosion. Poor battery cable connections can cause starting problems and decreased engine performance.
2   Check the cable-to-terminal connections at the ends of the cables for cracks, loose wire strands and corrosion. The presence of white, fluffy deposits under the insulation at the cable terminal connection is a sign that the cable is corroded and should be replaced. Check the terminals for distortion, missing mounting bolts and corrosion.
3   When replacing the cables, always disconnect the negative cable first and hook it up last or the battery may be shorted by the tool used to loosen the cable clamps. Even if only the positive cable is being replaced, be sure to disconnect the negative cable from the battery first.
4   Disconnect and remove the cable **(see illustrations)**. Make sure the replacement

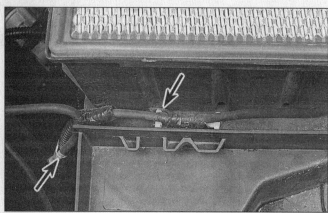

4.4a Detach any battery cable ties or retaining clips (arrows)

4.4b The positive battery cable is fastened at the engine compartment fuse box (arrow) . . .

4.4c . . . and at the starter (arrow)

4.4d The negative battery cable is fastened to the radiator support (arrow) . . .

4.4e . . . and to the transaxle bellhousing (arrow) next to the starter

cable is the same length and diameter.

5   Clean the threads of the relay or ground connection with a wire brush to remove rust and corrosion. Apply a light coat of petroleum jelly to the threads to prevent future corrosion.

6   Attach the cable to the relay or ground connection and tighten the mounting nut/bolt securely.

7   Before connecting the new cable to the battery, make sure that it reaches the battery post without having to be stretched. Clean the battery posts thoroughly and apply a light coat of petroleum jelly to prevent corrosion (see Chapter 1).

8   Connect the positive cable first, followed by the negative cable.

5   Ignition system - general information

Refer to illustrations 5.1a and 5.1b

Electronic Integrated (EI) Ignition system

1   The Electronic Integrated (EI) ignition system is a complete electronically controlled ignition system that does not incorporate a distributor or rotor and cap. The EI system consists of a crankshaft position sensor (CKP), Ignition Control Module (ICM) (except 1996 and later 3.8L engines and 1998 and later 3.0L engines), ignition coil pack, an EEC-V module (PCM), the spark plug wires and the spark plugs **(see illustrations)**. The EI system features a waste-spark method of spark distribution. Each cylinder is paired with its companion cylinder in the firing order (1-5, 4-3, 2-6) so one cylinder under compression fires simultaneously with its opposing cylinder, where the piston is on the

5.1a Ignition system components - 3.0L engine (1995 through 1998 models)

1	*Powertrain Control Module (PCM)*	*5*	*Spark plugs (left bank)*
2	*Crankshaft position (CKP) sensor (not visible)*	*6*	*Ignition Control Module (ICM) (1995 through 1997 models only)*
3	*Ignition coil pack*	*7*	*SPOUT connector (in wiring harness from PCM to ICM)*
4	*Spark plug wires*		

5

5.1b Ignition system components - 3.8L engine (1996 through 1998 models)

1   Powertrain Control Module (PCM) (under cowl)
2   Shorting bar (in PCM wiring harness)
3   Crankshaft position (CKP) sensor (not visible)

4   Ignition coil pack
5   Spark plug wires
6   Spark plugs (left bank)

exhaust stroke. Since the cylinder on the exhaust stroke requires very little of the available voltage to fire its plug, the majority of the voltage is used to fire the plug under compression. The ignition coil pack contains three separate coils. When a coil is triggered it supplies ignition voltage to two (companion) cylinders at the same time.

2   This ignition system does not have any moving parts (no distributor) and all engine timing and spark distribution is handled electronically. This system has fewer parts that require replacement and provides more accurate spark timing than conventional distributor type ignition systems. During engine operation, the Ignition Control Module receives signals from the crankshaft position sensor and the PCM to determine the proper spark advance and the turn on and firing time of the ignition coils. **Note:** *1996 and later 3.8L and 1998 and later 3.0L EI systems are not equipped with an Ignition Control Module. On these systems the PCM functions as the overall controller of the ignition system by receiving engine speed, camshaft and crankshaft position signals and determining the correct ignition timing and injector on time (rich/lean) for the fuel system. The PCM also functions as the controller of the ignition coil(s) primary circuit which was basically the job of the Ignition Control Module on earlier models.*

3   The CKP sensor is mounted on the engine front cover and triggered by a trigger wheel on the front of the crankshaft. The trigger wheel has 35 evenly spaced teeth and one gap where a 36th tooth would be. The gap lets the CKP sensor signal the PCM when the crankshaft is 90-degrees before TDC for cylinders 1 and 5. the PCM then computes actual TDC or any number of degrees before or after TDC. The CKP sensor is a magnetic pickup, or variable reluctance, sensor that produces a direct-current (dc) sine wave voltage signal. The sensor does the job of a distributor pickup in a distributor-type electronic ignition. Besides providing information on crankshaft position, the CKP sensor delivers the engine speed signal to the PCM. Refer to Chapter 6 for more information on the PCM and the CKP sensor.

6   Ignition system - general operating check

Refer to illustrations 6.3, 6.6 and 6.9
Warning: *Because of the very high secondary (spark plug) voltage generated by the ignition system, extreme care must be taken when this check is done.*
Note: *Beginning in 1994, Ford Motor Company began to manufacture a second genera-*

tion self-diagnosis system specified by EPA regulations called On Board Diagnosis (OBD) II. This system incorporates a series of diagnostic monitors that detect and identify emissions systems faults and store the information in the computer memory. This updated system also tests sensors and output actuators, diagnoses drive cycles, freezes data and clears codes. This powerful diagnostic computer must be accessed using the new OBD II SCAN tool and 16 pin Data Link Connector (DTC) located under the driver's dash area. All engines and powertrain combinations described in this manual are equipped with the On Board Diagnosis II (OBD-II) system. Refer to Chapter 6 for additional information on the OBD II system and its diagnostic capabilities.

1   A "calibrated ignition tester" is spark plug body with a shortened center electrode and a small clamp welded to the metal shell. They are sold at most auto parts stores, and two kinds are available. One kind of spark tester has a short center electrode extending from the insulator and is intended for use on breaker-point and older electronic ignitions. The other kind does not have a visible center electrode and is for use on EI and other high-voltage ignition systems. Be sure to use the high-voltage ignition tester on these systems.

2   If the engine cranks but won't start, disconnect the spark plug wire from any spark

6.3 To use a calibrated ignition tester, simply disconnect a spark plug wire, clip the tester to a convenient ground (like a valve cover bolt) and operate the starter - if there is enough power to fire the plug, sparks will be visible between the electrode tip and the tester body

6.6 Disconnect the electrical connector from the ignition coil and check for battery voltage to the coil with the ignition key on

6.9 Connect an LED test light to the positive battery terminal and the coil negative (-) terminals on the ignition coil harness connectors and watch for a blinking light when the engine is cranked

plug and attach it to the calibrated ignition tester.

3   Clip the tester to a bolt or metal bracket on the engine **(see illustration)**, crank the engine and watch the end of the tester to see if consistent, bright, well-defined sparks occur.

4   If sparks occur, sufficient voltage is reaching the plug to fire it (repeat the check at the remaining plug wires to verify that the distributor cap and rotor are OK). However, the plugs themselves may be fouled, so remove and check them as described in Chapter 1 or install new ones.

5   If no sparks or intermittent sparks occur, check for a bad spark plug wire by swapping wires.

6   If the problem isn't caused by the spark plug wire, check for battery voltage to the ignition coil with the ignition key ON (engine not running). Attach a 12 volt test light to the battery negative (-) terminal or other good ground. Disconnect the coil electrical connector and check for power at the positive (+) terminal **(see illustration)**. Battery voltage should be available. If there is no battery voltage, check the 25 amp fuse that protects the ignition circuit (see Chapter 12 for additional information on the fuses and the wiring schematics).

7   Check the primary and secondary resistances of the ignition coils (see Section 7).

8   Check the ignition coil electrical connectors for dirt, corrosion and damage.

9   If battery voltage is available to the ignition coils, attach an LED test light to the battery positive (+) terminal and each negative (-) terminal to the coil (on the vehicle harness side) **(see illustration)**, then crank the engine (be sure to check each negative terminal, one at a time). Confirm that the test light flashes. This test checks for the trigger signal (ground) from the computer or Ignition Control Module (if equipped). If a trigger signal is

present at the coil the computer and or the Ignition Control Module are functioning properly. **Caution:** *Use only an LED test light to avoid damaging the PCM.*

10   If the test light does not flash, check the crankshaft position sensor (see Chapter 6) and the Ignition Control Module (if equipped) (see Section 8). If the crankshaft sensor and the Ignition Control Module (if equipped) check out OK, have the PCM checked by a dealer service department or other qualified automotive repair shop.

7   Ignition coil - check and replacement

Check

Refer to illustrations 7.1 and 7.2

1   With the ignition off, disconnect the electrical connector(s) from the coil. Connect an ohmmeter across the coil positive (+) terminal and each negative (-) terminal **(see illustration)**. The resistance should be as listed in this Chapter's Specifications. If not,

7.1 To check the primary resistance of the ignition coil, connect the probes to the positive (+) terminal and each negative (-) terminal of the coil. The resistance should be the same for each check

replace the coil.

2   Connect an ohmmeter between the secondary terminals **(see illustration)** (the one that the spark plug wires connect to) of each coil pack. The resistance should be as listed in this Chapter's Specifications. If not, replace the coil. **Note:** *The ignition coil pack contains three separate coils. Each coil is paired according to its companion cylinders. Be sure to check resistance across the secondary terminals for each coil (1-5, 4-3, 2-6).*

Replacement

Refer to illustration 7.6

3   Disconnect the cable from the negative terminal of the battery.

4   Disconnect the ignition coil pack electrical connector **(see illustrations 5.1a and 5.1b)**. On 1995 3.8L engines, the coil pack is mounted to the front of the engine on the passenger side of the engine compartment.

5   Disconnect all the spark plug wires by squeezing the locking tabs and twisting while pulling. Do not pull on the wires.

5

7.2 Check the coil secondary resistance by probing the paired companion cylinders (1-5, 2-6, 3-4)

7.6 Remove the coil pack mounting screws (arrows) and lift it from the engine

6   Remove the four screws that secure the coil pack to the engine and disconnect the radio interference capacitor **(see illustration)**.
7   Installation is the reverse of the removal procedure with the following additions:

a) *Before installing the spark plug wire connector into the ignition coil, coat the entire interior of the rubber boot with silicone dielectric compound.*

b) *Insert each spark plug wire into the proper terminal of the ignition coil. Push the wire into the terminal and make sure the boots are fully seated and both locking tabs are engaged properly.*

c) *Reconnect the radio interference capacitor.*

8   Ignition Control Module (ICM) - check and replacement

Caution: *The Ignition Control Module is a delicate and relatively expensive electronic component. Failure to follow the step-by-step procedures could result in damage to the module and/or other electronic devices, including the PCM. Additionally, all devices under computer control are protected by a Federally mandated extended warranty. Check with your dealer concerning this warranty before attempting to diagnose and replace the module yourself.*
Note 1: *This procedure applies to 1995 models with a 3.8L engine and 1995 through 1997 models with a 3.0L engine.*
Note 2: *This test assumes that the coil trigger signal has been checked as described in Section 6 and is not present at the ignition coil harness side connector.*

Check

Refer to illustration 8.2

1   The Ignition Control Module is mounted to the passenger side fenderwell in between the right strut tower and the coolant reservoir **(see illustration 5.1a).**
2   Check for power to the ICM. Using a

voltmeter, probe terminal number 8 (red wire [+]) and check for battery voltage **(see illustration)**. With the ignition ON (engine not running), there should be battery voltage. If battery voltage is not present, refer to the wiring diagrams at the end of Chapter 12 and check the PCM power relay and related wiring for an open or shorted circuit.
3   Turn the ignition OFF. Using an ohmmeter, check for continuity between terminal 9 of the ICM connector and battery ground. Continuity should exist. If no continuity exists, refer to the wiring diagrams at the end of Chapter 12 and check the ground circuit for an open.
4   Check the resistance of the crankshaft position (CKP) sensor on terminals number 5 and number 6. It should be between 1.0 and 1.5 ohms.
5   Check the circuit from the ignition module to the SPOUT connector on terminal number 3 (pink wire [+]) for a complete circuit. Continuity should exist.
6   If the test results are correct, replace the ignition module with a new one.

Replacement

7   Disconnect the cable from the negative terminal of the battery.
8   Disconnect the electrical connector from the ignition module.
9   Remove the screws securing the Ignition Control Module to the passenger side fenderwell.
10   Installation is the reverse of the removal.

9   Ignition timing - check

Note 1: *This ignition timing procedure only checks the base timing setting specified by the factory. Timing cannot be adjusted, therefore the purpose of this check is to verify that the computer is controlling the ignition timing and that the base setting is correct. In most cases, the ignition system can be checked (see Section 6) but if the base setting remains incorrect, the PCM (computer) is defective. Take the vehicle to a dealer service department or other qualified repair shop to verify and repair the ignition system problem(s).*
Note 2: *1995 through 1997 3.0L engines and 1995 3.8L engines are equipped with a SPOUT connector in between the Ignition Control Module and the PCM* **(see Illustration 5.1a).** *The SPOUT controls the ignition signal from the computer. Disconnecting the SPOUT will display only the base timing settings without any changes from the computer. Do not remove the SPOUT connector except for checking ignition base timing.*
Note 3: *1996 and later 3.8L engines and 1998 and later 3.0L engines are equipped with a "shorting bar." This harness disconnect is used to remove the computer from the ignition timing control functions. Removal of the bar from the harness will retard the timing 2 to 3 degrees. The shorting bar is located in the right rear corner of the engine compart-*

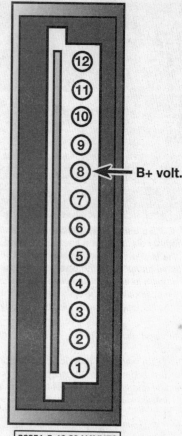

◄── B+ volt.

36051-5-10.28 HAYNES

8.2 Ignition Control Module (ICM) terminal guide

1 *Profile Ignition Pick-up (PIP) signal*
2 *Ignition Diagnostic Module (IDM) signal*
3 *Spark Output (SPOUT) input*
4 *Ignition ground*
5 *Crankshaft Position (CKP) sensor - negative*
6 *Crankshaft Position (CKP) sensor - positive*
7 *Shield ground*
8 *Battery positive (from PCM power relay)*
9 *Battery ground*
10 *Ignition coil No. 1*
11 *Ignition coil No. 3*
12 *Ignition coil No. 2*

ment in the PCM wiring harness **(see Illustration 5.1b)**. *Do not remove the Shorting Bar except for checking ignition base timing.*
1   Apply the parking brake and block the wheels.
2   Start the engine and warm it up. Once it has reached operating temperature, turn it off.
3   Unplug the SPOUT connector or shorting bar located in the wiring harness in the right rear corner of the engine compartment.
4   Connect an inductive timing light and a tachometer in accordance with the manufacturer's instructions. **Caution:** *Make sure that*

11.3 To measure charging voltage, attach the voltmeter leads to the battery terminals, start the engine and record the voltage reading

11.7 Alternator terminal identification

the timing light and tachometer wires don't hang anywhere near the cooling fan or they may become entangled in the fan blades when the fan begins to rotate.

5   Locate the timing marks on the crankshaft pulley (see Chapter 2).

6   Start the engine again. Place the transmission in DRIVE (parking brake applied). Turn off all accessories (heater, air conditioner, etc.).

7   Point the timing light at the pulley timing marks and note whether the specified timing mark is aligned with the timing pointer on the front of the timing chain cover. Refer to the Specifications listed in this Chapter.

8   If the proper mark isn't aligned with the stationary pointer, have the PCM checked by a dealer service department or other qualified repair shop.

9   Turn off the engine.

10   Insert the SPOUT connector or shorting bar into its electrical connector.

11   Remove the timing light and tachometer from the engine compartment.

10   Charging system - general information and precautions

The charging system includes the alternator, a voltage regulator (mounted on the backside of the alternator), a charge indicator or warning light, the battery, a large fuse (called a mega fuse) or fusible link (on some models) and the wiring between all the components. The charging system supplies electrical power for the ignition system, the lights, the radio, etc. The alternator is driven by a drivebelt at the front of the engine.

The purpose of the voltage regulator is to limit the alternator's voltage to a preset value. This prevents power surges, circuit overloads, etc., during peak voltage output. On integral voltage regulator systems, a solid state regulator is housed inside a plastic module mounted on the alternator itself.

These models are equipped a Motorcraft a 130 amp output rated alternator. The voltage regulator can be removed from the

backside of the alternator but the alternator must be removed from the engine first (see Section 12).

The charging system is protected by a series of large fuses (MEGA fuses) or fusible links which are located at the engine compartment fuse box. In the event of charging system problems, check the fuses and fusible links for damage or broken contacts.

The charging system doesn't ordinarily require periodic maintenance. However, the drivebelt, battery and wires and connections should be inspected at the intervals outlined in Chapter 1.

Be very careful when making electrical circuit connections to a vehicle equipped with an alternator and note the following:

a)   *When reconnecting wires to the alternator from the battery, be sure to note the polarity.*

b)   *Before using arc welding equipment to repair any part of the vehicle, disconnect the wires from the alternator and the battery terminals.*

c)   *Never start the engine with a battery charger connected.*

d)   *Always disconnect both battery cables before using a battery charger (negative cable first, positive cable last).*

11   Charging system - general operating check

Refer to illustrations 11.3 and 11.7

1   If a malfunction occurs in the charging circuit, do not immediately assume that the alternator is causing the problem. First, check the following items:

a)   *The battery cables where they connect to the battery. Make sure the connections are clean and tight.*

b)   *The battery electrolyte specific gravity. If it is low, charge the battery.*

c)   *Check the external alternator wiring and connections.*

d)   *Check the drivebelt condition and tension (see Chapter 1).*

e)   *Check the alternator mounting bolts for tightness.*

f)   *Run the engine and check the alternator for abnormal noise.*

g)   *Check the fusible links (if equipped) exiting the engine compartment fuse box (see Chapter 12). If they're burned, determine the cause and repair the circuit*

h)   *Refer to wiring diagrams in Chapter 12 and check all the fuses in series with the charging system. The location of the fuses may vary from year to year but the designations are the same.*

2   Using a voltmeter, check the battery voltage with the engine off. It should be approximately 12-volts.

3   Start the engine and check the battery voltage again. It should now be approximately 14 to 15-volts **(see illustration)**.

4   Turn ON the headlights. The voltage should drop and then come back up if the charging system is working properly.

5   If the voltage reading is more than the specified charging voltage, replace the voltage regulator (see Section 13).

6   If the voltage reading is less than the specified charging voltage, check the alternator as follows:

7   Using a voltmeter and working on the backside of the alternator, backprobe the "B+" terminal. There should be 12 volts present with the ignition key OFF **(see illustration)**.

8   With the ignition key ON (engine not running), backprobe each terminal. There should be 12 volts at the "A" terminal, one volt at the "I" terminal and 12 volts at the "B+" terminal.

9   Start the engine, then raise the engine to 2000 RPM, and backprobe each terminal again. There should be 14.0 to 14.7 volts at the "A" terminal and "B+" terminal and 13.0 to 14.0 volts at the "I" terminal.

10   If the voltages are not as specified, check the wiring harness. If the wiring harness is not defective, replace the alternator.

11   If you suspect that there is a voltage drain on the battery while the vehicle is sitting in the driveway, remove the battery positive

5

12.4a On 3.0L engines, detach the
alternator support bracket bolts (arrows)
(3.0L engine shown)

12.4b Alternator mounting bolts -
3.0L engine

12.4c Alternator mounting bolts -
3.8L engine

13.3 To detach the voltage regulator/brush holder assembly,
remove the four screws

13.4 Lift the assembly from the alternator

terminal, install a test light between the battery terminal and the cable (series connection) and observe the test light. If the light is
brightly illuminated, there is a circuit that continues to be activated. **Note:** *The test light
will glow dimly because of the parasitic drain
of the ECM, radio, clock, etc.*
12   Carefully remove the fuses one-by-one
that govern accessories such as radio,
blower motor, trunk lights, etc. until the test
light goes out. Track down the short circuit in
the particular fused circuit and repair the
problem. Recheck the electrical system as
described.
13   If all the fuses are pulled out and the test
light remains lit, remove the output cable at
the rear of the alternator then unplug all the
connectors from the backside of the alternator. If the test light goes out, then there is an
internal drain in the alternator or voltage regulator. Replace the alternator.

12 Alternator - removal and installation

Refer to illustrations 12.4a, 12.4b and 12.4c
1   Detach the cable from the negative terminal of the battery.

2   Unplug the electrical connectors from
the alternator.
3   Remove the drivebelt (see Chapter 1).
4   Remove the bolts and separate the
alternator from the engine **(see illustrations)**.
5   Installation is the reverse of removal.
6   After the alternator is installed, install the
drivebelt and reconnect the cable to the negative terminal of the battery.

13 Voltage regulator/alternator brushes - replacement

Refer to illustrations 13.3, 13.4, 13.5 and 13.9
NOTE: *On 1999 and later models, the alternator is a sealed unit and the voltage regula-
tor/alternator brushes can no longer be
replaced. If either of these components is
faulty, replace the alternator.*
1   Remove the alternator (see Section 12).
2   Set the alternator on a clean workbench.
3   Remove the four voltage regulator
mounting screws **(see illustration)**.
4   Detach the voltage regulator **(see illustration)**.
5   Detach the rubber plugs and remove the
brush lead retaining screws and nuts to separate the brush leads from the holder **(see**

illustration). Note that the screws have Torx
heads and require a special screwdriver.
6   After noting the relationship of the
brushes to the brush holder assembly,
remove both brushes. Don't lose the springs.
7   If you're installing a new voltage regulator, insert the old brushes into the brush

13.5 To remove the brushes from the
voltage regulator/brush holder assembly,
detach the rubber plugs from the two
brush lead screws and remove
both screws (arrows)

13.9 Before installing the voltage regulator/brush holder assembly, insert a paper clip as shown to hold the brushes in place during installation - after installation, simply pull the paper clip out

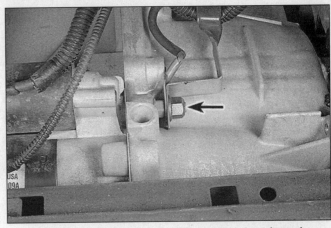

16.4 Remove the starter mounting bolt and nut (arrow) and separate the assembly from the transmission bellhousing (lower bolt not visible in this photo)

holder of the new regulator. If you're installing new brushes, insert them into the brush holder of the old regulator. Make sure the springs are properly compressed and the brushes are properly inserted into the recesses in the brush holder.

8   Install the brush lead retaining screws and nuts.

9   Insert a short section of wire, like a paper clip, through the hole in the voltage regulator **(see illustration)** to hold the brushes in the retracted position during regulator installation.

10   Carefully install the regulator. Make sure the brushes don't hang up on the rotor.

11   Install the voltage regulator screws and tighten them securely.

12   Remove the wire or paper clip.

13   Install the alternator (see Section 12).

14   Starting system - general information and precautions

1   The function of the starting system is to crank the engine fast enough to start it. The system is composed of the starter motor, starter relay, battery, switch and connecting wires.

2   Turning the ignition key to the Start position actuates the starter relay through the starter control circuit. The starter relay then connects the battery to the starter solenoid.

3   These models are equipped with a starter/solenoid assembly that is mounted to the transmission bellhousing.

4   All vehicles are equipped with a Transmission Range sensor in the starter control circuit, which prevents operation of the starter unless the shift lever is in Neutral or Park.

5   The starter circuit is equipped with a starter relay. This relay is located in the passenger compartment fuse box (see Chapter 12).

6   Never operate the starter motor for more than 15 seconds at a time without pausing to allow it to cool for at least two minutes.

Excessive cranking can cause overheating, which can seriously damage the starter.

15   Starter motor and circuit - in-vehicle check

Note: *Before diagnosing starter problems, make sure the battery is fully charged.*

1   If the starter motor doesn't turn at all when the switch is operated, make sure the shift lever is in Neutral or Park.

2   Make sure the battery is charged and that all cables at the battery and starter solenoid terminals are secure.

3   If the starter motor spins but the engine doesn't turn over, then the drive assembly in the starter motor is slipping and the starter motor must be replaced (see Section 16).

4   If, when the switch is actuated, the starter motor doesn't operate at all but the starter solenoid operates (clicks), then the problem lies with either the battery, the starter solenoid contacts or the starter motor connections.

5   If the starter solenoid doesn't click when the ignition switch is actuated, either the starter solenoid circuit is open or the solenoid itself is defective. Check the starter solenoid circuit (see the wiring diagrams at the end of this book) or replace the solenoid (see Section 17).

6   To check the starter solenoid circuit, remove the push-on connector from the solenoid wire. Make sure that the connection is clean and secure. If the connections are good, check the operation of the solenoid with a jumper wire. To do this, place the transmission in Park. Remove the push-on connector from the solenoid. Connect a jumper wire between the battery positive terminal and the exposed terminal on the solenoid. If the starter motor now operates, the starter solenoid is okay. The problem is in the ignition switch, Transmission Range sensor or in the starting circuit wiring (look for open or loose connections).

7   If the starter motor still doesn't operate, replace the starter solenoid (see Section 16).

8   If the starter motor cranks the engine at an abnormally slow speed, first make sure the battery is fully charged and all terminal connections are clean and tight. Also check the connections at the starter solenoid and battery ground. Eyelet terminals should not be easily rotated by hand. Also check for a short to ground. If the engine is partially seized, or has the wrong viscosity oil in it, it will crank slowly.

9   Check the starter circuit. Consult the wiring diagrams at the end of Chapter 12. Check the condition of the Transmission Range sensor (see Chapter 6).

10   Check the operation of the starter relay. Check for battery voltage to the relay and correct operation of the relay. Refer to Chapter 12 for additional information on the locations of relays and how to test them.

11   If the vehicle is equipped with an anti-theft alarm, check the circuit and the control module for shorts or damaged components (see Chapter 12).

16   Starter motor - removal and installation

Refer to illustration 16.4

1   Detach the cable from the negative terminal of the battery.

2   Raise the vehicle and support it securely on jackstands. **Warning:** *Some models covered by this manual are equipped with self leveling suspension systems. Always disconnect electrical power to the suspension system before lifting or towing the vehicle (see Chapter 10). Failure to perform this procedure may result in unexpected shifting or movement of the vehicle which could cause personal injury.*

3   Disconnect the large cable from the terminal on the starter motor and the solenoid terminal connections.

4   Remove the starter motor mounting bolt

and nut **(see illustration)** and detach the starter from the engine.

5   Installation is the reverse of removal.

17   Starter solenoid - replacement

Refer to illustrations 17.3a and 17.3b

1   Remove the starter assembly from the engine compartment (see Section 16).

2   Remove the electrical connector from the solenoid M terminal.

3   Remove the solenoid mounting bolts and separate the solenoid from the starter body **(see illustrations)**.

4   Installation is the reverse of removal.

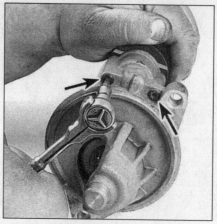

17.3a Remove the solenoid mounting bolts (arrows)

17.3b Separate the solenoid from the starter

Chapter 6
Emissions and engine control systems

Contents

1 General information

Refer to illustrations 1.1a, 1.1b and 1.7

To prevent pollution of the atmosphere from incompletely burned and evaporating gases, and to maintain good driveability and fuel economy, a number of emission control systems are incorporated **(see illustrations)**.

They include the:

Electronic Engine Control system (EEC-V) OBD-II
Evaporative Emission Control System (EVAP)
Positive Crankcase Ventilation (PCV) system
Exhaust Gas Recirculation (EGR) system
Catalytic converter

All of these systems are linked, directly or indirectly, to the emission control system.

The Sections in this Chapter include general descriptions, checking procedures within the scope of the home mechanic and component replacement procedures (when possible) for each of the systems listed above.

Before assuming that an emissions con-

1.1a Typical emission and engine control system components - 3.0L engine (1995 through 1998 models)

1	Powertrain Control Module (PCM) (under cowl)	6	Throttle Position Sensor (TPS)
2	PCV valve (on rear valve cover)	7	Mass Airflow (MAF) sensor
3	Exhaust Gas Recirculation (EGR) valve (not visible in photo)	8	Power distribution box
4	Camshaft position (CMP) sensor	9	Intake Air Temperature (IAT) sensor
5	Idle Air Control (IAC) valve	10	Engine Coolant Temperature (ECT) sensor

6

1.1b Typical emission and engine control system components - 3.8L engine (1996 through 1998)

1   Powertrain Control Module (PCM) (under cowl)
2   Camshaft Position (CKP) sensor (not visible - behind water pump)
3   Idle Air Control (IAC) valve
4   Throttle Position Sensor (TPS)

5   Intake Air Temperature (IAT) sensor
6   Mass Airflow (MAF) sensor
7   Power distribution box
8   EGR backpressure sensor
9   EGR valve

trol system is malfunctioning, check the fuel and ignition systems carefully. The diagnosis of some emission control devices requires specialized tools, equipment and training. If checking and servicing become too difficult or if a procedure is beyond your ability, consult a dealer service department. Remember, the most frequent cause of emissions problems is simply a loose or broken vacuum hose or wire, so always check the hose and wiring connections first.

This doesn't mean, however, that emission control systems are particularly difficult to maintain and repair. You can quickly and easily perform many checks and do most of the regular maintenance at home with common tune-up and hand tools. **Note:** *Because of a Federally mandated extended warranty which covers the emission control system components, check with your dealer about warranty coverage before working on any emissions-related systems. Once the warranty has expired, you may wish to perform some of the component checks and/or replacement procedures in this Chapter to save money.*

Pay close attention to any special precautions outlined in this Chapter. It should be noted that the illustrations of the various systems may not exactly match the system

installed on the vehicle you're working on because of changes made by the manufacturer during production or from year-to-year.

A Vehicle Emissions Control Information (VECI) label is located in the engine compartment **(see illustration)**. This label contains important emissions specifications and adjustment information, as well as a vacuum hose schematic with emissions components identified. When servicing the engine or emissions systems, the VECI label in your particular vehicle should always be checked for up-to-date information.

2   On Board Diagnosis (OBD) system and trouble codes

Note: *The diagnostic system and trouble codes are only accessible using expensive specialized equipment. The codes indicated in the text are designed and mandated by the EPA for all 1994 and later OBD-II vehicles produced by automobile manufacturers. These generic trouble codes do not include the manufacturer's specific trouble codes and because the OBD-II system requires a*

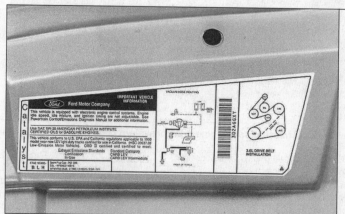

1.7 The Vehicle Emission Control Information (VECI) label is located in the engine compartment and contains information on the emission devices on your vehicle, vacuum line routing, etc. (3.0L engine shown)

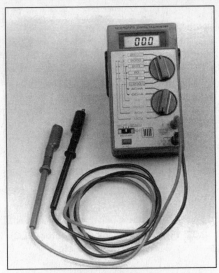

2.1 Digital multimeters can be used for testing all types of circuits; because of their high impedance, they are much more accurate than analog meters for measuring low-voltage computer circuits

2.2 Scanners like the Actron Scantool and the AutoXray XP240 are powerful diagnostic aids - programmed with comprehensive diagnostic information, they can tell you just about anything you want to know about your engine management system

special SCAN tool to access the trouble codes, have the vehicle diagnosed by a dealer service department or other qualified automotive repair facility if the proper SCAN tool is not available.

Diagnostic tool information

Refer to illustrations 2.1 and 2.2

1   A digital multimeter is necessary for checking fuel injection and emission related components **(see illustration)**. A digital volt-ohmmeter is preferred over the older style analog multimeter for several reasons. The analog multimeter cannot display the volts-ohms or amps measurement in hundredths and thousandths increments. When working with electronic circuits which are often very low voltage, this accurate reading is most important. Another good reason for the digital multimeter is the high impedance circuit. The digital multimeter is equipped with a high resistance internal circuitry (10 million ohms). Because a voltmeter is hooked up in parallel with the circuit when testing, it is vital that none of the voltage being measured should be allowed to travel the parallel path set up by the meter itself. This dilemma does not show itself when measuring larger amounts of voltage (9 to 12 volt circuits) but if you are measuring a low voltage circuit such as the oxygen sensor signal voltage, a fraction of a volt may be a significant amount when diagnosing a problem. Obtaining the diagnostic trouble codes is one exception where using an analog voltmeter is necessary.

2   Hand-held scanners are the most powerful and versatile tools for analyzing engine management systems used on later model vehicles **(see illustration)**. Early model scanners handle codes and some diagnostics for many OBD I systems. Each brand scan tool must be examined carefully to match the

year, make and model of the vehicle you are working on. Often, interchangeable cartridges are available to access the particular manufacturer (Ford, GM, Chrysler, etc.). Some manufacturers will specify by continent (Asia, Europe, USA, etc.).

3   With the arrival of the Federally mandated emission control system (OBD-II), a specially designed scanner must be used. At this time, several manufacturers plan to release OBD-II scan tools for the home mechanic. Ask the parts salesman at a local auto parts store for additional information concerning dates and costs.

OBD system general description

4   Beginning in 1994, Ford Motor Company began to manufacture a second generation self diagnosis system specified by the CARB and EPA regulations called On Board Diagnosis-II (OBD-II). This system incorporates a series of diagnostic monitors that detect and identify emissions systems faults and store the information in the computer memory. This updated system also tests sensors and output actuators, diagnoses drive cycles, freezes data and clears codes.

5   This powerful diagnostic computer must be accessed using the new OBD-II SCAN tool and 16 pin Data Link Connector (DLC) located under the driver's dash area. All engines and powertrain combinations described in this manual are equipped with the On Board Diagnosis II (OBD-II) system. This system consists of an onboard computer, known as the Powertrain Control Module (PCM), and information sensors, which monitor various functions of the engine and send data to the PCM. Based on the data and the information programmed into the computer's memory, the PCM generates output signals to control various engine func-

tions via control relays, solenoids and other output actuators.

6   The PCM is the "brain" of the EEC-V system. It receives data from a number of sensors and other electronic components (switches, relays, etc.). Based on the information it receives, the PCM generates output signals to control various relays, solenoids and other actuators. The PCM is specifically calibrated to optimize the emissions, fuel economy and driveability of the vehicle.

7   Because of a Federally mandated extended warranty which covers the EEC-V system components and because any owner-induced damage to the PCM, the sensors and/or the control devices may void the warranty, it isn't a good idea to attempt diagnosis or replacement of the PCM at home while the vehicle is under warranty. Take the vehicle to a dealer service department if the PCM or a system component malfunctions.

Information sensors

8   **Heated Oxygen sensors (HO2S)** - The HO2S generates a voltage signal that varies with the difference between the oxygen content of the exhaust and the oxygen in the surrounding air.

9   **Crankshaft position (CKP) sensor** - The CKP sensor provides information on crankshaft position and the engine speed signal to the PCM.

10   **Camshaft position (CMP) sensor** - The CMP sensor produces a signal in which the PCM uses to identify number 1 cylinder and to time the sequential fuel injection.

11   **Engine coolant temperature (ECT) sensor** - The ECT monitors engine coolant temperature and sends the PCM a voltage signal that affects PCM control of the fuel mixture, ignition timing, and EGR operation.

12   **Intake Air Temperature (IAT) sensor** - The IAT provides the PCM with intake air temperature information. The PCM uses this information to control fuel flow, ignition timing, and EGR system operation.

13   **Throttle Position Sensor (TPS)** - The TPS senses throttle movement and position, then transmits a voltage signal to the PCM. This signal enables the PCM to determine when the throttle is closed, in a cruise position, or wide open.

14   **Mass airflow (MAF) sensor** - The MAF sensor measures the molecular mass of the intake airflow entering the engine. The MAF sensor, along with the IAT sensor, provide mass airflow and air temperature information for the most precise fuel metering.

15   **Vehicle speed sensor (VSS)** - The vehicle speed sensor provides information to the PCM to indicate vehicle speed.

16   **EGR backpressure sensor** - The EGR backpressure sensor is used to monitor the rate and flow of exhaust gas recirculation into the intake system.

17   **Fuel tank pressure sensor** - The fuel tank pressure sensor is part of the evaporative emission control system and is used to monitor vapor pressure in the fuel tank. The

6

PCM uses this information to turn on and off the purge valves and solenoids of the evaporative emission system.

18   Power steering pressure (PSP) switch - The PSP sensor is used to increase transaxle hydraulic line pressure during low-speed vehicle maneuvers.

19   Brake On-Off (BOO) switch - This switch is used when the driver applies the brakes, to disengage the cruise control and change fuel metering and spark advance during deceleration.

20   Transaxle sensors - In addition to the vehicle speed sensor, the PCM receives input signals from the following sensors inside the transaxle or connected to it: (a) the turbine shaft speed sensor, (b) the transmission fluid temperature sensor, and (c) the transmission range sensor.

21   A/C clutch control switch - When battery voltage is applied to the air conditioning compressor solenoid, a signal is sent to the PCM, which interprets the signal as an added load created by the compressor and increases engine idle speed accordingly to compensate.

Output actuators

22   PCM power relay - The main PCM power relay is activated by the ignition switch and supplies battery power to the PCM and the EEC-V system when the switch is in the Start or Run position. On 1995 through 1997 3.0L engines and 1995 3.8L engines, the PCM power relay is part of the Constant Control Relay Module (CCRM). On 1998 models, the power relay is in the power distribution box in the engine compartment. Refer to Chapter 12 or your owner's manual for more information on relay location.

23   Fuel pump relay - The fuel pump relay is activated by the PCM with the ignition switch in the Start or Run position. When the ignition switch is turned on, the relay is activated to supply initial line pressure to the system. On 1995 through 1997 3.0L engines and 1995 3.8L engines, the fuel pump relay is part of the Constant Control Relay Module (CCRM). On 1998 models, the fuel pump relay is in the power distribution box in the engine compartment. For more information on fuel pump check and replacement, refer to Chapter 4.

24   Fuel injectors - The PCM opens the fuel injectors individually in firing order sequence. The PCM also controls the time the injector is open, called the "pulse width." The pulse width of the injector (measured in milliseconds) determines the amount of fuel delivered. For more information on the fuel delivery system and the fuel injectors, including injector replacement, refer to Chapter 4.

25   Ignition Control Module (ICM) - The ICM triggers the ignition coils and determines proper spark advance based on inputs from the PCM. An externally mounted ICM is used on all 1995 through 1997 3.0L engines and 1995 3.8L engines. All other models use an ignition module which is incorporated into the PCM. Refer to Chapter 5 for more information on the Ignition Control Module.

26   Idle air control (IAC) valve - The IAC valve controls the amount of air to bypass the throttle plate when the throttle valve is closed or at idle position. The IAC valve opening and the resulting airflow is controlled by the PCM. Refer to Chapter 4 for more information on the IAC valve.

27   EGR vacuum solenoid - The EGR vacuum solenoid is controlled by the PCM to regulate the opening of the vacuum-operated EGR valve.

28   Secondary air injection pump - The secondary air injection system supplies air to the exhaust manifolds to aid catalytic converter warm-up and operation. The secondary air pump is an electric pump, controlled by the PCM.

29   Canister purge valve - The evaporative emission canister purge valve is a solenoid valve, operated by the PCM to purge the fuel vapor canister and route fuel vapor to the intake manifold for combustion.

30   Canister vent solenoid - The evaporative emission canister vent solenoid is operated by the PCM during the OBD-II evaporative emission monitor (see Section 3) and during an emission test of the evaporative system.

Obtaining OBD-II system codes

Refer to illustration 2.32

31   On OBD-II systems, the PCM will illuminate the Malfunction Indicator Light on the

2.32 Typical Data Link Connector (DLC) on an OBD-II vehicle

dash if it recognizes a component fault for two consecutive drive cycles. It will continue to set the light until the PCM does not detect any malfunction for three or more consecutive drive cycles. Because the OBD-II system requires a SCAN tool to reset the light, if the tool is not available for diagnostics, have the system checked by a dealer service department or other qualified repair facility.

32   The diagnostic codes for the EEC-V (OBD-II) systems can be extracted from the PCM using a special SCAN tool that is programmed to interface with this new system by plugging into the DLC **(see illustration)**. If the tool is not available, have the vehicle checked at a dealer service department or other qualified repair shop.

Clearing codes

33   To clear the codes from the PCM memory, install the OBD-II SCAN tool, scroll the menu for the function that describes "CLEARING CODES" and follow the prescribed method for that particular SCAN tool. If necessary, have the codes cleared by a dealer service department or other qualified repair facility. **Caution:** *Do not disconnect the battery from the vehicle to clear the codes. This will erase stored operating parameters from the memory and cause the engine to run rough for a period of time while the computer relearns the information.*

OBD-II Trouble Codes

Code	Probable cause
P0102	Mass Airflow (MAF) sensor circuit low input
P0103	Mass Airflow (MAF) sensor circuit high input
P0112	Intake Air Temperature (IAT) sensor circuit low input
P0113	Intake Air Temperature (IAT) sensor circuit high input
P0117	Electronic Coolant Temperature (ECT) sensor circuit low input
P0118	Electronic Coolant Temperature (ECT) sensor circuit high input
P0121	In range Throttle Position Sensor (TPS) fault
P0122	Throttle Position Sensor (TPS) circuit low input
P0123	Throttle Position Sensor (TPS) circuit high input
P0131	Upstream heated O2 sensor circuit low voltage (Bank 1)

Code	Probable cause
P0133	Upstream heated O2 sensor circuit slow response (Bank 1)
P0135	Upstream heated O2 sensor heater circuit fault (Bank 1)
P0136	Downstream heated O2 sensor fault (Bank 1)
P0141	Downstream heated O2 sensor heater circuit fault (Bank 1)
P0151	Upstream heated O2 sensor circuit low voltage (Bank 2)
P0153	Upstream heated O2 sensor circuit slow response (Bank 2)
P0155	Upstream heated O2 sensor heater circuit fault (Bank 2)
P0156	Downstream heated O2 sensor fault (Bank 2)
P0161	Downstream heated O2 sensor heater circuit fault (Bank 2)
P0171	System Adaptive fuel too lean (Bank 1)
P0172	System Adaptive fuel too rich (Bank 1)
P0174	System Adaptive fuel too lean (Bank 2)
P0175	System Adaptive fuel too rich (Bank 2)
P0191	Injector Pressure sensor system performance
P0192	Injector Pressure sensor circuit low input
P0193	Injector Pressure sensor circuit high input
P0301	Cylinder no. 1 misfire detected
P0302	Cylinder no. 2 misfire detected
P0303	Cylinder no. 3 misfire detected
P0304	Cylinder no. 4 misfire detected
P0305	Cylinder no. 5 misfire detected
P0306	Cylinder no. 6 misfire detected
P0325	Knock sensor circuit fault
P0326	Knock sensor circuit performance
P0351	Ignition coil no. 1 primary circuit fault
P0352	Ignition coil no. 2 primary circuit fault
P0353	Ignition coil no. 3 primary circuit fault
P0354	Ignition coil no. 4 primary circuit fault
P0355	Ignition coil no. 5 primary circuit fault
P0356	Ignition coil no. 6 primary circuit fault
P0400	EGR flow fault
P0401	EGR insufficient flow detected
P0402	EGR excessive flow detected
P0420	Catalyst system efficiency below threshold (Bank 1)
P0421	Catalyst system efficiency below threshold (Bank 1)
P0430	Catalyst system efficiency below threshold (Bank 2)
P0431	Catalyst system efficiency below threshold (Bank 2)
P0442	EVAP small leak detected
P0443	EVAP VMV circuit fault
P0452	EVAP fuel tank pressure sensor low input
P0453	EVAP fuel tank pressure sensor high input
P0500	VSS fault
P0505	IAC valve system fault
P0603	PCM Keep Alive Memory test error
P0605	PCM Read Only Memory test error

6

3   Powertrain Control Module (PCM) - replacement

Refer to illustrations 3.4 and 3.5
Warning: *The models covered by this manual are equipped with airbags. Always disconnect the negative battery cable, then the positive battery cable and wait two minutes before working in the vicinity of the impact sensors, steering column or instrument panel* to avoid the possibility of accidental deployment of the airbag, which could cause personal injury (see Chapter 12).
Note: *Because of a Federally mandated extended warranty which covers the emission control system components, check with your dealer about warranty coverage before working on any emissions-related systems. Once the warranty has expired, you may wish to perform some of the component checks and/or replacement procedures in this Chap-* ter to save money.

1   Disconnect the cable from the negative terminal of the battery, then the positive cable and wait at least two minutes before proceeding.
2   Remove the cowl cover (see Chapter 11).
3   Working in the right rear corner of the engine compartment, remove the bolt that retains the electrical connector to the PCM, then unplug the connector.

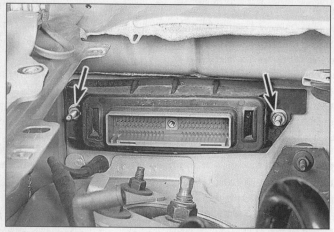

3.4 Remove the PCM bracket mounting nuts (arrows)

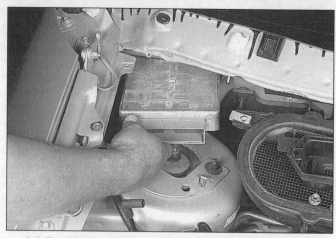

3.5 The PCM is removed through the engine compartment

4.1 Location of the upstream (A) and the downstream (B) oxygen sensors

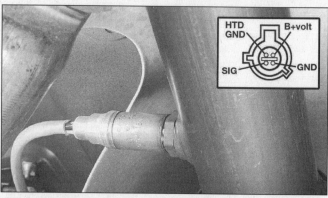

4.6 Using pins or paper clips, backprobe the SIG (+) and GND (-) terminals and check for a varying millivolt signal as the system adjusts the air/fuel ratio. Voltage should remain steady as the engine warms up (open loop) and then vary from 0.10 volts (100 millivolts) to 0.9 volts (900 millivolts) if the system is operating properly (closed loop)

4   Remove the PCM bracket retaining nuts **(see illustration)**.

5   Carefully slide the PCM out **(see illustration)**. **Note:** *Avoid any static electricity damage to the computer by using gloves and a special anti-static pad to store the PCM on once it is removed.*

6   Installation is the reverse of removal.

---

4   Information sensors - general information and testing

Note: *If any of the following checks indicate that a sensor is good (and not the cause of the driveability problem or trouble code), check the wiring harness and electrical connectors between the sensor and the PCM for an open or short-circuit condition. If no problems are found, have the vehicle checked by a dealer service department or other qualified repair shop.*

Heated oxygen sensor (HO2S)

Refer to illustrations 4.1 and 4.6

1   Because of OBD-II catalyst monitoring requirements, the EEC-V system on these models has four heated oxygen sensors (HO2S), two in each exhaust pipe at the inlet and in each catalytic converter. The oxygen in the exhaust reacts with the HO2S to produce a voltage output that varies from 0.1 volt (high oxygen, lean mixture) to 0.9 volt (low oxygen, rich mixture). The upstream HO2S in each branch of the exhaust system provides a feedback signal to the PCM that indicates the amount of leftover oxygen in the exhaust **(see illustration)**. The PCM monitors this variable voltage continuously to determine the required fuel injector pulse width and to control the engine air/fuel ratio. A mixture ratio of 14.7 parts air to 1 part fuel is the ideal ratio for minimum exhaust emissions, as well as the best combination of fuel economy and engine performance. Based on HO2S signals, the PCM tries to maintain this air/fuel ratio of 14.7:1 at all times.

2   The downstream HO2S in each branch of the exhaust system has no effect on PCM control of the air/fuel ratio. These sensors are identical to the upstream sensors and operate in the same way. The PCM uses their signals, however, for the OBD-II catalyst monitor. A downstream HO2S will produce a more slowly

fluctuating voltage signal that reflects the lower oxygen content in the postcatalyst exhaust.

3   An HO2S produces no voltage when it is below its normal operating temperature of about 600-degrees F. During this warm-up period, the PCM operates in an open-loop fuel control mode. It does not use the HO2S signal as a feedback indication of residual oxygen in the exhaust. Instead, the PCM controls fuel metering based on the inputs of other sensors and its own programs.

4   Proper operation of an HO2S depends on four conditions:

a) *Electrical - The low voltages generated by the sensor require good, clean connections which should be checked whenever a sensor problem is suspected or indicated.*

b) *Outside air supply - The sensor needs air circulation to the internal portion of the sensor. Whenever the sensor is installed, make sure the air passages are not restricted.*

c) *Proper operating temperature - The PCM will not react to the sensor signal until the sensor reaches approximately 600-degrees F. This factor must be con-*

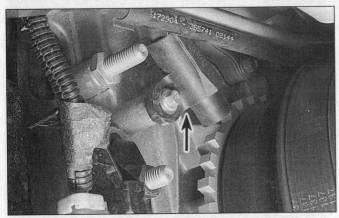

4.9 Location of the crankshaft position sensor (arrow) (3.0L shown, 3.8L similar)

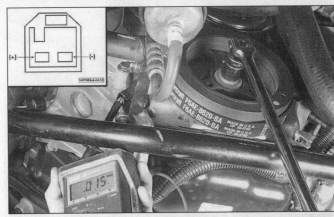

4.12 With the voltmeter set on the AC scale, check for a pulsing voltage signal between 0 and 0.05 volts as the engine is slowly rotated

sidered when evaluating the performance of the sensor.

d) **Unleaded fuel** - Unleaded fuel is essential for proper operation of the sensor.

5   The OBD-II system can detect several different HO2S problems and set DTC's to indicate the specific fault. If an OBD-II scan tool is not available, have the codes read by a dealer service department or other qualified driveability technician with the necessary equipment. When an HO2S fault occurs that sets a DTC, the PCM will disregard the HO2S signal voltage and revert to open-loop fuel control as described previously. **Note:** Refer to the wiring diagrams in Chapter 12 to identify circuit functions by wire color coding for the following tests. **Caution:** The HO2S is very sensitive to excessive circuit loads and circuit damage of any kind. For safest testing, install jumper wires in the HO2S connector to connect your voltmeter. If jumper wires aren't available, carefully backprobe the wires in the connector shell with straight pins or similar devices. Do not puncture the HO2S wires or try to backprobe the sensor itself. Use only a digital voltmeter to test an HO2S.

6   Turn the ignition On but do not start the engine. Connect your voltmeter negative (-) lead to a good ground and the positive (+) lead to the SIG wire at the HO2S connector **(see illustration)**. The meter should read approximately 400 to 450 millivolts (0.40 to 0.45 volt). If it doesn't, trace and repair the circuit from the sensor to the PCM.

7   Start the engine and let it warm up to normal operating temperature; again check the HO2S signal voltage.

a) Voltage from an upstream sensor should range from 100 to 900 millivolts (0.1 to 0.9 volt) and switch actively between high and low readings.

b) Voltage from a downstream sensor should also read between 100 to 900 millivolts (0.1 to 0.9 volt) but it should not switch actively. The downstream HO2S voltage may stay toward the center of its range (about 400 millivolts) or stay for relatively longer periods of time at the upper or lower limits of the range.

8   Also check the battery voltage supply to the HO2S heating circuits. Move the voltmeter negative (-) lead to the HTR GND terminal and the positive (+) lead to the B+Volt terminal of the sensor connector **(see illustration 4.6)**. With the ignition On, the meter should read more than 10 volts. Battery voltage is supplied to the sensors through a relay for only about three seconds when the engine is not running. Have an assistant turn the ignition on while you read the voltmeter. Refer to the wiring diagrams in Chapter 12 for more information on the circuits and relays.

Crankshaft position (CKP) sensor

Refer to illustrations 4.9 and 4.12

9   The crankshaft position (CKP) sensor is a variable reluctance type sensor which is mounted on the engine front cover, adjacent to the crankshaft trigger wheel **(see illustration)**. The trigger wheel has 35 evenly spaced teeth and one gap where a 36th tooth would be. The gap lets the CKP sensor signal the PCM when the crankshaft is 60-degrees before TDC for cylinders 1 and 5. The PCM then computes actual TDC or any number of degrees before or after TDC and uses this information to control ignition spark advance. The CKP sensor also provides the engine speed signal to the PCM and is part of the

misfire monitor circuit described in Section 3.

10   The OBD-II system can detect different CKP sensor problems and set DTC's to indicate the specific fault. If an OBD-II scan tool is not available, have the codes read by a dealer service department or other qualified driveability technician with the necessary equipment.

11   Disconnect the CKP sensor electrical connector and with the ignition key ON (engine not running), check for battery voltage to the CKP sensor.

12   Connect a voltmeter to the crankshaft sensor and, using the AC scale, check the voltage pulses as the gear is slowly rotated **(see illustration)**. Use a large socket and breaker bar to rotate the crankshaft pulley.

13   If no pulsing voltage signal is produced, replace the crankshaft sensor.

Camshaft position (CMP) sensor

Refer to illustration 4.15

14   The camshaft position (CMP) sensor is a Hall-effect switch that sends a voltage signal pulse to the PCM to indicate when the number 1 piston is approaching TDC on the compression stroke. The PCM uses this signal to synchronize and sequence the fuel injectors.

15   The CMP sensor is mounted at the rear of the engine block where the distributor was

6

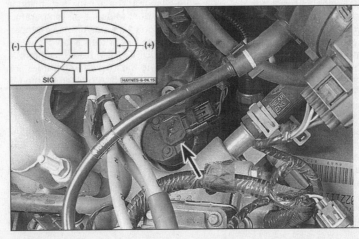

4.15 The camshaft position sensor (arrow) is located at the rear (left side) of the engine block (3.0L shown, 3.8L similar)

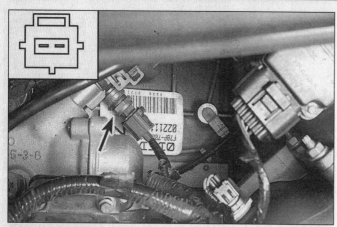

4.20 The ECT sensor (arrow) is located next to the thermostat housing on the lower intake manifold (3.0L shown, 3.8L similar) - check the resistance of the coolant temperature sensor with the engine completely cold and then with the engine at operating temperature. Resistance should decrease as temperature increases

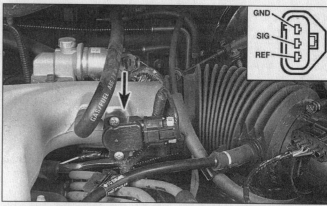

4.28 The TPS (arrow) is located on the side of the throttle body facing the front of the engine compartment (3.0L shown, 3.8L similar) - backprobe the TPS using straight pins or other suitable probes on SIG (+) and GND (-). Check the signal voltage, it should be 0.5 to 1.0 volt at closed throttle. Rotate the accelerator completely to wide open throttle and confirm the voltage increases steadily to 4.5 to 5.0 volts

installed on earlier versions of this engine **(see illustration)**.

16   Turn the ignition On but do not start the engine. Identify the battery positive (B+) terminal of the CMP connector **(see illustration 4.15)**. Backprobe the wire for this terminal with your voltmeter positive (+) lead; the reading should be more than 10 volts. If it isn't, trace and repair the circuit between the CMP sensor and the PCM.

17   Move the voltmeter positive (+) lead to the sensor signal terminal of the connector and crank the engine. The meter should show a pulsing DC-voltage reading of approximately 5 volts, once for each sensor revolution.

18   If the CMP sensor doesn't produce a pulsating voltage signal but battery voltage is present in Step 16, refer to the wiring diagrams and check the sensor ground connection. If the ground is OK, then replace the sensor.

Engine Coolant Temperature (ECT) sensor

Refer to illustration 4.20

19   The coolant temperature sensor is a thermistor, which is a variable resistor that changes its resistance as temperature changes. The sensor is installed in the engine cooling system to sense coolant temperature. As coolant temperature increases, sensor resistance decreases and vice versa. The PCM uses this information to compute the engine operating temperature. A problem in the ECT sensor circuit will set a trouble code. The fault may be in the circuit wiring or connections or in the sensor itself.

20   Vehicles equipped with a digital instrument cluster have two almost identical coolant temperature sensor units. One is the sender for the instrument panel temperature gauge, the other is the ECT sensor for the EEC-V system The temperature sender for the instrument panel gauge has a single-wire

connector and has a tan or brown plastic body, the ECT sensor has a two-wire connector and the sensor itself has a gray plastic body **(see illustration)**. **Note:** *Before condemning an ECT sensor, check the coolant level in the system.*

21   Disconnect the ECT sensor and use an ohmmeter to measure resistance across the two terminals of the sensor. At 65-degrees F, resistance should be approximately 40,500 ohms **(see illustration 4.20)**.

22   Next, start the engine and warm it up until it reaches operating temperature. The resistance should be lower. For example, at 180 to 220-degrees F resistance should be 3,800 to 1,840 ohms.

23   If the resistance values of the sensor are correct, refer to the wiring diagrams in Chapter 12 and check the voltage from the PCM to the disconnected sensor connector **(see illustration)**. The open-circuit voltage should be approximately 5 volts.

Intake Air Temperature (IAT) sensor

24   Like the ECT sensor, the IAT sensor is a thermistor that changes resistance as temperature changes. The sensor is installed in the intake air duct to sense air temperature **(see illustrations 1.1a and 1.1b)**. As temperature increases, sensor resistance decreases and vice versa. The PCM uses this information to compute the intake temperature and fine-tune fuel metering. A problem in the IAT sensor circuit will set a trouble code. The fault may be in the circuit wiring or connections or in the sensor itself.

25   With the engine cool, disconnect the IAT sensor and use an ohmmeter to measure resistance across the two terminals of the sensor. For example, at 68-degrees F the resistance should be approximately 37,300 ohms.

26   Next, start the engine and warm it up until it reaches operating temperature. Turn

the engine off, disconnect the sensor and measure the resistance again. It should be lower. If the sensor resistance doesn't change as described, replace it.

27   If the resistance values of the sensor are correct, refer to the wiring diagrams in Chapter 12 and check the voltage from the PCM to the disconnected sensor. The open-circuit voltage should be approximately 5 volts.

Throttle Position Sensor (TPS)

Refer to illustration 4.28

28   The Throttle Position Sensor (TPS) is a variable-resistance potentiometer, mounted on the side of the throttle body and connected to the throttle shaft **(see illustration)**. It senses throttle movement and position, then transmits a voltage signal to the PCM. This signal enables the PCM to determine when the throttle is closed, in a cruise position, or wide open. A defective TPS can cause surging, stalling, rough idle and other driveability problems because the PCM thinks the throttle is moving when it is not. The OBD-II system can detect several different TPS problems and set trouble codes to indicate the specific fault. If an OBD-II scan tool is not available, have the codes read by a dealer service department or other qualified repair shop.

29   Backprobe the signal terminal of the sensor connector with the positive (+) lead of your voltmeter **(see illustration 4.28)**. Backprobe the ground terminal with the meter negative (-) lead. Do not disconnect the sensor connector from the sensor for these tests.

30   Turn the ignition On but do not start the engine. The meter should read less than 1.0 volt with the throttle closed.

31   Open the throttle (or have a helper depress the accelerator) until the throttle is wide open. The voltmeter reading should increase smoothly and steadily to approximately 5.0 volts.

32   Also, check the TPS reference voltage.

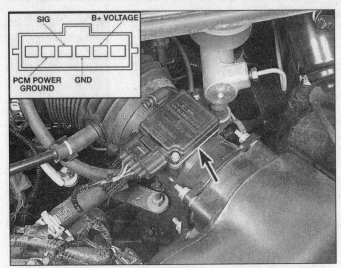

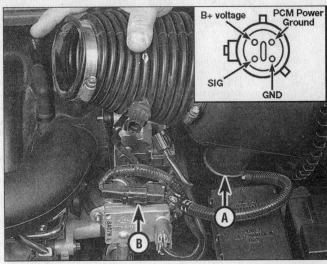

4.33a The MAF sensor (arrow) on 3.0L engines and 1995 3.8L engines is located in the air intake duct - check for battery voltage to the B+ terminal on the MAF sensor (key ON engine not running)

4.33b The MAF sensor (A) on 1996 and later 3.8L engines is located in the air cleaner housing - unplug the connector (B) and check for battery voltage to the B+ terminal on the MAF sensor (key ON engine not running)

Insert the voltmeter positive (+) probe into the reference voltage terminal of the connector and the negative (-) probe into the ground terminal. With the ignition On but the engine not running, the meter should read 5.0 ± 0.1 volts. **Note:** *If TPS-related driveability problems continue but these general tests don't indicate a TPS fault, have the sensor tested with an oscilloscope by an experienced driveability technician. A TPS often develops a voltage signal dropout of such short duration that it can't be seen on a voltmeter. The PCM can see such a signal fault and a driveability problem will result.*

Mass Airflow (MAF) sensor

Refer to illustrations 4.33a and 4.33b

33 The Mass Airflow (MAF) sensor is installed in the air intake duct on 3.0L engines and 1995 3.8L engines, and in the air cleaner housing on 1996 and later 3.8L engines **(see illustrations)**. This sensor uses a hot-wire sensing element to measure the molecular mass (or weight) of air entering the engine. The air passing over the hot wire causes it to cool, and the sensor converts this temperature change into an analog voltage signal to the PCM. The PCM in turn calculates the required fuel injector pulse width to obtain the necessary air/fuel ratio. A defective MAF sensor can cause surging, stalling, rough idle and other driveability problems. The OBD-II system can detect several different MAF sensor problems and set trouble codes to indicate the specific fault. If an OBD-II scan tool is not available, have the codes read by a dealer service department or other qualified repair shop.

34 To check for power to the MAF sensor, disconnect the MAF sensor electrical connector.

35 Refer to illustrations 4.33a and 4.33b and connect the positive (+) lead of your voltmeter to the B+ terminal of the harness con-

nector; connect the meter negative (-) lead to the sensor connector ground terminal.

36 Turn the ignition On but do not start the engine. The meter should read more than 10 volts, or close to battery voltage.

37 Reconnect the electrical connector and use straight pins or other suitable probes to backprobe the MAF signal (+) and ground (-) terminals with the voltmeter. Start the engine and check the voltage, it should be 0.5 to 0.7 volts at idle.

38 Increase the engine rpm. The MAF signal voltage should increase to about 1.5 to 3.0 volts. It is impossible to simulate driving conditions in the driveway, but it is necessary to watch the voltmeter for an increase in signal voltage as the engine speed is raised. The engine is not under load, but signal voltage should vary slightly.

39 If you suspect a defective MAF sensor, stop the engine and disconnect the MAF harness connector. Using an ohmmeter, probe the MAF signal (+) and ground (-) terminals. If the hot-wire element inside the sensor has been damaged, the ohmmeter will show an open circuit (infinite resistance).

40 If the voltage readings are correct, refer to the wiring diagrams and check the wiring harness for open circuits or a damaged harness. **Note:** *If MAF-related driveability problems continue but these general tests don't indicate a MAF fault, have the sensor tested by an experienced driveability technician. A MAF sensor can develop voltage signal problems that can't be seen on a voltmeter. The PCM can see such signal faults and a driveability problem will result.*

Vehicle Speed Sensor (VSS)

Refer to illustrations 4.41 and 4.45

41 The vehicle speed sensor (VSS) is a pickup coil (variable-reluctance) sensor mounted on the transaxle case above the right driveaxle **(see illustration)**. It produces

an AC voltage sine wave, the frequency of which is proportional to vehicle speed. The PCM uses the sensor input signal for several different engine and transmission control functions. The VSS signal also drives the speedometer on the instrument panel. A defective VSS can cause various driveability and transmission problems. The OBD-II system can detect sensor problems and set trouble codes to indicate specific faults. If an OBD-II scan tool is not available, have the codes read by a dealer service department or other qualified repair shop.

42 Refer to the wiring diagrams in Chapter 12 to identify the functions of connector terminals.

43 Disconnect the VSS connector and turn the ignition On but do not start the engine. Use a voltmeter to check for voltage between the sensor connector and ground **(see illustration 4.41)**. Approximately 1.5 volts should be present on one of the sensor wires with the key on and the engine off.

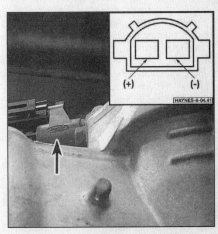

4.41 Disconnect the VSS harness connector and check for voltage at the connector

6

4.45 Remove the VSS and check for a pulsing AC voltage signal as the VSS gear is slowly turned

44   Remove the VSS from the transaxle as described in Section 5.

45   Connect a voltmeter to the VSS, set the meter on the AC scale, and check for voltage pulses as you spin the sensor drive gear **(see illustration)**.

46   If no pulsing voltage signal is produced, replace the sensor.

Power Steering Pressure (PSP) switch

47   The power steering pressure (PSP) switch is a normally closed switch, mounted on the auxiliary actuator of the steering gear. When steering system pressure reaches a high-pressure setpoint, the PSP switch opens and sends a signal to the PCM that the PCM uses to maintain engine idle speed during parking maneuvers. The OBD-II system can detect switch problems and set trouble codes to indicate specific faults. If an OBD-II scan tool is not available, have the codes read by a dealer service department or other qualified repair shop.

48   Check the operation of the PSP switch if the engine stalls during parking or if the engine idles continuously at high rpm.

49   Refer to the wiring diagrams at the end of this manual to identify the functions of the connector terminals.

50   Disconnect the PSP switch connector and connect an ohmmeter to the terminals on the switch body.

51   Start the engine and let it idle.

52   Turn the steering wheel to point the front wheels straight ahead and read the ohmmeter. It should indicate continuity of close to zero ohms.

53   Turn the steering wheel to either side and watch the ohmmeter. The PSP switch should open as the wheel nears the steering stop on either side, and the meter should indicate an open circuit (infinite resistance).

54   If the switch fails either test, replace it. If the switch is OK, troubleshoot the engine idle control operation if high idle speed or stalling problems continue.

Brake On-Off (BOO) switch

55   The brake on-off switch (also called the brake pedal position switch) tells the PCM when the brakes are being applied. The switch closes when brakes are applied and opens when the brakes are released. The switch is mounted on the brake pedal.

56   The brake light circuit is controlled by this switch, and burned-out bulbs or other circuit problems will cause the engine to idle roughly. Therefore, check the BOO switch operation when troubleshooting any rough-idle problems.

57   Refer to the wiring diagrams at the end of this manual to identify the functions of the connector terminals.

58   Disconnect the switch connector and connect your voltmeter positive (+) lead to the connector terminal that provides battery positive (B+) voltage to the switch. Connect the negative (-) meter lead to a good ground.

59   Turn the ignition On and read the meter. It should indicate more than 10 volts, or close to battery voltage.

60   Connect an ohmmeter to the switch terminals and manually open and close the switch. The meter should alternate from continuity to an open-circuit reading.

61   Also check continuity from the switch to the brake light bulbs. Replace any burned-out bulbs or damaged wire looms.

62   Replacement of the switch is covered in Chapter 9.

Transmission Range (TR) sensor

Refer to illustration 4.63

63   The Transmission Range (TR) sensor is mounted on top of the transaxle bellhousing **(see illustration)** and senses the position of the gear selector as chosen by the driver. The TR sensor contains a series of resistors and switch contacts. Depending on the gear selector position, the sensor contacts route current through different combinations of resistors and produce voltage signals of different levels. The sensor sends these signals to the PCM, which uses them for a number of engine and transmission control operations.

64   The TR sensor input affects operation of the EGR system, idle speed control, and transaxle torque converter lockup. The sensor also takes the place of the neutral safety switch used on older automatic transmissions. If the gear selector is not in Park or Neutral, the TR sensor will not let the starter motor operate. The OBD-II system can detect sensor problems and set trouble codes to indicate specific faults. If an OBD-II scan tool is not available, have the codes read by a dealer service department or other qualified repair shop.

65   If you suspect a problem with the TR sensor, check the connector for looseness and damaged terminals and wires.

66   Refer to illustration 4.63 to identify the functions of the connector terminals.

67   Disconnect the TR sensor connector and connect your voltmeter positive (+) lead

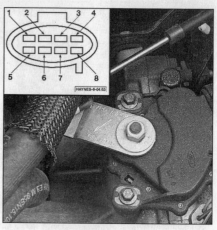

4.63 Transaxle range sensor terminal guide

1 Battery positive	6 Back-up lights
2 SIG ground	7 Accessory feed
3 SIG positive	8 Starter control
4 Liftgate release	
5 Starter control to interlock	

to the connector terminal that provides battery positive (B+) voltage to the sensor. Connect the negative (-) meter lead to a good engine ground.

68   Turn the ignition On and read the meter. It should indicate more than 10 volts, or close to battery voltage.

69   Disconnect the electrical connector from the ignition coil pack to disable the ignition.

70   Crank the engine and use the voltmeter to verify that at least 9 volts is present at the appropriate connector terminal as shown on the wiring diagrams.

71   The preceding checks indicate whether or not voltage is available to the sensor with the ignition On and during cranking. If the sensor passes these tests but problems continue, have the system diagnosed by an experienced driveability technician.

Air conditioning clutch control

72   During air conditioning operation, the PCM monitors the application of the air conditioning compressor clutch. The PCM commands the IAC valve to adjust the idle speed of the engine to compensate for the additional load.

73   First, check for battery voltage to the air conditioning clutch with the engine running, the air conditioning system activated and the air conditioning clutch harness connector disconnected. Battery voltage should be available. If not, check the air conditioning clutch solenoid.

74   Use a jumper wire from the battery positive terminal (+) and apply voltage to the air conditioning clutch. There should be a definite "click" when the clutch is activated.

75   Check for battery voltage to the air conditioning clutch solenoid. Battery voltage should be present with the air conditioning system properly charged and the air conditioning selected. **Note:** *If no power exists,*

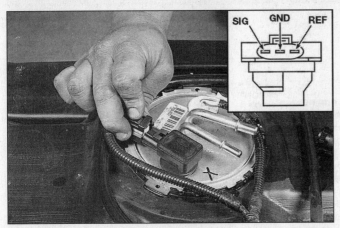

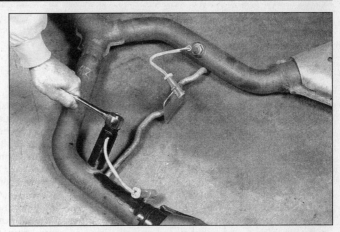

4.78 The fuel tank pressure sensor is located on the top of the fuel tank (1998 model shown)

5.4 Removing an oxygen sensor using a special slotted socket (exhaust system removed for clarity)

then check the air conditioning high pressure cut-out fan switch, the air conditioning clutch cycling pressure switch, the air conditioning/heater control assembly (see Chapter 3) and the 15 amp fuse that protects the circuit (see Chapter 12).

76   In most cases, if the air conditioning does not function, the problem is probably related to the air conditioning system relays and switches and not the PCM. Refer to Chapter 3 for additional information on the air conditioning system and diagnostics.

Fuel Tank Pressure (FTP) sensor

Refer to illustration 4.78

77   The Fuel Tank Pressure (FTP) sensor is used to monitor the fuel tank pressure or vacuum during the OBD-II test portion for emissions integrity. This test scans various sensors and output actuators to detect abnormal amounts of fuel vapors that may not be purging into the canister and/or the intake system for recycling. The FTP sensor helps the PCM monitor this pressure differential (pressure vs. vacuum) inside the fuel tank.

78   With the ignition key ON (engine not running), check for REF voltage to the fuel tank pressure sensor **(see illustration)**. Voltage should be available. It may be difficult to access the harness with the fuel tank in place. Find a location in the harness near the tank to check for voltage without removing the FTP sensor and fuel tank from the vehicle.

79   If voltage is available, the remaining checks must be performed with a specialized SCAN tool. Have the FTP sensor and EVAP system checked by a dealer service department or other qualified repair facility.

---

5   Information sensors - replacement

Heated oxygen sensor (HO2S)

Refer to illustration 5.4

1   The exhaust pipe contracts when cool,

and the HO2S may be hard to loosen when the engine is cold. To make sensor removal easier, start and run the engine for a minute or two; then shut it off. Be careful not to burn yourself during the following procedure. Also observe these guidelines when replacing an HO2S.

a) *The sensor has a permanently attached pigtail and electrical connector which should not be removed from the sensor. Damage or removal of the pigtail or electrical connector can harm operation of the sensor.*

b) *Keep grease, dirt and other contaminants away from the electrical connector and the louvered end of the sensor.*

c) *Do not use cleaning solvents of any kind on the oxygen sensor.*

d) *Do not drop or roughly handle the sensor.*

2   Raise the vehicle and place it securely on jackstands. **Warning:** *Some models covered by this manual are equipped with self leveling suspension systems. Always disconnect electrical power to the suspension system before lifting or towing the vehicle (see Chapter 10). Failure to perform this procedure may result in unexpected shifting or*

5.9 Crankshaft position sensor retaining bolts (arrows) (3.0L shown, 3.8L similar)

movement of the vehicle which could cause personal injury.

3   Disconnect the electrical connector from the sensor.

4   Using a suitable wrench or specialized O2 sensor socket, unscrew the sensor from the exhaust manifold **(see illustration)**.

5   Anti-seize compound must be used on the threads of the sensor to aid future removal. The threads of most new sensors will be coated with this compound. If not, be sure to apply anti-seize compound before installing the sensor.

6   Install the sensor and tighten it securely.

7   Lower the vehicle and reconnect the electrical connector for the sensor.

Crankshaft position (CKP) sensor

Refer to illustration 5.9

8   Be sure the ignition is off and disconnect the sensor electrical connector.

9   Remove the retaining bolts, sensor shield (if equipped) and remove the sensor from the engine cover **(see illustration)**.

10   Installation is the reverse of removal.

Camshaft position (CMP) sensor

Refer to illustrations 5.15, 5.19a and 5.19b

11   The CMP sensor is mounted on a drive unit. If you are replacing only the CMP sensor, remove the sensor screws, detach it from the drive assembly, and install the new sensor. You do not have to remove the drive assembly from the engine. Many engine repair procedures, however, require removal of the drive assembly. In these cases, you must time the drive assembly when you reinstall it. This procedure requires a Ford special tool to align the sensor properly. Read the entire procedure and obtain the necessary tool before beginning.

12   Refer to Chapter 2A and position the number 1 piston at TDC.

13   Disconnect the battery ground (negative) cable.

6

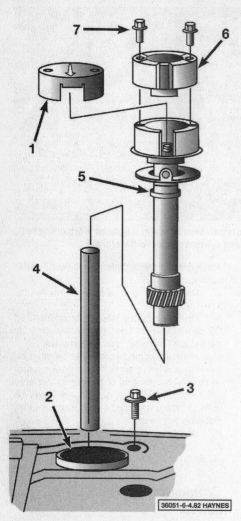

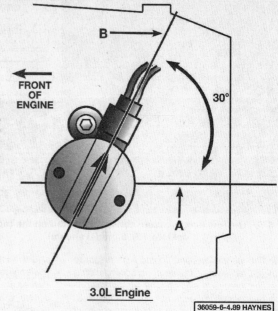

5.19a On 3.0L engines, with the housing seated against the engine block, on 1985-1998 models the arrow on the tool (B) should point 30-degrees from the center line of the engine (A). On 1999 and later models, the arrow on the tool (B) should point 38-degrees from the center line of the engine (A)

3.0L Engine

36059-6-4.89 HAYNES

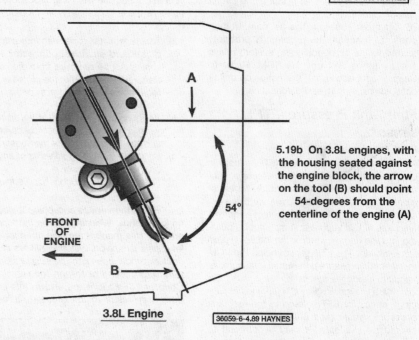

5.19b On 3.8L engines, with the housing seated against the engine block, the arrow on the tool (B) should point 54-degrees from the centerline of the engine (A)

3.8L Engine

36059-6-4.89 HAYNES

36051-6-4.82 HAYNES

5.15 Exploded view of the camshaft position sensor and synchronizer assembly (3.8L engine shown, 3.0L engine similar)

1   *Synchro positioning tool*
2   *Timing chain cover*
3   *Clamp*
4   *Oil pump intermediate shaft*
5   *Synchronizer*
6   *Camshaft Position sensor*
7   *Bolt*

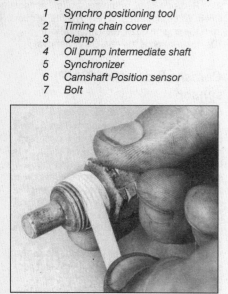

5.24 To prevent leakage, wrap the threads of the coolant temperature sensor with Teflon tape before installing it

14   Mark the relative position of the CMP sensor electrical connector so the assembly can be oriented properly during installation. **Note:** *This is necessary only if the drive assembly will be removed.* Disconnect the electrical connector from the CMP sensor. Remove the screws and remove the sensor from the drive assembly.

15   To remove the drive assembly, remove the bolt and hold-down clamp and lift the drive out of the engine. Remove the oil pump intermediate shaft along with the CMP sensor drive **(see illustration)**.

16   Place the alignment tool onto the drive assembly and align the vane of the synchronizer with the radial slot in the tool.

17   Turn the tool on the drive assembly until the boss on the tool engages the notch on the drive housing.

18   Transfer the oil pump intermediate shaft onto the drive assembly. Lubricate the gear, the thrust washer, and the lower bearing of the drive with clean engine oil.

19   Insert the assembly into the engine so that the drive gear engages the camshaft gear and the oil pump shaft engages the pump. Then rotate the sensor assembly so that the arrow on the tool is aligned as indicated **(see illustrations)**. This should be the orientation point for the connector that you marked before removing the sensor drive assembly.

20   Check the position of the electrical connector on the sensor to make sure it is aligned with the mark you made during removal. If it

5.29 Remove the screws (arrows) and disengage the TPS from the throttle shaft

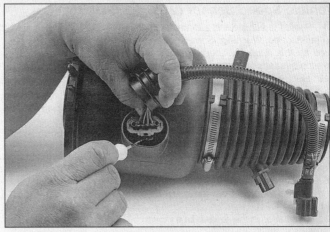

5.31a On 1996 and later 3.8L engines, unplug the connector from the sensor . . .

isn't oriented correctly, do not rotate the drive assembly to reposition it. Doing so will result in the fuel system being out of time with the engine and possible engine damage. If the connector is not oriented properly, repeat the installation procedure.

21   Install the hold-down clamp and bolt and tighten it securely. Remove the positioning tool.

22   Install the CMP sensor and tighten the screws securely.

23   Connect the sensor electrical connector and reconnect the battery ground cable.

Engine Coolant Temperature (ECT) sensor

Refer to illustration 5.24

Warning: *Wait until the engine is completely cool before performing this procedure.*

24   Before installing the new sensor, wrap the threads with Teflon sealing tape to prevent leakage and thread corrosion (see illustration). Remove the radiator cap to release any residual pressure in the cooling system, then reinstall the cap.

25   Unscrew the ECT sensor from the engine. Install the new sensor as quickly as possible to minimize coolant loss. Tighten the sensor securely and reconnect the electrical connector.

26   Check the coolant level as described in Chapter 1, adding some, if necessary. Start the engine and allow it to reach normal operating temperature, then check for coolant leaks. Check the coolant level in the expansion tank after the engine has warmed up and then cooled down again.

Intake Air Temperature (IAT) sensor

27   Disconnect the electrical connector and carefully remove the IAT sensor from the air intake duct or air cleaner housing (see illustrations 1.1a and 1.1b). Slight prying may be necessary, but be careful not to damage any of the plastic parts.

28   Install and connect the new sensor.

5.31b . . . then separate the MAF sensor from the air cleaner housing

Throttle Position Sensor (TPS)

Refer to illustration 5.29

29   Unplug the electrical connector, remove the two retaining screws and remove the TPS from the throttle body (see illustration).

30   Install the new sensor, making sure it engages the throttle shaft correctly. Reconnect the electrical connector. The TPS is not adjustable.

Mass Airflow (MAF) sensor

Refer to illustrations 5.31a, 5.31b, 5.32a and 5.32b

Note: *The plastic MAF sensor body and the metal air duct on which it is mounted are an assembly that must be replaced as a unit. Do not try to separate the sensor body from the metal duct.*

31   Disconnect the electrical connector from the MAF sensor and remove the air intake duct. On 1996 and later engines, remove the air cleaner housing, then detach the MAF sensor and the air cleaner housing (see illustrations).

32   Remove the four nuts that secure the sensor to the air cleaner housing and remove

5.32a On 3.0L engines and 1995 3.8L engines, disconnect the air inlet duct, remove the nuts (arrows) and separate the MAF sensor from the air cleaner housing

5.32b MAF sensor retaining nuts (arrows) on a 1996 and later 3.8L engine

the clamp that secures the sensor to the intake air duct (see illustrations).

33   Install and connect the new sensor.

6

Vehicle Speed Sensor (VSS)

34   Raise the vehicle and support it securely on jackstands. **Warning:** *Some models covered by this manual are equipped with self leveling suspension systems. Always disconnect electrical power to the suspension system before lifting or towing the vehicle (see Chapter 10). Failure to perform this procedure may result in unexpected shifting or movement of the vehicle which could cause personal injury.*

35   Disconnect the electrical connector from the VSS.

36   Remove the hold-down bolt and clamp and remove the VSS from the transaxle **(see illustration 4.41)**.

37   Installation is the reverse of removal.

Power Steering Pressure (PSP) switch

38   Raise the vehicle and support it securely on jackstands. **Warning:** *Some models covered by this manual are equipped with self leveling suspension systems. Always disconnect electrical power to the suspension system before lifting or towing the vehicle (see Chapter 10). Failure to perform this procedure may result in unexpected shifting or movement of the vehicle which could cause personal injury.*

39   Disconnect the electrical connector from the switch and unscrew the switch from the auxiliary actuator on the steering gear.

40   Install and connect the new switch and lower the vehicle to the ground.

41   Refer to Chapter 10 and bleed air from the power steering system. Add fluid as necessary (see Chapter 1).

Brake On-Off (BOO) switch

42   Replacement of the switch is covered in Chapter 9.

Transmission Range (TR) sensor

43   Replacement of the switch is covered in Chapter 7

---

6   Exhaust Gas Recirculation (EGR) system

General description

Refer to illustration 6.2

1   The EGR system is used to lower oxides of nitrogen (NOx) emission levels caused by high combustion temperatures. The EGR recirculates a small amount of exhaust gases into the intake manifold. The additional mixture lowers the temperature of combustion thereby reducing the formation of NOx compounds.

2   The EGR flow rate is determined by monitoring the pressure across a fixed metering orifice as exhaust gasses pass through it. This system is called the Differential Pressure Feedback (DPFE) system **(see illustration)**.

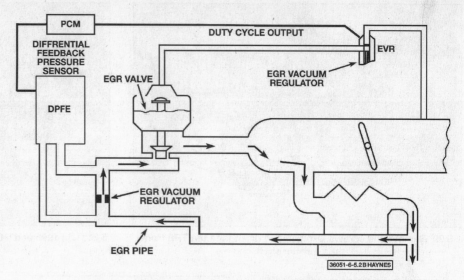

6.2 Typical Differential Pressure Feedback EGR (DPFE) system

The pressure sensor monitors upstream (before) and downstream (after) exhaust backpressure. This backpressure coefficient is relayed to the PCM and the correct amount of EGR (duty cycle) is applied to the EGR vacuum regulator control (EVR). By calculating the difference between the two pressures, the PCM determines exactly the EGR flow rate at all driving conditions. The DPFE is more accurate than early systems in that the computer does not have to guess at the upstream pressure coefficient to determine EGR flow rate as the engine drives through various road conditions such as hard acceleration, downshifting, engine misfire, poor fuel combustion, etc. All these conditions will cause the exhaust backpressure to vary and requires more strict and responsive EGR control to limit NOx emission levels.

3   Here is a list of the various components on the DPFE EGR system: the EGR valve, the EGR vacuum regulator, differential feedback pressure sensor, the PCM, the EGR pipe and the various vacuum and pressure lines for the EGR system.

4   The OBD-II system can detect a variety of different EGR system problems and set codes to indicate the specific trouble area. If an OBD-II SCAN tool is not available, have the codes extracted from the PCM by a dealer service department or other qualified automotive repair shop.

Check

5   Too much EGR flow tends to weaken combustion, causing the engine to run rough or stop. When EGR flow is excessive, the engine can stop after a cold start or at idle after deceleration, the vehicle can surge at cruising speeds or the idle may be rough. If the EGR valve remains constantly open, the engine may not idle at all.

6   Too little or no EGR flow allows combustion temperatures to get too high during acceleration and load conditions. This can cause spark knock (detonation), engine overheating or emission test failure.

7   The following checks will help you pinpoint problems in the EGR system. Where the procedure says to lift up on the EGR valve diaphragm, it's a good idea to wear a heat-resistant glove to prevent burns.

EGR valve

8   The EGR valve is controlled by a normally open EGR vacuum regulator (EVR) which allows vacuum to pass when energized. The PCM energizes the EVR to turn on the EGR. The PCM controls the EGR when three conditions are present: engine coolant is above 113-degrees F, the TPS is at part throttle and the MAF sensor is in its mid-range.

9   Make sure the vacuum hoses are in good condition and hooked up correctly.

10   To perform a leakage test, hook up a vacuum pump to the EGR valve. Apply a vacuum of 5 to 6 in-Hg to the valve. The vacuum pump should hold vacuum.

11   If access is possible, position your fingertip under the vacuum diaphragm and apply vacuum to the EGR valve. You should feel movement of the EGR diaphragm. **Warning:** *The EGR valve becomes very hot during engine operation - it's a good idea to wear a glove when performing this check.*

12   Remove the EGR valve (see Step 20) and clean the inlet and outlet ports with a wire brush or scraper. Do not sandblast the valve or clean it with gasoline or solvents. These liquids will destroy the EGR valve diaphragm.

13   If the specified conditions are not met, replace the EGR valve.

EGR control system

Refer to illustrations 6.16, 6.17 and 6.18

14   If a code is displayed there are several

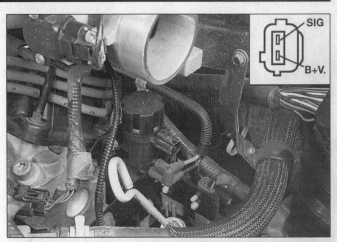

6.16 Working on the harness side of the Electronic Vacuum Regulator (arrow) electrical connector, check for battery voltage

6.17 Check the resistance of the EGR vacuum regulator. It should be approximately 30 to 70 ohms

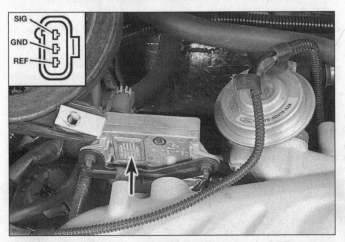

6.18 Check for the correct reference voltage to the DPFE sensor (arrow) with the ignition key ON (engine not running)

6.25 EGR mounting bolts (arrows) on a 3.0L engine

possibilities for EGR failure. Engine coolant temperature (ECT) sensor, TPS, MAF sensor, TCC system and the engine rpm govern the parameters the EGR system use for distinguishing the correct ON time.

15   These systems use an Electronic Vacuum Regulator (EVR) to control the amount of exhaust gas through the EGR valve. The valve is normally open (engine at operating temperature) and the vacuum source is a ported signal. The PCM uses a controlled "pulse width" or electronic signal to turn the EGR ON and OFF (the "duty cycle"). The duty cycle should be zero percent (no EGR) when in Park or Neutral, when the TPS input is below the specified value or when Wide Open Throttle (WOT) is indicated.

16   To check the EGR vacuum regulator, disconnect the electrical connector to the EGR vacuum regulator, turn the ignition key ON (engine not running) and check for battery voltage to the solenoid (see illustration). Battery voltage should be present.

17   Next, use an ohmmeter and check the resistance of the EGR vacuum regulator. It should be between 30 and 70 ohms (see illustration).

18   Check for reference voltage to the DPFE

sensor. With the ignition key on (engine not running), check for voltage on the harness side of the electrical connector (see illustration) on terminal VREF. It should be between 4.0 and 6.0 volts. If the test results are incorrect, replace the DPFE sensor.

19   Check the operation of the Differential Pressure Feedback (DPFE) sensor. Note: The DPFE sensor on the Differential Pressure Feedback EGR systems have two exhaust lines hooked into the EGR tube. Check for signal voltage to the sensor. Backprobe the correct terminals and check for a voltage signal while the engine is running first at cold temperatures and then at warm operating temperatures. With the engine cold there should be no EGR, therefore the voltage should be approximately 0.20 to 0.70 volts. As the engine starts to warm and EGR is signaled by the computer, voltage values should increase to approximately 4.0 to 6.0 volts.

Component replacement

EGR valve

Refer to illustration 6.25

20   When buying a new EGR valve, make sure that you have the right EGR valve. Use

the stamped code located on the top of the EGR valve.

21   Detach the cable from the negative terminal of the battery.

22   Remove the air cleaner housing assembly (see Chapter 4).

23   Detach the vacuum line from the EGR valve.

24   Remove the EGR pipe from the exhaust manifold.

25   Remove the bolts securing the EGR valve to the intake manifold/air intake plenum (see illustration).

26   Remove the EGR valve and gasket from the manifold. Discard the gasket.

27   With a wire wheel, buff the exhaust deposits from the EGR valve mounting surface on the manifold and, if you plan to use the same valve, the mounting surface of the valve itself. Look for exhaust deposits in the valve outlet. Remove deposit build-up with a screwdriver. Caution: *Never wash the valve in solvents or degreaser - both agents will damage the diaphragm. Sandblasting is also not recommended because it will affect the operation of the valve.*

28   If the EGR passage contains an excessive build-up of deposits, clean it out with a

6

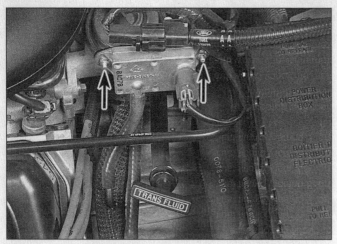

6.38 Remove the DPFE sensor mounting nuts

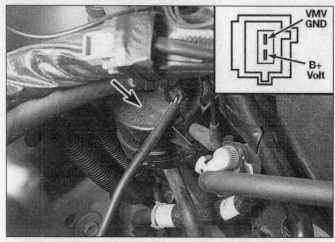

7.15 Check for battery voltage to the Vapor Management valve (arrow) with the ignition key ON (engine not running)

wire wheel. Make sure that all loose particles are completely removed to prevent them from clogging the EGR valve or from being ingested into the engine.
29   Installation is the reverse of removal.

EGR vacuum regulator

30   Detach the cable from the negative terminal of the battery.
31   Unplug the electrical connector from the solenoid.
32   Clearly label and detach both vacuum hoses.
33   Remove the solenoid mounting screw and remove the solenoid.
34   Installation is the reverse of removal.

DPFE sensor

Refer to illustration 6.38

35   Detach the cable from the negative terminal of the battery.
36   Unplug the electrical connector from the sensor.
37   Clearly label and detach both vacuum hoses.
38   Remove the sensor mounting nuts **(see illustration)** and remove the assembly.
39   Installation is the reverse of removal.

---

7   Evaporative Emissions Control System (EVAP)

General description

1   This system is designed to trap and store fuel vapors that evaporate from the fuel tank, throttle body and intake manifold during non-operation or idling, store them in the charcoal canister and then route them into the combustion chamber to be burned during engine operation.
2   The Evaporative Emission Control System (EVAP) consists of a charcoal-filled canister and the lines connecting the canister to the fuel tank, fuel vapor management valve (VMV), a fuel tank pressure sensor, fuel filler

cap, fuel vapor valve, ported vacuum and intake manifold vacuum. **Note 1:** *The evaporative charcoal canister is mounted on the frame near the fuel tank on all models.* **Note 2:** *The fuel tank pressure sensor location and checks are covered in Section 4.*
3   Fuel vapors are transferred from the fuel tank, throttle body and intake manifold to a canister where they are stored when the engine is not operating. When the engine is running, the fuel vapors are purged from the canister by a vapor management valve (VMV) which is PCM controlled, and consumed in the normal combustion process. The fuel tank pressure sensor relays the inside fuel tank pressure to the PCM which in turn regulates the EVAP system purge control system.
4   The OBD-II system can detect a variety of different EVAP system problems and set codes to indicate the specific trouble area. If an OBD-II SCAN tool is not available, have the codes extracted from the PCM by a dealer service department or other qualified automotive repair facility.

Check

General system checks

5   Poor idle, stalling and poor driveability can be caused by an inoperative vapor management valve, a damaged canister, split or cracked hoses or hoses connected to the wrong ports.
6   Evidence of fuel loss or fuel odor can be caused by fuel leaking from fuel lines or the throttle body, a cracked or damaged canister, an inoperative vapor management valve (VMV), disconnected, misrouted, kinked, deteriorated or damaged vapor or control hoses or an improperly seated air cleaner or air cleaner gasket.
7   Inspect each hose attached to the canister for kinks, leaks and breaks along its entire length. Repair or replace as necessary.
8   Inspect the canister. If it is cracked or damaged, replace it.
9   Look for fuel leaking from the bottom of the canister. If fuel is leaking, replace the

canister and check the hoses and hose routing.

Excessive pressure in fuel tank

10   The easiest way to check for excess fuel vapor pressure in the fuel tank is simply remove the fuel filler cap and listen for the sound of pressure release, similar to a flat tire or air compressor discharge. If the weather is extremely hot, take into account for the extra pressure from the heated molecules. The most accurate test is using the OBD-II SCAN tool. This will run a series of checks using the fuel tank pressure sensor and other output actuators to detect excess pressure. Have the vehicle diagnosed by a dealer service department or other qualified automotive repair facility.
11   If excess pressure is detected, check the canister fuel vapor hose and inlet port for blockage or collapsed hoses. Also inspect the hoses near the VMV and between the fuel tank and body for kinks and damage.
12   Also, check the evaporative emission valve. Remove the fuel tank (see Chapter 4) and check the evaporative emission valve to make sure the passage through the orifice is open to atmospheric pressure. If it is plugged, replace the valve.
13   Check the charcoal canister for tightness. Remove the close-off line near the canister and install a hand-held pressure pump. Pump the canister to approximately 2.5 psi and confirm that pressure vents through the close-off line. Now install the pump directly at the canister and apply 2.5 psi. The pressure should hold steady.

Fuel vapor odor in engine compartment

Refer to illustration 7.15

14   Check the hoses around the VMV for damage or incorrectly routed lines. Correct if necessary.
15   Check the VMV for battery voltage. With the ignition key ON (engine not running), check for battery voltage on the B+ VOLT terminal on the computer side of the VMV har-

7.17 Remove the charcoal canister mounting bolts (1995 through 1998 models)

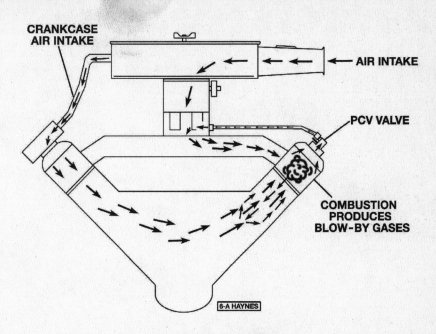

8.1 Gas flow in a typical PCV system

ness connector **(see illustration)**. If there is no battery voltage available, have the PCM diagnosed by a dealer service department or other qualified automotive repair facility. If reference voltage is available to the VMV, have the VMV checked using an OBD-II SCAN tool at a dealer service department or other qualified repair facility.

Component replacement

Refer to illustration 7.17

16   Clearly label, then detach, all vacuum lines from the canister.

17   On 1995 through 1998 models, loosen the canister mounting bolts and pull the canister out **(see illustration)**.

18   On 1999 and later models, remove the four mounting bolts and lower the canister assembly. Disconnect the dust separator and remove the fuel vent valve. Remove the canister from the holder.

19   Installation is the reverse of removal.

8   Positive Crankcase Ventilation (PCV) system

Refer to illustration 8.1

1   The Positive Crankcase Ventilation (PCV) system reduces hydrocarbon emissions by scavenging crankcase vapors. It does this by circulating fresh air from the air cleaner through the crankcase, where it mixes with blow-by gases and is then re-routed through a PCV valve to the intake manifold **(see illustration)**.

2   The main components of the PCV system are the PCV valve, a fresh air filtered inlet and the vacuum hoses connecting these two components with the engine and the EECS system.

3   To maintain idle quality, the PCV valve restricts the flow when the intake manifold vacuum is high. If abnormal operating conditions arise, the system is designed to allow excessive amounts of blow-by gases to flow back through the crankcase vent tube into the air cleaner to be consumed by normal combustion.

4   Checking and replacement of the PCV valve and filter is covered in Chapter 1.

9   Catalytic converter

General description

1   The catalytic converter is an emission control device added to the exhaust system to reduce pollutants from the exhaust gas stream. A single-bed converter design is used in combination with a three-way (reduction) catalyst. The catalytic coating on the three-way catalyst contains platinum and rhodium, which lowers the levels of oxides of nitrogen (NOx) as well as hydrocarbons (HC) and carbon monoxide (CO).

Check

2   The test equipment for a catalytic converter is expensive and highly sophisticated. If you suspect that the converter on your vehicle is malfunctioning, take it to a dealer or authorized emissions inspection facility for diagnosis and repair.

3   Whenever the vehicle is raised for servicing of underbody components, check the converter for leaks, corrosion and other damage. If damage is discovered, the converter should be replaced.

Replacement

4   Because the converter is part of the exhaust system, converter replacement requires removal of the exhaust pipe assembly (see Chapter 4). Take the vehicle, or the exhaust system, to a dealer service department or a muffler shop.

6

Notes

Chapter 7
Automatic transaxle

Contents

Specifications

General

Fluid type and capacity .... See Chapter 1

Torque specifications

	Ft-lbs (unless otherwise specified)
Transaxle Range Sensor-to-case bolts	80 to 106 in-lbs
Shift control assembly nuts	18
Shift cable bracket bolts (at transaxle)	18
Speed sensor bolt	80 to 106 in-lbs
Transaxle-to-engine bolts	39 to 53
Torque converter-to-driveplate bolts	20 to 33
Torque converter cover bolts	80 to 106 in-lbs
Transaxle mount tube-to-transaxle bolts	56 to 75
Transaxle mount	
Mount-to-transaxle mount tube nuts	65 to 87
Mount-to-subframe bolts	56 to 75

1   General information

The vehicles covered by this manual are equipped with the AX4S electronic four-speed automatic transaxle, equipped with Overdrive and very similar to the Ford AXOD that has been in use for many years.

Due to the complexity of the clutches and the hydraulic control system, and because of the special tools and expertise required to perform an automatic transaxle overhaul, it should not be undertaken by the home mechanic. Therefore, the procedures in this Chapter are limited to general diagnosis, routine maintenance, adjustment and transaxle removal and installation.

If the transaxle requires major repair work, it should be left to a dealer service department or an automotive or transaxle repair shop.

Replacement and adjustment procedures the home mechanic can perform include those involving the indicator cable, speed sensor and the shift linkage. Removal of the transaxle requires a vehicle hoist, a transaxle jack and a three-bar engine support fixture. The engine stays in the vehicle while the front suspension subframe is removed and the transaxle lowered with a special transaxle jack. It is a job not recommended for the home mechanic.

Caution: *Never tow a disabled vehicle with an automatic transaxle at speeds greater than 35 mph or for distances over 50 miles if the front wheels are on the ground.*

2   Diagnosis - general

Note: *Automatic transaxle malfunctions may be caused by five general conditions: poor*

engine performance, improper adjustments, hydraulic malfunctions, mechanical malfunctions or malfunctions in the computer or its signal network. Diagnosis of these problems should always begin with a check of the easily repaired items: fluid level and condition (see Chapter 1) and shift linkage adjustment. Next, perform a road test to determine if the problem has been corrected or if more diagnosis is necessary. If the problem persists after the preliminary tests and corrections are completed, additional diagnosis should be done by a dealer service department or transaxle repair shop using a scan tool to diagnose the electronics of the transaxle. Refer to the Troubleshooting section at the front of this manual for information on symptoms of transaxle problems.

Preliminary checks

1   Drive the vehicle to warm the transaxle to normal operating temperature.

7

2   Check the fluid level as described in Chapter 1:

a) *If the fluid level is unusually low, add enough fluid to bring the level within the designated area of the dipstick, then check for external leaks (see below).*

b) *If the fluid level is abnormally high, drain off the excess, then check the drained fluid for contamination by coolant. The presence of engine coolant in the automatic transaxle fluid indicates that a failure has occurred in the internal radiator walls that separate the coolant from the transaxle fluid (see Chapter 3).*

c) *If the fluid is foaming, drain it and refill the transaxle, then check for coolant in the fluid or a high fluid level.*

3   Check the engine idle speed. **Note:** *If the engine is malfunctioning, do not proceed with the preliminary checks until it has been repaired and runs normally.*

4   Check the shift cable (see Section 5). Make sure it's properly adjusted and operates smoothly.

Fluid leak diagnosis

5   Most fluid leaks are easy to locate visually. Repair usually consists of replacing a seal or gasket. If a leak is difficult to find, the following procedure may help.

6   Identify the fluid. Make sure it's transaxle fluid and not engine oil or brake fluid (automatic transaxle fluid is a deep red color).

7   Try to pinpoint the source of the leak. Drive the vehicle several miles, then park it over a large sheet of cardboard. After a minute or two, you should be able to locate the leak by determining the source of the fluid dripping onto the cardboard.

8   Make a careful visual inspection of the suspected component and the area immediately around it. Pay particular attention to gasket mating surfaces. A mirror is often helpful for finding leaks in areas that are hard to see.

9   If the leak still cannot be found, clean the suspected area thoroughly with a degreaser or solvent, then dry it.

10   Drive the vehicle for several miles at normal operating temperature and varying speeds. After driving the vehicle, visually inspect the suspected component again.

11   Once the leak has been located, the cause must be determined before it can be properly repaired. If a gasket is replaced but the sealing flange is bent, the new gasket will not stop the leak. The bent flange must be straightened.

12   Before attempting to repair a leak, check to make sure that the following conditions are corrected or they may cause another leak. **Note:** *Some of the following conditions cannot be fixed without highly specialized tools and expertise. Such problems must be referred to a transaxle repair shop or a dealer service department. A factory technician with the proper scan tool can quickly diagnose*

many transaxle problems with an electronic transaxle such as the AX4S.

Gasket leaks

13   Check the pan periodically. Make sure the bolts are tight, no bolts are missing, the gasket is in good condition and the pan is flat (dents in the pan may indicate damage to the valve body inside).

14   If the pan gasket is leaking, the fluid level or the fluid pressure may be too high, the vent may be plugged, the pan bolts may be too tight, the pan sealing flange may be warped, the sealing surface of the transaxle housing may be damaged, the gasket may be damaged or the transaxle casting may be cracked or porous. If sealant instead of gasket material has been used to form a seal between the pan and the transaxle housing, it may be the wrong type of sealant.

Seal leaks

15   If a transaxle seal is leaking, the fluid level or pressure may be too high, the vent may be plugged, the seal bore may be damaged, the seal itself may be damaged or improperly installed, the surface of the shaft protruding through the seal may be damaged or a loose bearing may be causing excessive shaft movement.

16   Make sure the dipstick tube seal is in good condition and the tube is properly seated. Periodically check the area around the speedometer gear or speed sensor for leakage. If transaxle fluid is evident, check the O-ring for damage.

Case leaks

17   If the case itself appears to be leaking, the casting is porous and will have to be repaired or replaced.

18   Make sure the oil cooler hose fittings are tight and in good condition.

Fluid comes out vent pipe or fill tube

19   If this condition occurs, the transaxle is overfilled, there is coolant in the fluid, the case is porous, the dipstick is incorrect, the vent is plugged or the drain-back holes are plugged.

---

3   Driveaxle oil seals - replacement

Refer to illustration 3.3

1   Raise the vehicle and support it securely on jackstands. **Warning:** *On models equipped with an air suspension system, always disconnect the electrical power to the air suspension system before raising the vehicle (see Chapter 10). Failure to do so may result in unexpected shifting or movement of the vehicle, which could cause personal injury.*

2   Remove the driveaxle(s) (see Chapter 8).

3   The seal is a two-piece design, with an outer metal ring and an inner rubber seal. Use a hammer and hook tool to pry up the outer lip of the seal to dislodge it so it can be pried out of the housing **(see illustration)**.

3.3 Dislodge the differential seal by working around the outer edge with a hook tool and hammer

4   Compare the new seal to the old one to make sure they're the same.

5   Coat the lips of the new seal with transaxle fluid.

6   Place the new seal in position and tap it into the bore with a hammer and a large socket or a piece of pipe that's the same diameter as the outside metal edge of the seal.

7   Reinstall the various components in the reverse order of removal.

---

4   Vehicle Speed Sensor (VSS) - replacement

Refer to illustrations 4.4, 4.5 and 4.8

Note: *Any time the speed sensor is removed, you MUST install a new O-ring.*

1   Disconnect the negative battery cable.

2   The speed sensor is located on the right (passenger) side of the extension housing. To determine if the O-ring is leaking, look for transaxle fluid around the sensor.

3   Refer to Chapter 4 and remove the exhaust system flex pipe, being careful not to bend either of the two flex portions, then remove the exhaust Y-pipe.

4   Remove the transaxle heat shield **(see illustration)**. It is retained by three push-nuts.

5   Unplug the sensor electrical connector **(see illustration)**.

6   On some models, there is a speedometer cable mounted to the end of the speed sensor. Disconnect the cable from the speed sensor.

7   Remove the sensor hold-down bolt and remove the sensor **(see illustration 4.5)**.

8   Remove the old O-ring **(see illustration)** and install a new O-ring on the sensor. **Note:** *If a new speed sensor is being installed, remove the clip and swap the plastic drive gear onto the new sensor and secure it with the clip.*

9   Installation is the reverse of removal. Tighten the sensor hold-down bolt securely.

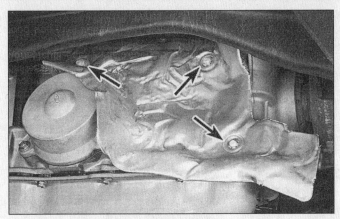

4.4 Remove the transaxle heat shield, retained by three push-nuts (arrows)

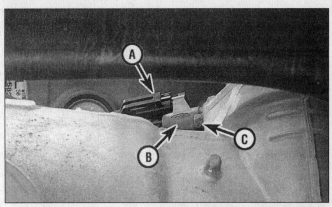

4.5 Disconnect the electrical connector (A) from the speed sensor (B), then remove the hold-down bolt (C) and sensor (this view is through the fenderwell)

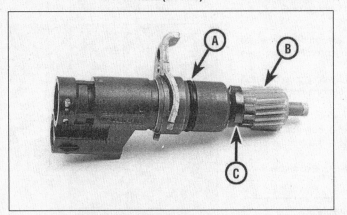

4.8 Before reinstalling the speed sensor, replace the O-ring (A) with a new one - if the gear (B) is to be changed, remove the clip (C)

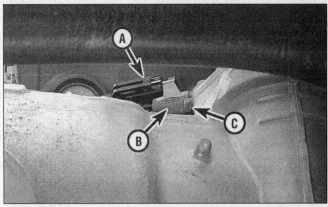

5.3 Use a small screwdriver to release the shift cable terminal lock (A) until the "window" (B) shows black, then pull the terminal from the transaxle lever (C)

5   Shift cable - removal and installation

Refer to illustrations 5.3, 5.4 and 5.6

Warning: *The models covered by this manual are equipped with airbags. Always disable the airbag system before working in the vicinity of the impact sensors, steering column or instrument panel to avoid the possibility of accidental deployment of the airbag(s), which* could cause personal injury (see Chapter 12). *The yellow wires and connectors routed through the console are for this system. Do not use electrical test equipment on these yellow wires or tamper with them in any way while working around the dash or console.*

1   Disconnect the cable from the negative battery terminal, see Warning above.
2   Refer to Chapter 4 and remove the air cleaner and duct.
3   Disconnect the shift cable from the shift lever on the transaxle **(see illustration)**.

4   Remove the bolts holding the shift cable bracket to the transaxle **(see illustration)**. **Caution:** *Do not remove the cable from the bracket. If the cable is removed from the bracket, the cable and its retaining clip must be replaced.*
5   Refer to Chapter 11 and remove the driver's side knee bolster.
6   Disconnect the cable end from the shift control lever, and remove the cable and bracket from the U-shaped mount on the steering column **(see illustration)**.

7

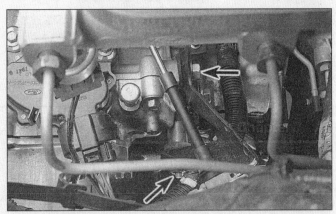

5.4 Remove the bolts (arrows) holding the shift cable bracket to the transaxle - do not remove the cable from the bracket!

5.6 Pry off the cable end (A) at the steering column, then remove the cable and bracket from the channel (B)

6.4a Pry up the plastic boot from the pin (arrow) at the top of the column . . .

6.4b . . . then pull the boot out of the slot at the bottom right (arrow)

6.5 Use a small punch to drive out the shift lever pin (arrow)

7   Pull the cable through the hole in the firewall to the engine side.

8   Installation is the reverse of removal. **Note:** *When attaching the cable end to the transaxle lever, make sure that the window in the terminal lock shows yellow (locked position).*

6   Shift lever assembly - removal and installation

Refer to illustrations 6.4a, 6.4b and 6.5
Warning: *The models covered by this manual are equipped with airbags. Always disable the airbag system before working in the vicinity of the impact sensors, steering column or instrument panel to avoid the possibility of accidental deployment of the airbag(s), which could cause personal injury (see Chapter 12). The yellow wires and connectors routed through the dash and steering column are for this system. Do not use electrical test equipment on these yellow wires or tamper with them in any way while working around the dash or steering column.*

1   Disconnect the negative cable from the battery, see Warning above.

2   Remove the driver's side knee bolster and upper and lower steering column covers

(see Chapter 11).

3   Disconnect the electrical connector from the transmission control switch (see Section 9).

4   Remove the boot by prying up one end and pulling the bottom out of a slot in the column **(see illustrations)**.

5   Drive the retaining pin from the shift lever base with a small punch and hammer **(see illustration)**. Remove the shift lever and the clip.

6   Installation is the reverse of removal.

7   Shift indicator cable - replacement and adjustment

Warning: *The models covered by this manual are equipped with airbags. Always disable the airbag system before working in the vicinity of the impact sensors, steering column or instrument panel to avoid the possibility of accidental deployment of the airbag(s), which could cause personal injury (see Chapter 12). The yellow wires and connectors routed through the dash and steering column are for this system. Do not use electrical test equipment on these yellow wires or tamper with them in any way while working around the dash or steering column.*

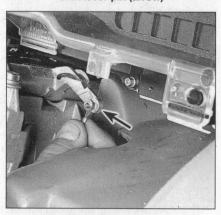

7.4a Pry off the washer and remove the cable loop from the pin (arrow)

Replacement

Refer to illustrations 7.4a, 7.4b and 7.5

1   Disconnect the negative battery cable, see **Warning** above.

2   Refer to Chapter 11 and remove the driver's side knee bolster and the upper and lower steering column covers.

3   Remove the ignition lock cylinder (see Chapter 12).

4   The shift indicator cable attaches to the shift control lever on the top side of the steering column **(see illustrations)**.

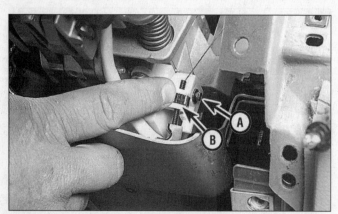

7.4b Remove the bolt (A) and detach the cable end from the column - B is the adjustment thumbwheel

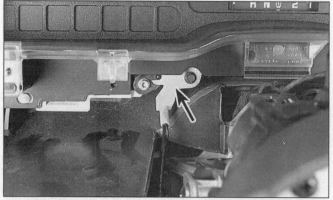

7.5 Remove the indicator cable mount (arrow) at the instrument panel

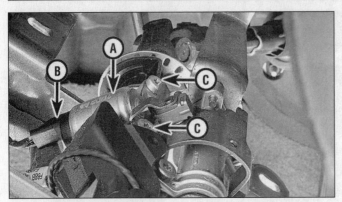

8.2 The interlock solenoid (A) on column-shift models is located near the bottom of the steering column (shown with column dropped from dash) - (B) is the electrical connector and (C) are the mounting screws

9.2 Pry the cover off the end of the shift lever top expose the TCS

5   Follow the cable routing from there to the instrument panel, removing any retaining clips along the way, and disconnect the upper end from the shift indicator in the instrument panel **(see illustration)**.

6   Replacement is the reverse of the removal process.

Adjustment

7   Position the shift lever in the Overdrive position, which is two clicks counterclockwise from the most clockwise position. Observe the indicator, which has a "flag" around the indicated position (with standard instrument panels) or a pointer (electronic instrument panel).

8   Put a few pounds of pressure downward (clockwise) on the shifter. If the flag or pointer isn't correctly centered on the Overdrive symbol, adjust the thumbwheel on the shift indicator cable until it is centered **(see illustration 7.4b)**.

9   After adjustment, try the shift lever in all positions to make sure the indicator matches all the shift lever detents.

8   Brake/transaxle shift interlock system - description, check and solenoid replacement

Warning: *The models covered by this manual are equipped with airbags. Always disable the airbag system before working in the vicinity of the impact sensors, steering column or instrument panel to avoid the possibility of accidental deployment of the airbag(s), which could cause personal injury (see Chapter 12). The yellow wires and connectors routed through the dash and steering column are for this system. Do not use electrical test equipment on these yellow wires or tamper with them in any way while working around the dash or steering column.*

Description

1   The brake transaxle interlock system prevents shift lever from being moved out of Park unless the brake pedal is depressed simultaneously. When the car is started, a solenoid is energized, locking the shift lever in Park; when the brake pedal is depressed, the solenoid is de-energized, unlocking the shift lever so that it can be moved out of Park. **Note:** *Before making the following checks, refer to Chapter 9 and verify that the brake light switch is functioning properly, because it is part of the interlock circuit. Also check the 10-amp fuse in position 21 in the fuse panel below the instrument panel (see Chapter 12). This fuse serves the interlock solenoid.*

Check

Refer to illustration 8.2

2   Remove the driver's side knee bolster and upper and lower steering column covers (see Chapter 11). Using a flashlight, locate the brake/transaxle shift interlock solenoid **(see illustration)**.

3   Verify that the brake/transaxle shift interlock solenoid operates as follows:

a) *When the ignition key is in the Lock position, the solenoid plunger should be in and you should not be able to move the shift lever, even with the brake pedal applied.*

b) *With the ignition key is turned to the Off position, the solenoid plunger should be popped out and you should be able to move the shift lever to any gear position without applying the brake pedal.*

c) *When the ignition key is turned to the Run position, the solenoid should go back into the solenoid and you should not be able to move the shift lever; except with the brake pedal applied, the solenoid plunger should be released (pop out) and you should be able to move the shift lever out of Park into any gear.*

d) *Place the key in the Lock position, then turn the key to the Acc (accessory) position. The solenoid plunger should go in a little further than it does when the key is turned to Lock, and you should not be able to move the shift lever (even with the brake applied).*

4   If the solenoid doesn't operate as described above, unplug the electrical connector from the solenoid, apply battery voltage to the solenoid and verify that it "clicks" on, then open the circuit and verify that the solenoid clicks off. If the solenoid doesn't operate as described, replace it.

5   Check for battery voltage (10 or more volts) at the connector at the terminal for the wire from the fuse panel (red/white wire on 1995 models, pink/black on 1996 and later).

6   Step on the brake pedal and check for voltage at the wire from the brake light switch (red/green). There should be battery voltage when the pedal is depressed, but less than one volt with the pedal released.

7   Using an ohmmeter, check the resistance at the black wire on the solenoid connector. If the resistance is greater than 5 ohms, there is a problem in the ground circuit.

Solenoid replacement

8   Disable the airbag system (see Chapter 12). Remove the driver's side knee bolster and upper and lower steering column covers (see Chapter 11).

9   Lower the steering column and support it (see Chapter 10).

10 Unplug the electrical connector from the solenoid, and remove the solenoid mounting screws **(see illustration 8.2)**. Remove the solenoid.

11 Installation is the reverse of the removal procedure.

9   Transaxle Control Switch (TCS) - check and replacement

1   The Transmission Control Switch located on the shift lever allows the driver to turn the Overdrive capability On or Off. In normal driving the Overdrive is always turned On.

Check

Refer to illustrations 9.2 and 9.7

2   Pull the cover (on the end of the shift lever) from the TCS **(see illustration)**.

3   Check the 10-amp fuse (designated "V") in the engine compartment fuse panel.

7

9.7 The TCS simply pulls out of the shift lever - check the pins for continuity

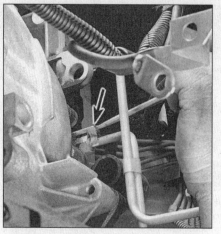

10.2a Remove the retaining clip (arrow) from the cooler line at the transaxle

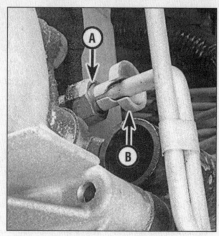

10.2b Push the tool (A) into the fitting with the slot lined up with the flat on the integral retainer (B) - hold the tool in firmly while pulling the tube out

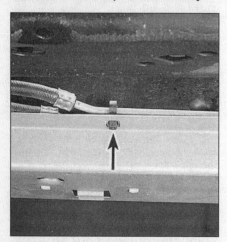

10.3 Remove the clip (arrow) from the line clamp at the radiator support

10.4 Disconnect the lower cooling line from the right side of the radiator, using a spring-lock coupling tool (arrow)

4   Use a volt-ohmmeter to check for battery voltage at one of the pin sockets in the end of the shift lever, with the ignition key in the On position (engine not running).

5   With a self-powered test light, check the continuity of the three pins on the TCS. Across two of the pins there should be continuity only when the switch is depressed.

6   Check for continuity between the TCS, the TCS indicator light and the circuit between the TCS and the PCM (see the wiring diagrams in Chapter 12). If the checks are not correct, repair the circuit or replace the TCS switch.

7   The TCS simply pulls out of the end of the shift lever. To replace it, push the new switch into the lever, making sure the pins are aligned with the sockets. Press the cover back on **(see illustration)**.

10   Auxiliary oil cooler - removal and installation

Refer to illustrations 10.2a, 10.2b, 10.3 and 10.4

1   On 3.8L models with trailer-towing

option, an auxiliary oil cooler is provided for the automatic transaxle fluid. The cooler looks like as small radiator and is mounted in front of the engine radiator and air-conditioning condenser, just behind the grille. Transaxle fluid flow comes from the transaxle to the auxiliary cooler through the transaxle fluid output line, and then from the auxiliary cooler to the standard cooler in the right-hand tank of the engine radiator. **Note:** *Do not confuse the auxiliary transaxle oil cooler with the similar-looking power steering fluid cooler, which is mounted to the left (facing the front of the radiator) of the transaxle cooler. They are both mounted with the same bolt.*

2   To remove the cooler, Place a suitable drain pan under the transaxle to catch the fluid and use a fuel-line disconnect tool to remove the lower fluid line from the transmission **(see illustrations)**. Plug the line after the fluid stops draining.

3   Remove the screws holding the two tubing clamps to the left side of the radiator support **(see illustration)**.

4   At the transaxle cooler at the right side of the radiator, disconnect the lower tube fit-

ting **(see illustration)**.

5   Remove the bolt **(see illustration 10.1)** and remove the cooler with the fluid lines. **Caution:** *If hot transaxle fluid spills on painted body surfaces, clean it off immediately to avoid damaging the finish.*

6   Installation is the reverse of the removal procedure. **Caution:** *Do not tighten the mounting bolts until you are sure the cooling lines are properly connected, and there are no kinks or leaks.*

11   Transaxle range sensor - check and replacement

Check

Refer to illustrations 11.1 and 11.2

1   With the ignition key Off, remove the electrical connector from the TRS **(see illustration)**. **Caution:** *Disconnect the connector by hand, do not use a pry tool.*

2   Use small mirror to examine the end of the connector for loose, corroded or bent

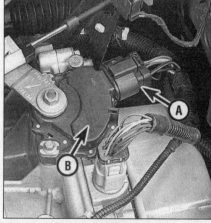

11.1 Electrical connector (A) on the side of the TRS switch (B)

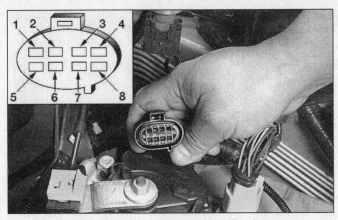

11.2 TRS sensor terminal identification

1	Battery power	5	Starter control to interlock
2	Signal return	6	To backup lamps
3	TRS to PCM	7	Fused accessory feed
4	To liftgate release	8	To starter control

11.13 Mark and remove the lever (A), then remove the two mounting bolts (B) and take off the TRS sensor

pins **(see illustration)**.

3   Major testing of the TRS sensor and harness should be conducted using a factory test tool. However, you can check for basic continuity between the terminals on the sensor with an ohmmeter or continuity tester.

4   Check for continuity between the white/pink terminal and the red/light blue terminal. There should be continuity only when in Park or Neutral (key Off).

5   To check the backup light portion of the sensor, check for continuity between the purple/orange terminal and the black/pink terminal. There should be continuity only when the transaxle is shifted to Reverse.

6   Further testing should be done with the factory tool.

Replacement

Refer to illustration 11.13

7   Disconnect the negative cable from the battery.

8   Shift the transaxle into Neutral.

9   Raise and support the front of the vehicle securely on jackstands.

10   Refer to Section 5 and remove the shift cable from the shift lever and the cable bracket.

11   Make a paint or scribe mark on the transaxle shift lever, relative to its position on the splined shaft from the transaxle when in Neutral, then remove the nut and take off the shift lever.

12   Disconnect the electrical connectors from the TRS sensor **(see illustration 11.1)**.

13   Remove the bolts and detach the sensor **(see illustration)**.

14   To install the sensor (transaxle shaft still in its Neutral position), line up the flats on the switch with the flats in the shaft and lower the sensor onto the shaft.

15   Install the bolts. The alignment of the sensor is made by moving the switch slightly when tightening the two mounting bolts. A factory tool makes this easier, but the tool isn't absolutely necessary. If you are reinstalling your original switch, the bolts will have made a scratched circle on the mount-

ing bosses. If you bolt it down with the bolts aligned exactly over these circles, the switch will be positioned correctly.

16   The remainder of installation is the reverse of removal.

17   Connect the negative battery cable, apply the parking brake and verify that the engine will start only in Neutral or Park.

12 Transaxle mount - check and replacement

Refer to illustrations 12.1, 12.5a and 12.5b

1   Insert a large screwdriver or prybar into the space between the transaxle bracket and the mount and try to pry the transaxle up slightly **(see illustration)**. The transaxle bracket should not move away from the insulator much at all.

2   To replace the mount, raise and suitably support the vehicle on jackstands. **Warning:** *On models equipped with an air suspension system, always disconnect the electrical power to the air suspension system before raising the vehicle (see Chapter 10). Failure to do so may result in unexpected shifting or movement of the vehicle, which could cause*

personal injury.

3   Remove the left front wheel and the inner splash shield (see Chapter 11).

4   Support the engine with a floorjack and a block of wood, and raise the engine/transaxle enough to take some tension off the transaxle mount.

5   Remove the nuts attaching the insulator to the front subframe and the nuts attaching the insulator to the transaxle **(see illustrations)**.

12.1 Pry on the transaxle mount to check for a broken insulator

7

12.5a Remove the mount-to-chassis bolts (A), then the mount-to-transaxle-bracket nuts (B) - 3.0L models

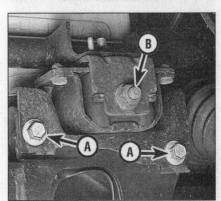

12.5b Remove the mount-to-chassis bolts (A), then the mount-to-transaxle-bracket throughbolt (B) - 3.8L models

13.5 Make paint marks (arrows) on the driveplate and torque converter stud to align them in later assembly

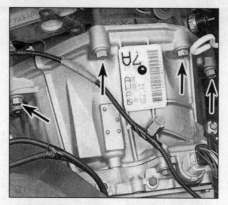

13.14 Remove the upper transaxle bolts (arrows)

13.18 Remove the four bolts (arrows) holding the subframe to the vehicle

6   Raise the engine/transaxle slightly with the jack and remove the insulator, noting which holes are used in the support for proper alignment during installation.

7   Installation is the reverse of the removal procedure. Be sure to tighten the nuts/bolts to the torque listed in this Chapter's Specifications.

13 Automatic transaxle - removal and installation

Caution: *This is a difficult procedure for the home mechanic, requiring the use of several specialized tools, including a transaxle jack and a three-bar engine support fixture. Such tools can be rented, but the job is still difficult to do without a hydraulic lift. Safely raising the vehicle enough for the transaxle and subframe to be pulled out from underneath is a problem without a hoist.*

Removal

Refer to illustrations 13.5, 13.14, 13.18 and 13.19

1   Disconnect the negative cable from the battery. Refer to Chapter 5 and remove the battery and battery tray.

2   Raise the vehicle and support it securely on a vehicle hoist. **Warning:** *On models equipped with an air suspension system, always disconnect the electrical power to the air suspension system before raising the vehicle (see Chapter 10). Failure to do so may result in unexpected shifting or movement of the vehicle, which could cause personal injury.*

3   Drain the transaxle fluid (Chapter 1).

4   Remove the torque converter cover.

5   Mark the torque converter-to-driveplate relationship with white paint so they can be installed in the same position **(see illustration)**.

6   Remove the torque converter-to-driveplate nuts. Turn the crankshaft pulley bolt for access to each nut.

7   Remove the starter motor (see Chapter 5).

8   Remove the driveaxles from the transaxle (see Chapter 8).

9   Disconnect the speed sensor (see Section 4).

10   Disconnect the electrical connectors from the transaxle.

11   Remove any exhaust components which will interfere with transaxle removal (see Chapter 4).

12   Disconnect the shift linkage from the transaxle (see Section 5).

13   Support the engine using a three-bar support fixture to retain the engine in the body (Refer to Chapter 2). **Note:** *You will probably have to remove the hood and upper cowl panel to install the three-bar fixture (see Chapter 11).*

14   Remove the upper transaxle-to-engine bolts and set aside the wiring harness attached to one of the studs on some models **(see illustration)**.

15   Disconnect and plug the transaxle cooler lines (see Section 10). Make sure all of the fluid has drained out into a suitable container.

16   Remove the dipstick tube.

17   Refer to Chapter 10 and disconnect the tie rod ends, steering column lower coupler, lower balljoints and stabilizer links. **Warning:** *Lock the steering wheel in the straight-ahead position before disconnecting the steering column lower coupler and DO NOT allow the steering wheel to rotate or damage to the airbag clockspring could occur.*

18   Refer to Chapter 2 and disconnect the engine mount nuts at the chassis, then support the front subframe assembly with a jack and remove the four bolts holding the subframe to the vehicle **(see illustration)**. **Warning:** *Never put any part of your body under the front subframe while it is unbolted from the vehicle.*

19   Lower the front subframe and roll it out of the way on the subframe jack. Position a transmission jack under the transaxle and remove the lower transaxle bolts **(see illustration)**.

20   Move the transaxle back to disengage it from the engine block dowel pins and make sure the torque converter is detached from the driveplate. Secure the torque converter to the transaxle so it will not fall out during removal. Lower the transaxle with the transaxle jack.

13.19 Remove the lower transaxle-to-engine bolts (arrow indicates one above the starter location, there is another on the opposite side of the oil pan)

Installation

21   Prior to installation, make sure that the torque converter hub is securely engaged in the pump. Lubricate the torque converter hub with multi-purpose grease.

22   With the transaxle secured to the jack, raise it evenly into position. Be sure to keep it level so the torque converter does not slide out.

23   Turn the torque converter to line up the studs with the holes in the driveplate. The white paint mark on the torque converter and the driveplate made in Step 5 must line up.

24   Move the transaxle forward carefully until the dowel pins and the torque converter are engaged.

25   Install the transaxle-to-engine bolts. Tighten them securely.

26   Install the torque converter-to-driveplate nuts. Tighten the bolts to the specified torque.

27   Install the front subframe and reconnect the suspension and chassis components which were removed. Tighten the bolts and nuts to the torque specified in Chapter 10.

28   The remainder of the installation is the reverse of the removal procedure.

29   Fill the transaxle with new fluid (Chapter 1), run the vehicle and check for fluid leaks and proper shifting. Verify before driving that the vehicle will only start in Park or Neutral.

Chapter 8   Driveaxles

Contents

Specifications

Driveaaxle length

Left side .... 19.7 inches
Right side .... 25.1 inches

Torque specifications   **Ft-lbs**

Driveaxle hub nut .... 157 to 212
Lower arm to steering knuckle nut .... 40 to 55
Wheel lug nuts .... 85 to 105

1   Driveaxles - general information and inspection

Power is transmitted from the transaxle to the front wheels by two driveaxles, which consist of splined solid axles with constant velocity (CV) joints at each end.

There are three types of CV joints used on the vehicles covered by this manual. The outer joint, on all models, is a fixed ball-and-cage design which allows for a wide range of angular movement. The inner CV joint is either a tripod design (1995 models) or a double offset ball-and-cage design (1996 and later models). The inner CV joint not only allows angular movement, but permits the driveaxle to vary in length by sliding in-and-out of the housing as the suspension travels up-and-down.

The CV joints are protected by rubber boots, which are retained by clamps so the joints are protected from water and dirt. The boots should be inspected periodically (see Chapter 1). The inner boots have very small breather holes which may leak a small amount of lubricant under some circumstances, such as when the joint is compressed during removal. Damaged CV joint boots must be replaced immediately or the joints can be damaged. Boot replacement involves removing the driveaxles (Section 2). It's a good idea to disassemble, clean, inspect and repack the CV joint whenever replacing a CV joint boot to make sure the joint isn't contaminated with moisture or dirt, which would cause premature failure of the CV joint (see Section 3).

The most common symptom of worn or damaged CV joints, besides lubricant leaks, are a clicking noise in turns, a clunk when accelerating from a coasting condition or vibration at highway speeds.

Some specialized tools and procedures are required to disassemble and overhaul the CV joints in a driveaxle. The outer joint is not rebuildable, and is sold only as an assembly, with the shaft, outer joint, ABS sensor ring and front wheel dust shield. The inner joint on all models can be rebuilt using a repair kit.

Warning 1: *Since many of the procedures covered in this Chapter involve working under the vehicle, make sure it's securely supported on sturdy jackstands or on a hoist where the vehicle can easily be raised and lowered.*

Warning 2: *On models equipped with electronic rear suspension, the system's switch must be turned OFF before the vehicle is towed or raised on a jack or hoist (see Chapter 10). Failure to do so could result in a sudden inflation or deflation of the air springs, causing instability of the vehicle while it's off the ground.*

Warning 3: *Whenever the driveaxle hub is loosened or removed it must be replaced with a new hub nut. Whenever the driveaxle is removed, the driveaxle circlips must be replaced with new circlips. Whenever any of the suspension fasteners are loosened or removed they must be inspected and if necessary replaced with new fasteners. All fasteners must be replaced with new fasteners of the same part number or of the original equipment quality and design. Torque specifications must be followed for proper assembly and component retention.*

2   Driveaxles - removal and installation

Refer to illustrations 2.1, 2.2, 2.5, 2.7a,2.7b, 2.9 and 2.12

Warning: *On models equipped with an air suspension system, always disconnect the electrical power to the air suspension system before raising the vehicle (see Chapter 10). Failure to do so may result in unexpected shifting or movement of the vehicle, which could cause personal injury.*

Removal

1   Remove the wheel cover and loosen the hub nut **(see illustration)**. Loosen the wheel lug nuts, raise the front of the vehicle and

2.1 Before raising the vehicle, remove the wheel center cap and use a breaker bar and socket to loosen the hub nut

8

support it securely on jackstands. Apply the parking brake and block the rear wheels to keep the vehicle from rolling off the jackstands. Remove the front wheel.

2   Remove and discard the driveaxle hub nut. Purchase a new hub nut for installation **(see illustration).**

3   Refer to Chapter 9 and disconnect the electrical connector from the ABS wheel sensor, then remove the wheel sensor and wire it out of the way.

4   Remove and discard the nut on the pinch bolt at the lower balljoint steering knuckle (see Chapter 10). Drive the bolt out with a punch (the bolt must be replaced with a new one), then pry the pinch opening open a little with a large screwdriver. Use a large prybar to force the lower arm away from the knuckle.

5   When removing the left-side driveaxle, insert a large prybar through the lower A-arm and against the frame. Position the point of the prybar behind the flange of the inner CV joint and pry the driveaxle out of the transaxle **(see illustration). Caution:** *Do not tear the inner joint's boot with the prybar.*

6   Once the inner joint is loose from the transaxle, support the axle with a heavy wire, but do not allow the wire to contact the inner boot, or the boot could be damaged.

7   To disengage the right-side driveaxle's

2.2   Use a prybar to hold the caliper while removing the hub nut

inner joint from the transaxle, use a hook adapter on a slide-hammer to get behind the inner joint housing and extract it from the transaxle **(see illustrations).** There generally isn't room to get a prybar in on that side.

8   Refer to Chapter 10 and remove the stabilizer bar link at the bar. Refer to Chapter 9 for removal of the brake caliper and disc.

9   Use a two-jaw puller over the hub flange, with the center bolt pushing on the axleshaft to push the outer joint and axle out

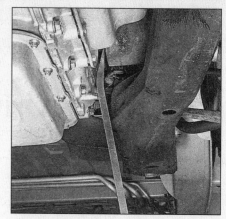

2.5   A large prybar (arrow) can be used to pry the inner joint of the left driveaxle outward from the transaxle

of the knuckle **(see illustration). Caution:** *Do not use a hammer to force the driveaxle out of the hub.*

10   Support the CV joints and carefully remove the driveaxle from the vehicle. **Caution:** *Do not tear the inner or outer boots on any sharp components as you withdraw the axle. Do not nick the ABS exciter ring on the outer joint of each driveaxle.*

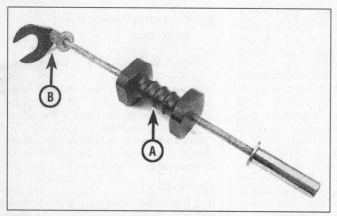

2.7a   Use a slide-hammer (A) with an adapter (B) to pull the right-side driveaxle out

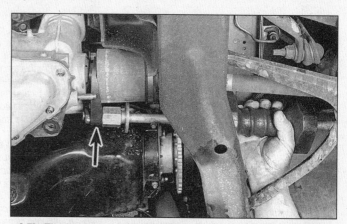

2.7b   The hook tool (arrow) can help extract an axle when there isn't a convenient place to position a prybar

2.9   A puller is required to separate the axle from the hub - a prybar keeps the hub from rotating during the procedure

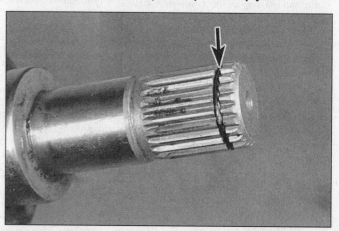

2.12   Always install a new circlip (arrow) when installing the inner CV joint to the transaxle

3.3 Cut the old boot clamps off and discard them (if you're planning to reuse the boot, make sure you don't accidentally cut the boot)

3.4 Pull the boot back and slide the tripod from the housing

3.5 Expand the stop-ring and move it back on the shaft

Installation

11 If the drixeaxle was replaced with a new or rebuilt unit, it should be fully assembled.

12 Install a new circlip on the stubshaft of the inner CV joint (see illustration). Lubricate the differential seal with multi-purpose grease, raise the driveaxle into position and through the steering knuckle. While supporting the CV joints, insert the splined end of the inner CV joint into the differential side gear. Seat the shaft in the side gear by positioning the end of a screwdriver in the groove in the CV joint and tapping it into position with a hammer until you feel the circlip engage. Grasp the inner joint housing and pull outward (away from the transaxle) to ensure that the circlip is engaged.

13 Insert the outer axleshaft through the hub, making sure the splines are aligned. Use two lug nuts and washers to retain the rotor to the hub and install the axle washer and new hub nut. **Warning:** The axle hub nut must be NEW, do not reuse the old nut. Keep the rotor from turning (see illustration 2.2), and tighten the hub nut. Final tightening to the torque listed in this Chapter's Specifications should be done with the wheel/tire

installed and the vehicle on the ground (see illustration 2.1).

3 Driveaxle boot - replacement

Note: *If the CV joints exhibit wear indicating the need for an overhaul (usually due to torn boots), explore all options before beginning the job. Complete rebuilt driveaxles are available on an exchange basis, which eliminates a lot of time and work. Whatever is decided, check on the cost and availability of parts before disassembling the vehicle.*

1 Remove the driveaxle (see Section 2).

2 Place the driveaxle in a vise lined with soft jaws to avoid damage to the shaft.

Inner CV joint

1995 models

Refer to illustrations 3.3, 3.4, 3.5, 3.6, 3.10, 3.11, 3.14a and 3.14b

3 Cut off the boot seal retaining clamps and slide the boot towards the center of the driveaxle (see illustration). Mark the housing and driveaxle so they can be reinstalled in the same relative positions.

4 Withdraw the tripod bearing assembly

from the housing, using tape or a cloth wrapped around the bearing assembly to retain the bearings during removal and installation (see illustration).

5 Expand the stop ring with snap-ring pliers and move it back on the shaft (see illustration).

6 Push the inner race back enough to remove the circlip (see illustration). Discard the circlip; a new circlip must be used on installation.

7 Remove the tripod bearing assembly from the axle, remove the stop ring, then slide the boot off the axle.

8 Clean all of the old grease out of the housing and the bearing assembly.

9 Pack the housing with half of the grease furnished with the new boot and place the remainder in the boot. **Note:** The inner joint of the left-hand axle requires 8.8 ounces of hi-temp CV joint grease, while the right-hand inner joint requires only 7.1 ounces.

10 Wrap the driveaxle splines with tape to avoid damaging the boot, then slide the new boot onto the axle until the small end fits into the groove on the axle (see illustration).

11 Install the stop ring, pushing it back far enough to allow circlip installation. Slip the bearing assembly onto the shaft, making sure

3.6 Push the inner race back to expose and remove the circlip with pliers or a screwdriver

3.10 Before installing the CV joint boot, wrap axle splines with electrical tape to prevent damage to the boot

3.11 When installing the bearing assembly on the driveaxle, make sure the chamfered side is facing the stop ring

8

3.14a Equalize the pressure inside the boot by inserting a small, dull screwdriver between the boot and the housing

3.14b You'll need a special boot clamp installation tool (available at auto parts stores) to tighten the clamps - follow the instructions provided by the manufacturer

3.16 Use pliers or a screwdriver to remove the inner retaining ring

the chamfered side of the bearing assembly faces the stop ring **(see illustration)**.

12   Install the new circlip, push the bearing assembly towards the circlip, and move the stop-ring back to its groove.

13   Carefully insert the assembly into the greased housing. **Note:** *On the left-hand axle inner joint only, there is an insert that snaps over the housing before the bearing assembly is inserted.*

14   Seat the boot in the housing and axle-shaft grooves and install the retaining clamps **(see illustrations)**. Install the driveaxle as described in Section 2.

1996 through 1998 models

Refer to illustrations 3.16, 3.17, 3.18, 3.19, 3.21 ,3.22, 3.23, 3.24a, 3.24b, 3.26, 3.29,3.30, 3.31, 3.32 and 3.34

15   Cut off the boot seal retaining clamps and slide the boot towards the center of the driveaxle **(see illustration 3.3)**. Mark the housing and driveaxle so they can be rein-stalled in the same relative positions.

16   Remove the front retaining ring **(see illustration)**.

17   Withdraw the ball-and-cage assembly from the housing **(see illustration)**.

3.17 With the retainer removed, the housing can be pulled off the ball-and-cage assembly

18   Expand the stop-ring and move it back on the axle **(see illustration)**.

19   Push the inner race back enough to remove the circlip **(see illustration)**.

20   Slide the inner bearing assembly off the axleshaft.

21   Mark the inner race and cage to ensure

3.18 Spread the stop ring and slide it back on the axle

that they are reassembled with the correct sides facing out **(see illustration)**.

22   Using a screwdriver or piece of wood, pry the balls from the cage **(see illustration)**. Be careful not to scratch the inner race, the balls or the cage.

23   Rotate the inner race 90-degrees, align the inner race lands with the cage windows

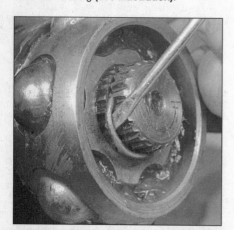

3.19 Push the inner race back enough to pry the circlip off the shaft

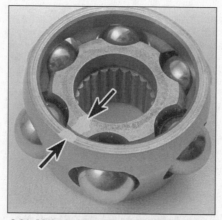

3.21 Make index marks on the inner race and cage so they'll both be facing the same way when reassembled

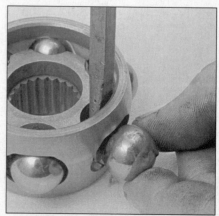

3.22 Pry the balls from the cage with a screwdriver (be careful not to nick or scratch them)

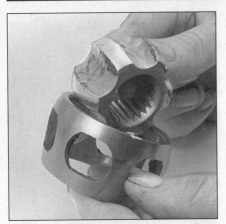

3.23 Tilt the inner race 90 degrees and rotate it out of the cage

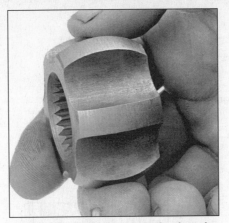

3.24a Check the inner race lands and grooves for pitting and score marks

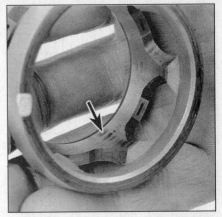

3.24b Check the cage for cracks, pitting and score marks (shiny spots are normal and don't affect operation)

3.26 Press the balls into the cage through the windows

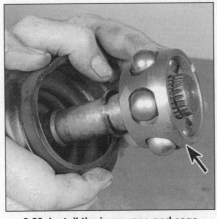

3.29 Install the inner race and cage assembly with the "bulge" (arrow) facing the axleshaft end

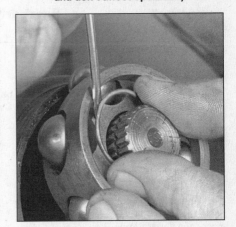

3.30 Install the circlip . . .

and rotate the race out of the cage (see illustration).

24 Clean all of the old grease out of the housing and the bearing assembly. Inspect the balls, cage, inner race and housing for scoring, pitting and other signs of abnormal wear. Apply a coat of CV joint grease to the inner bearing surfaces to hold the balls in place when reassembling the spider assembly. Shiny, polished spots are normal and will not adversely affect CV joint performance (see illustrations).

25 Begin reassembly by inserting the inner race into the cage. Verify that the match-marks are on the same side. However, it's not necessary for them to be in direct alignment with each other.

26 Press the balls into the cage windows with your thumbs (see illustration).

27 Wrap the axleshaft splines with tape to avoid damaging the boot. Slide the small boot clamp and boot onto the axleshaft, then remove the tape (see illustration 3.10).

28 Install a new stop ring on the axleshaft. Don't seat it in the groove at this time, but slide it past the splined area.

29 Install the inner race and cage assembly on the axleshaft with the larger diameter side or "bulge" of the cage facing the axleshaft end (see illustration).

30 Install the new circlip and slide the inner race and cage assembly out until the inner race contacts the circlip (see illustration).

31 Install the stop ring in the groove (see illustration). Make sure it's completely seated by pushing on the inner race and cage assembly.

32 Fill the outer race and boot with CV joint grease (normally included with the new boot

kit). Pack the inner race and cage assembly with grease, by hand, until grease is worked completely into the assembly (see illustration). Note: *The inner joint of the left-hand axle requires 8.8 ounces of hi-temp CV joint grease, while the right-hand inner joint requires only 7.1 ounces.*

33 Slide the outer race down onto the inner race and install the wire ring retainer. Wipe any excess grease from the axle boot groove

8

3.31 . . . then seat the stop ring in the groove

3.32 Pack grease into the bearing until it's completely full

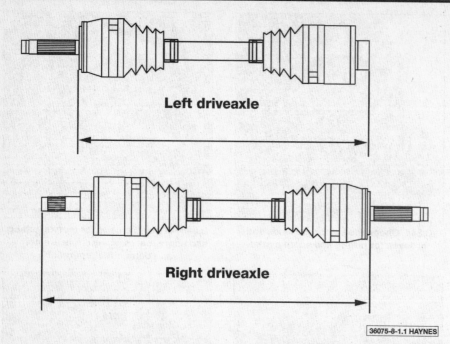

Left driveaxle

Right driveaxle

36075-8-1.1 HAYNES

3.34 The driveaxle standard length should be set to the dimension listed in this Chapter's Specifications before the inner CV joint boot clamps are tightened

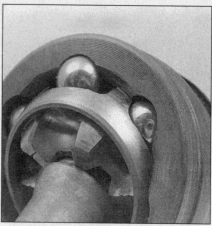

3.41 After the assembly has been cleaned and the solvent blown out with compressed air, rotate the joint through its full range of motion, inspecting the components for wear or damage

on the outer race. Seat the small diameter of the boot in the recessed area on the axleshaft and install the clamp. Push the other end of the boot onto the outer race.

34   Adjust the length of the driveaxle to the dimension listed in this Chapter's Specifications **(see illustration)**.

35   With the axle set to the proper length, equalize the pressure in the boot by inserting a dull screwdriver between the boot and the outer race **(see illustration 3.14a)**. Don't damage the boot with the tool.

36   Install the boot clamp. A pair of special clamp-crimping pliers are required. The pliers are available at most auto parts stores **(see illustration 3.14b)**.

37   Install a new circlip on the stub axle and install the driveaxle as described in Section 2.

1999 and later models

The inner CV joint is an integral part of the drive shaft and cannot be serviced.

Outer CV joint

Refer to illustration 3.41

38   The outer boot can only be replaced while the inner joint is off the shaft. Follow the procedure above and remove the inner joint and boot.

39   Cut the clamps off and slide the old outer boot off the driveaxle.

40   Clean the bearing assembly as best as possible. **Note:** *Because the outer joint can't be disassembled, it's difficult to wash away all the old grease, and to rid the bearings of solvent once they're cleaned. But it is imperative that the job be done thoroughly, so take*

your time and do it right.

41   Bend the outer CV joint at an angle to the driveaxle to expose the bearings, inner race and cage **(see illustration)**. If any of the components show excessive wear or damage, the joint must be replaced. New outer joints are available only with the axleshaft.

42   Lube the bearing assembly, using three ounces of hi-temp CV joint grease in the bearing assembly.

43   Tape the axle splines to protect the boot **(see illustration 3.10)**, and slide the new boot into place. When the small end of the boot is in position, pack three ounces of grease into the boot and position the large end of the boot over the housing until it seats into the groove. Install a new clamp on the outboard side and tighten both boot clamps **(see illustration 3.14b)**.

44   Assemble the inner CV joint and boot onto the axleshaft

45   Install a new circlip on the stub axle and install the driveaxle as described in Section 2.

Chapter 9 Brakes

Contents

Specifications

General
Brake fluid type.. See Chapter 1

Disc brakes
Brake pad lining minimum thickness................................ See Chapter 1
Disc
  Front
    Minimum thickness.. Refer to the marks cast into the disc
    Runout (maximum)... 0.003 inch
    Thickness variation limit...................................... 0.0003 inch
  Rear
    Minimum thickness.. Refer to the marks cast into the disc
    Runout (maximum)... 0.003 inch
    Thickness variation limit...................................... 0.0005 inch

Drum brakes
Shoe lining minimum thickness...................................... See Chapter 1
Maximum inner diameter.. Refer to the marks cast into the drum

Torque Specifications   **Ft-lbs** (unless otherwise indicated)
Bleeder valve
  Front.. 60 to 120 in-lbs
  Rear... 97 in-lbs
Caliper mounting bolts
  Front.. 25
  Rear... 132 to 168 in-lbs
Caliper mounting bracket bolts
  Front.. 65 to 87
  Rear (spindle bolts)... 48 to 59
Brake hose-to-caliper bolt... 35 to 46
Brake pressure control valve mounting bolt.................... 91 to 122 in-lbs
Master cylinder-to-power booster nuts........................... 22
Power booster-to-firewall nuts....................................... 15 to 20
Rear wheel cylinder mounting bolts (drum brakes)........ 108 to 156 in-lbs
Wheel lug nuts.. See Chapter 1

9

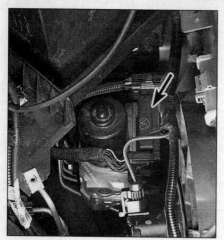

2.2 The Anti-lock Braking System (ABS) is controlled by the EHCU (arrow) mounted under the battery tray on the left front frame rail - it combines the hydraulic control unit and the microprocessor into one assembly

2.3a The front wheel ABS sensor (arrow) is mounted to the steering knuckle

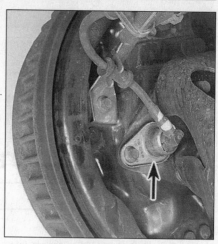

2.3b An ABS sensor (arrow) is also mounted at each rear wheel, behind the brake assembly

1   General information

The vehicles covered by this manual are equipped with hydraulically operated front and rear brake systems. All front brake systems are disc type, while the rear brakes are either disc or drum. All brakes are self adjusting. Front and rear disc brakes automatically compensate for pad wear. Rear drum brakes incorporate an adjusting mechanism which is activated any time the service brakes are applied.

Hydraulic system

The hydraulic system consists of two separate circuits. The master cylinder has separate reservoirs for the two circuits and in the event of a leak or failure in one hydraulic circuit, the other circuit will remain operative. A visual warning of circuit failure or air in the system is given by a warning light activated on the dash.

Brake pressure control valve

The brake pressure control valve is designed to provide better front-to-rear braking balance with heavy brake application. It is mounted on the driver's side frame rail near the fuel filter. It reduces the pressure allowed to the rear brakes under certain braking operations due to the fact the rear of the vehicle is lighter and does not require as much braking force.

Power brake booster

The power brake booster, utilizing engine manifold vacuum and atmospheric pressure to provide assistance to the hydraulically operated brakes, is mounted on the firewall in the engine compartment.

Parking brake

The parking brake operates the rear

brakes only, through cable actuation. It's activated by a handle assembly mounted between the front seats.

Service

After completing any operation involving disassembly of any part of the brake system, always test drive the vehicle to check for proper braking performance before resuming normal driving. When testing the brakes, perform the tests on a clean, dry flat surface. Conditions other than these can lead to inaccurate test results.

Test the brakes at various speeds with both light and heavy pedal pressure. The vehicle should stop evenly without pulling to one side or the other. Avoid locking the brakes because this slides the tires and diminishes braking efficiency and control of the vehicle.

Tires, vehicle load and front-end alignment are factors which also affect braking performance.

Warning: *On models with air suspension, electrical power to the air suspension system must be turned off before raising the vehicle (see Chapter 10). Failure to do so can result in a sudden inflation or deflation of the air springs, causing instability of the vehicle while it's off the ground.*

2   Anti-lock Brake System (ABS) - general information

Refer to illustrations 2.2, 2.3a and 2.3b

An optional four-wheel Anti-lock Brake System (ABS) maintains vehicle maneuverability, directional stability, and optimum deceleration under severe braking conditions on most road surfaces. It does so by monitoring the rotational speed of the wheels and controlling the brake line pressure to the wheels during braking. This prevents the wheels from locking up on slippery roads or during hard braking.

Hydraulic modulator/motor pack assembly

The Electronic Hydraulic Control Unit (EHCU) is located on the left front frame rail, under the battery tray **(see illustration)**. It consists of a hydraulic unit and an integrated microprocessor. The EHCU monitors the ABS system and controls the anti-lock valve solenoids. It accepts and processes information received from the brake switch and wheel speed sensors to control the hydraulic line pressure and avoid wheel lock up. It also monitors the system and stores fault codes which indicate specific problems.

Each sensor assembly consists of a variable reluctance sensor mounted adjacent to a "toothed ring" with an air gap between them **(see illustrations)**. A wheel speed sensor is mounted on the steering knuckle of each front wheel and at each rear wheel, in close proximity to a toothed ring. The air gap between the sensors and the rings is not adjustable, and the sensors themselves are not rebuildable. If a sensor malfunctions, the sensor can be replaced. If a front ring becomes damaged, the outer CV joint and axleshaft assembly must be replaced. If a rear ring becomes damaged, it can be pressed off the rear wheel bearing assembly, and a new one pressed on.

A wheel speed sensor measures wheel speed by monitoring the rotation of the toothed ring. As the teeth of the ring move through the magnetic field of the sensor, an AC voltage is generated. This signal frequency increases or decreases in proportion to the speed of the wheel. The EHCU monitors these three signals for changes in wheel speed; if it detects the sudden deceleration of a wheel, i.e. wheel lockup, the EHCU activates the ABS system.

Warning lights

The ABS system has self-diagnostic capabilities. Each time the vehicle is started, the EHCU runs a self-test. There are two

3.3 Spray the caliper and disc with brake cleaner before beginning any repairs

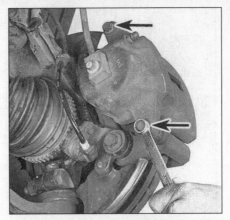

3.4a Remove the lower caliper mounting bolt, check it for thread damage and replace as necessary - the upper bolt doesn't have to be removed (unless you're removing the caliper)

3.4b Pivot the caliper up - it doesn't need to be removed completely unless the caliper or the disc is to be removed - arrow points to the anti-rattle clip

warning lights on the instrument panel, a red BRAKE light and an amber ABS light, each with their own functions. During starting, the red BRAKE warning light should come on briefly then go out. If the red BRAKE light stays on, it indicates a problem with the main braking system, such as low fluid level detected or the parking brake is still on. If the lights stays on after the parking brake is released, check the brake fluid level in the master cylinder reservoir (see Chapter 1).

The amber ABS light indicates a problem with the ABS system, not the main or basic brake system. If the light stays on steadily, it indicates that there is a problem with the ABS system, but the main system is still working. If you have a steady ABS light, drive to a dealer service department or other qualified repair shop for diagnosis and repair. If the EHCU senses that the ABS light is not working, the red BRAKE light will come on instead.

Checks

Although a special electronic tester is necessary to properly diagnose the system, the home mechanic can perform a few preliminary checks before taking the vehicle to a dealer service department (or other qualified repair shop) which is equipped with this tester:

a) Make sure the brake calipers/wheel cylinders are in good condition.
b) Check the electrical connector at the controller.
c) Check the fuses.
d) Follow the wiring harness to the speed sensors and brake light switch and make sure all connections are secure and the wiring isn't damaged.

If the above preliminary checks don't rectify the problem, the vehicle should be diagnosed by a dealer service department or other qualified repair shop.

3 Disc brake pads - replacement

Refer to illustrations 3.3, 3.4a through 3.4p and 3.4q through 3.4w

Warning: *Disc brake pads must be replaced on both wheels at the same time - never replace the pads on only one wheel. Also, brake system dust may contain asbestos, which is hazardous to your health. DO NOT blow it out with compressed air and DO NOT inhale it. An approved filtering mask should be worn when working on the brakes. DO NOT use gasoline or solvents to remove the*

dust. Use brake system cleaner only! **Note:** *This procedure applies to the front and rear disc brakes.*

1   If the vehicle is equipped with air suspension, turn off the air suspension system (see Chapter 10). **Warning:** *On models with air suspension, electrical power to the air suspension system must be turned off before raising the vehicle (see Chapter 10). Failure to do so can result in a sudden inflation or deflation of the air springs, causing instability of the vehicle while it's off the ground.*

2   Loosen the front wheel lug nuts, raise the front of the vehicle and support it securely on jackstands. Apply the parking brake. Remove the wheels.

3   Remove about two-thirds of the fluid from the master cylinder reservoir and discard it; as the pistons are pushed in for clearance to allow the pads to be removed, the fluid will be forced back into the reservoir. Position a drain pan under the brake assembly and clean the caliper and surrounding area with brake system cleaner **(see illustration)**.

4   To replace the front brake pads, follow the accompanying photos, beginning with **illustration 3.4a** for 1995 through 1998 mod-

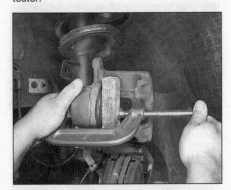

3.4c Use a large C-clamp and a block of wood to retract the piston - tighten the clamp to push the piston back, taking care not to damage the phenolic piston

3.4d Remove the outer brake pad

3.4e Remove the inner brake pad

9

3.4f Lubricate the caliper mounting bolt sleeve (arrow)

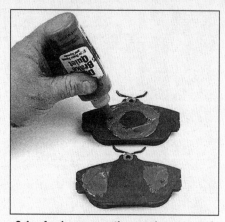

3.4g Apply some anti-squeal compound to the backing plates of the new pads - the upper one here is the inner pad that rides against the piston

3.4h Install the new outer brake pad; make sure both ends of the pad are properly seated in the mounting bracket

3.4i Install the inner pad, check for the anti-rattle clip (arrow), then lower the caliper over the pads, install the lower bolt and tighten it to the torque listed in this Chapter's Specifications

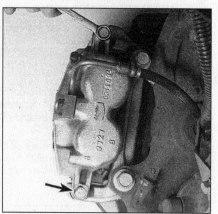
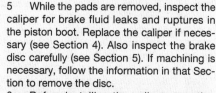

3.4j Remove the two caliper mounting bolts, check them for thread damage and replace as necessary

3.4k Pull off the caliper . . .

els or **illustration 3.4j** for 1999 and later models. To replace the rear brake pads, start with **illustration 3.4q.** Be sure to stay in order and read the caption under each illustration. Work on one brake assembly at a time so that you'll have something to refer to if necessary.

5   While the pads are removed, inspect the caliper for brake fluid leaks and ruptures in the piston boot. Replace the caliper if necessary (see Section 4). Also inspect the brake disc carefully (see Section 5). If machining is necessary, follow the information in that Section to remove the disc.

6   Before installing the caliper mounting

bolts, clean them and check them for corrosion and damage. If they're significantly corroded or damaged, replace them. Be sure to tighten the caliper mounting bolts to the torque listed in this Chapter's Specifications. **Note:** *Before bolting the caliper down over the new pads, make sure the large anti-rattle clip is in place* **(see illustration 3.4b).**

7   Install the brake pads on the opposite wheel, then install the wheels and lower the

3.4l . . . and hang it from the upper control arm with a piece of wire; DON'T allow the caliper to hang by the brake hose!

3.4m Remove the outer brake pad

3.4n Remove the inner brake pad

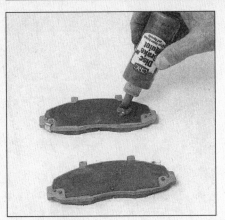

3.4o Apply some anti-squeal compound to the backing plates of the new pads

3.4p Apply high temperature grease to the caliper sliding studs (arrows)

3.4q Wash the disc and brake pads with brake cleaner to remove brake dust; DO NOT blow off brake dust with compressed air

3.4r Remove the two caliper mounting bolts (arrow indicates upper bolt), check them for thread damage and replace as necessary

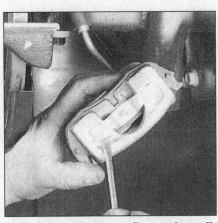

3.4s Remove the rear caliper and pry off the old outer brake pad

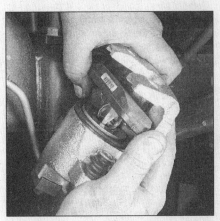

3.4t Pop the inner brake pad retainer out of the caliper piston face and remove the inner pad. Depress the piston into its bore (see illustration 3.4c)

vehicle. Tighten the lug nuts to the torque listed in the Chapter 1 Specifications.

8   Add brake fluid to the reservoir until it's full (see Chapter 1). Pump the brakes several

times to seat the pads against the disc, then check the fluid level again.

9   Check the operation of the brakes before driving the vehicle in traffic. Try to

avoid heavy brake applications until the brakes have been applied lightly several times to seat the pads.

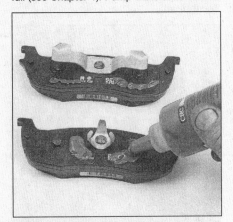

3.4u Apply some anti-squeal compound to the backing plates of the new pads

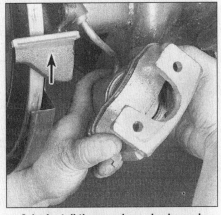

3.4v Install the new inner brake pad; make sure the retainer is seated all the way into the hole in the piston face - arrow indicates "slipper" on caliper mount; make sure both slippers are in good condition and in place

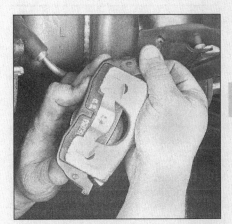

3.4w Install the new outer brake pad; make sure the retainer is fully seated on the caliper bridge. Install the caliper and tighten the caliper bolts to the torque listed in this Chapter's Specifications

9

4 Brake caliper - removal and installation

Warning: *Dust created by the brake system may contain asbestos, which is harmful to your health. Never blow it out with compressed air and don't inhale any of it. An approved filtering mask should be worn when working on the brakes. Do not, under any circumstances, use petroleum-based solvents to clean brake parts. Use brake system cleaner only!*

Note: *If replacement is indicated (usually because of fluid leakage) explore all options before beginning the job. New and factory-rebuilt calipers are available on an exchange basis, which makes this job quite easy.*

Removal

Refer to illustrations 4.1a and 4.1b

1   The removal procedure is the same as in Section 3, with the exception of disconnecting the hydraulic line from the caliper and, if you're working on a front caliper, removing the upper mounting bolt **(see illustrations)**.

Installation

2   The installation procedure is the same as in Section 3. It may be necessary to depress the piston into the bore slightly, to allow the pads to fit over the brake disc **(see illustration 3.4c)**.

3   When attaching the hydraulic line to the caliper, use new copper washers at the line-to-caliper connection (use a new washer on each side of the fitting block). Tighten the fitting bolt to the torque listed in this Chapter's Specifications.

4   Bleed the hydraulic system as described in Section 10.

5 Brake disc - inspection, removal and installation

Refer to illustrations 5.2, 5.3, 5.4a, 5.4b, 5.5a and 5.5b

Inspection

1   Loosen the wheel lug nuts, raise the vehicle and support it securely on jackstands. **Warning:** *On models with air suspension, electrical power to the air suspension system must be turned off before raising the vehicle (see Chapter 10). Failure to do so can result in a sudden inflation or deflation of the air springs, causing instability of the vehicle while it's off the ground.* Apply the parking brake and block the wheels to keep the vehicle from rolling off the jackstands. Remove the wheel and install the lug nuts to hold the disc in place. **Note:** *It may be necessary to place washers under the lug nuts to hold the disc securely against the hub.*

2   Remove the brake caliper as outlined in Section 4. You don't have to disconnect the brake hose. After removing the caliper bolts,

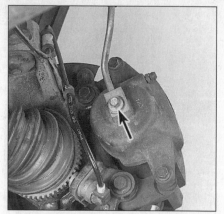

4.1a Remove the banjo bolt (arrow) to separate the hydraulic line from the caliper (front shown, rear caliper similar)

suspend the caliper out of the way with a piece of wire - DO NOT let it hang by the hose. If you're removing a front disc, remove the two bolts holding the caliper mounting bracket to the knuckle **(see illustration)**.

3   Visually inspect the disc surface for score marks and other damage. Light scratches and shallow grooves are normal and may not be detrimental to brake operation, but deep score marks - over 0.015-inch (0.38 mm) deep - require disc removal and refinishing by an automotive machine shop. Be sure to check both sides of the disc **(see illustration)**. If pulsating has been felt during application of the brakes, suspect excessive disc runout.

4   To check disc runout, mount a dial indicator with the stem resting at a point about 1/2-inch from the outer edge of the disc **(see illustration)**. Set the indicator to zero and turn the disc. The indicator reading should not exceed the specified allowable runout limit. If it does, the disc should be refinished by an automotive machine shop. **Note:** *The discs should be resurfaced, regardless of the dial indicator reading, to impart a smooth finish and ensure perfectly flat brake pad sur-*

5.3 The brake pads on this vehicle were obviously neglected, as they wore down to the rivets and cut deep grooves into the disc - wear this severe will require replacement of the disc

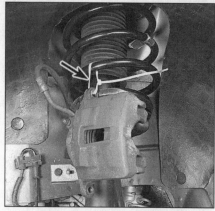

4.1b When the caliper is to be removed for other repair procedures, and the line is to remain connected, hang the caliper from the suspension with a wire hook or zip-tie (arrow) to avoid damage to the brake hose

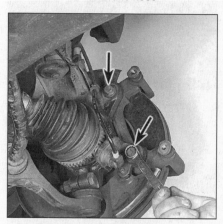

5.2 Remove the two bolts (arrows) and the mounting bracket

*faces which will eliminate pedal pulsations. At the very least, if you don't have the discs resurfaced, remove the glaze with sandpaper or emery cloth using a swirling motion **(see illustration)**.*

5.4a Check for runout with a dial indicator - mount the indicator with the plunger about 1/2-inch from the outer edge of the disc

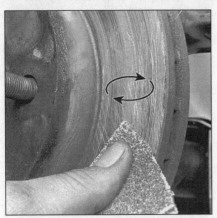

5.4b If you don't have the discs machined, at the very least be sure to remove the glaze from the disc surface with sandpaper or emery cloth (use a swirling motion as shown here)

5 The disc must not be machined to a thickness less than the minimum allowable thickness. The minimum thickness is cast into the disc (see illustration). The disc thickness can be checked with a micrometer (see illustration).

Removal

6 Remove the lug nuts that were put on to hold the disc in place, and remove the disc. If it is stuck, apply penetrating oil and hit the disc with a plastic mallet. Once you break it free, clean any rust from the hub with a wire brush.

Installation

7 Place the disc in position over the threaded studs. If you used penetrating oil to free the disc, clean it thoroughly with brake system cleaner before installation.

8 Install the caliper mounting bracket (front brake only), caliper and brake pad assembly (refer to Section 3 for the caliper

5.5a The minimum wear (or discard) thickness is cast into the disc

installation procedure, if necessary). Tighten the mounting bracket bolts and the caliper bolts to the torque listed in this Chapter's Specifications.

9 Install the wheel, then lower the vehicle to the ground. Tighten the lug nuts to the torque listed in the Chapter 1 Specifications. Depress the brake pedal a few times to bring the brake pads into contact with the disc. Bleeding of the system won't be necessary unless the brake hose was disconnected from the caliper. Check the operation of the brakes carefully before driving the vehicle in traffic.

---

6 Drum brake shoes - replacement

Warning: The brake shoes must be replaced on both rear wheels at the same time - never replace the shoes on only one wheel. Also, brake system dust may contain asbestos, which is harmful to your health. Never blow it out with compressed air and don't inhale any of it. Do not, under any circumstances, use petroleum-based solvents to clean brake

5.5b Measure the thickness of the disc at several points with a micrometer

parts. Use brake system cleaner only. Whenever the brake shoes are replaced, the return and hold-down springs should also be replaced. Due to the continuous heating/cooling cycle that the springs are subjected to, they lose their tension over a period of time and may allow the shoes to drag on the drum and wear at a much faster rate than normal.

Drum removal

Refer to illustrations 6.3 and 6.4

1 Remove about two-thirds of the brake fluid from the master cylinder reservoir.

2 Loosen the wheel lug nuts, raise the rear of the vehicle and support it on jackstands. Block the front wheels and remove the rear wheels from the vehicle. Warning: On models with air suspension, electrical power to the air suspension system must be turned off before raising the vehicle (see Chapter 10). Failure to do so can result in a sudden inflation or deflation of the air springs, causing instability of the vehicle while it's off the ground.

3 Remove the retainers holding the drum to the hub, if present (see illustration). Discard the retainer(s) - it isn't necessary to reinstall them.

4 Grasp the brake drum and pull it off. If the drum is stuck, remove the rubber plug behind the backing plate and loosen the brake adjuster star wheel (see illustration).

Inspection

Refer to illustrations 6.5a and 6.5b

5 Check the drum for cracks, score marks, deep grooves and signs of overheating of the shoe contact surface. If the drums have blue spots, indicating overheated areas, they should be replaced. Also, look for grease or brake fluid on the shoe contact surface. Grease and brake fluid can be removed from the drum with denatured alcohol or brake cleaner, but the brake shoes must be replaced if they are contaminated. Surface glazing, which is a glossy, highly polished finish, can be removed from the drum with sandpaper or emery cloth (see illustration).

6.3 If present, remove the retainers holding the brake drum to the hub

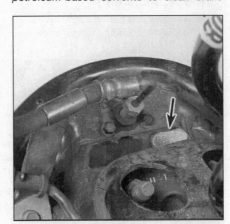

6.4 To retract the brake shoes, remove the rubber plug from the backing plate (arrow), then use a screwdriver to push the adjusting lever off the adjuster star wheel and another screwdriver to turn the star wheel

6.5a Remove glaze from the drum surfaces with sandpaper or emery cloth

6.5b Drum maximum diameter markings

6.8a Before doing any work on the rear brakes, wash the assembly with brake system cleaner

6.8b Use a pair of needle-nose pliers to remove the upper spring (arrow)

6.8c Remove the brake shoe adjusting lever (arrow)

6.8d Using a hold-down spring tool, remove the leading shoe's hold-down spring (arrow)

Note: *Professionals recommend resurfacing the drums whenever a brake job is done. Resurfacing will eliminate the possibility of out-of-round drums. If the drums are worn so much that they can't be surfaced without exceeding the maximum allowable diameter* **(see illustration)***, then new ones will be required.*

6   Inspect the wheel cylinder for fluid leak-

age as described in Chapter 1.
7   Inspect the surface of the brake shoes for cracks, contamination from grease, and wear and compare their thickness to the Chapter 1 Specifications.

Shoe replacement

Refer to illustrations 6.8a through 6.8q
8   Follow the accompanying photos **(see**

illustrations 6.8a through 6.8q) for the actual shoe replacement procedure. Be sure to stay in order and read the information in the caption under each illustration.
9   Once the new shoes are in place, turn the adjuster until the diameter of the shoe assembly is just smaller than the inner diameter of the brake drum. **Note:** *Work on only*

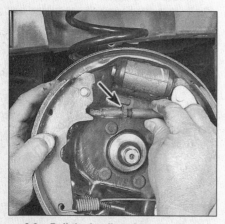

6.8e Pull the leading shoe away and remove the adjuster assembly (arrow)

6.8f Release the lower spring (arrow) and remove the leading shoe

6.8g Use the tool to remove the hold-down spring on the trailing shoe

6.8h Pull the parking brake cable end out of the end of the lever with pliers

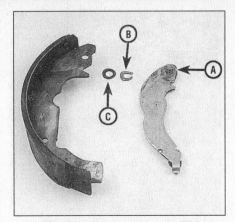

6.8i Pry the old horseshoe clip (B) from the old shoe - grease the pin (A) of the lever and assemble it to the new trailing shoe with the wave-washer (C) and a new horseshoe clip

6.8j Squeeze the new clip over the lever's pin

6.8k Clean the backing plate and apply high-temperature brake grease at the points indicated

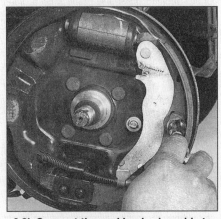

6.8l Connect the parking brake cable to the lever, then install the trailing shoe and hold-down spring

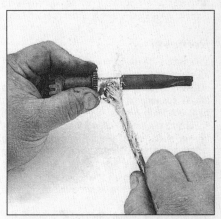

6.8m Clean the threads of the adjuster and lubricate it with high-temperature brake grease

one side of the vehicle at a time. The left and right adjusters are different, with a right-hand thread for the right side and left-hand thread for the left side. The socket ends are marked R and L, and the right-side nut has two machined grooves, while the left side nut has only one.

10 Install the brake drum. Insert a narrow screwdriver or brake adjusting tool through the adjustment hole and turn the star wheel until the brakes drag slightly as the drum is turned.

11 Turn the star wheel in the opposite direction until the drum turns freely. Keep the

adjuster lever from contacting the star wheel or it won't turn.

12 Repeat the adjustment on the opposite wheel.

13 Install the plug in the backing plate access hole.

14 Install the wheels and lower the vehicle.

6.8n Install the adjuster and the new leading shoe

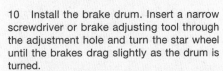

6.8o Secure the leading shoe with its hold-down spring

6.8p Install the adjuster lever (A) and hook the upper spring (B) into the hole in the top of the adjuster lever

9

Tighten the lug nuts to the torque listed in the Chapter 1 Specifications.

15   Pump the brake pedal several times and top off the master cylinder with brake fluid, if necessary (see Chapter 1). Check brake operation carefully before driving the vehicle in traffic.

7   Wheel cylinder - removal and installation

Refer to illustration 7.4

Note: *If replacement is indicated (usually because of fluid leakage or sticking brakes) explore all options before beginning the job. If the vehicle has high mileage, new wheel cylinders are highly recommended. New or factory-rebuilt wheel cylinders are available, which makes this job quite easy.*

Warning: *Never replace just one wheel cylinder. Replace both rear wheel cylinders to achieve balanced braking of the rear wheels.*

Removal

1   Raise the rear of the vehicle and support it securely on jackstands. **Warning:** *On models with air suspension, electrical power to the air suspension system must be turned off before raising the vehicle (see Chapter 10). Failure to do so can result in a sudden inflation or deflation of the air springs, causing instability of the vehicle while it's off the ground.* Block the front wheels to keep the vehicle from rolling off the jackstands.

2   Remove the brake shoe assembly (see Section 6).

3   Carefully clean the area around the wheel cylinder on both sides of the backing plate.

4   Unscrew the brake line fitting **(see illustration)**. Immediately plug the brake line to prevent fluid loss and contamination.

5   Remove the wheel cylinder retaining bolts **(see illustration 7.4)**.

6   Remove the wheel cylinder from the

6.8q Hook the lower spring into the hole in the trailing shoe from the backside (arrow), then into the hole in the leading shoe

brake backing plate and place it on a clean workbench.

Installation

7   Place the wheel cylinder in position and connect the brake line loosely.

8   Install the wheel cylinder bolts and tighten them to the torque listed in this Chapter's Specifications.

9   Tighten the brake line fitting securely.

10   Install the brake shoes (see Section 6).

11   Install the wheel and lug nuts, lower the vehicle and tighten the lug nuts to the torque listed in the Chapter 1 Specifications.

12   Bleed the brakes (see Section 10).

8   Master cylinder - removal and installation

Note: *New and rebuilt master cylinders are readily available at auto parts stores. Rebuilding of the master cylinder is not recommended.*

7.4 A flare-nut wrench should be used to disconnect the brake line - arrows point to the wheel cylinder mounting bolts

Removal

Refer to illustrations 8.2a and 8.2b

1   Detach the cable from the negative battery terminal.

2   Unplug the electrical connector from the fluid level sensor switch and the speed control pressure switch **(see illustrations)**.

3   Place rags under the line fittings and prepare caps or plastic bags to cover the ends of the lines once they're disconnected. **Caution:** *Brake fluid will damage paint. Cover all painted parts and be careful not to spill fluid during this procedure.*

4   Loosen the fittings at the ends of the brake lines where they enter the master cylinder **(see illustration 8.2b)**. To prevent rounding off the flats on the fittings, use a flare-nut wrench, which wraps around the hex.

5   Pull the brake lines away from the master cylinder and plug the ends to prevent contamination.

6   Remove the clip over the two mounting nuts, then remove the nuts **(see illustration 8.2b)** and detach the master cylinder from the power booster studs.

8.2a Unplug the fluid level sensor connector (arrow) from the reservoir

8.2b Disconnect the speed control pressure switch (A) if equipped, then disconnect the line fittings (B) and the mounting nuts (C)

7   Remove the reservoir cap, then discard any remaining fluid.

Installation

Note: *Whenever the master cylinder is removed, the complete hydraulic system must be bled (see Section 10). The time required to bleed the system can be reduced if the master cylinder is filled with fluid and bench bled before it's installed on the vehicle.*

8   Insert threaded plugs of the correct size into the brake line outlet holes and fill the reservoirs with brake fluid. The master cylinder should be supported so brake fluid won't spill during the bench bleeding procedure.

9   Loosen one plug at a time and push the piston assembly into the bore to force air from the master cylinder. To prevent air from being drawn back in, the appropriate plug must be replaced before allowing the piston to return to its original position.

10   Stroke the piston three or four times for each outlet to ensure that all air has been expelled.

11   Since high pressure isn't involved in the bench bleeding procedure, there is an alternative to the removal and replacement of the plugs with each stroke of the piston assembly. Before pushing in on the piston assembly, remove one of the plugs completely. Before releasing the piston, however, instead of replacing the plug, simply put your finger tightly over the hole to keep air from being drawn back into the master cylinder. Wait several seconds for the brake fluid to be drawn from the reservoir into the piston bore, then repeat the procedure. When you push down on the piston it'll force your finger off the hole, allowing the air inside to be expelled. When only brake fluid is being ejected from the hole, replace the plug and go on to the other port.

12   Refill the master cylinder reservoir and install the cap.

13   Install the master cylinder by reversing the removal steps, tightening the mounting nuts to the torque listed in this Chapter's Specifications, then bleed the brakes at each wheel (see Section 10).

9   Brake hoses and lines - inspection and replacement

Refer to illustration 9.2

Warning: *On ABS-equipped vehicles, do not replace any rubber brake hoses except with parts specifically designated for ABS models.*

1   About every six months, raise the vehicle and support it securely on jackstands, then check the flexible hoses that connect the steel brake lines to the front and rear brake assemblies. Look for cracks, chafing of the outer cover, leaks, blisters and other damage. The hoses are important and vulnerable parts of the brake system and the inspection should be thorough. A light and mirror will be helpful to see into restricted areas. If a hose exhibits any of the above conditions, replace it with a new one.

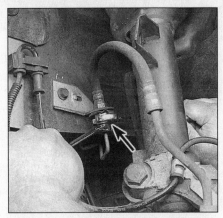

9.2 Loosen the tube nut (arrow) with a flare-nut wrench while holding the hose fitting with an open-end wrench, then remove the U-clip from the hose fitting

Front brake hose

2   Disconnect the brake line fitting from the hose fitting, being careful not to bend the frame bracket or brake line **(see illustration)**. It will probably be necessary to hold the hose fitting with an open-end wrench to prevent the bracket from bending or the line from twisting.

3   Use pliers to remove the U-clip from the female fitting at the bracket, then remove the hose from the bracket.

4   At the caliper end of the hose, remove the bolt from the fitting block, then remove the hose and the copper washers on either side of the fitting block.

5   When installing the hose, always use new copper washers on either side of the fitting block and lubricate all bolt threads with clean brake fluid.

6   With the fitting flange engaged with the caliper locating ledge, attach the hose to the caliper.

7   Without twisting the hose, install the female fitting in the hose bracket. It'll fit the bracket in only one position.

8   Install the U-clip retaining the female fitting to the frame bracket.

9   Attach the brake line to the hose fitting, holding the hose fitting with an open-end wrench, then tighten the fitting securely.

10   When the brake hose installation is complete, there shouldn't be any kinks in the hose. Make sure the hose doesn't contact any part of the suspension. Check it by turning the wheels to the extreme left and right positions. If the hose makes contact, remove the hose and correct the installation as necessary.

Rear brake hose

11   Using a flare-nut wrench on the line fitting and an open-end wrench on the hose fitting, disconnect the hose at both ends, being careful not to bend the bracket or steel lines.

12   Remove the U-clip with pliers and separate the female fittings from the brackets.

13   Unbolt the hose retaining clip and remove the hose.

14   Without twisting the hose, install the female ends in the frame bracket. It'll fit the bracket in only one position.

15   Install the U-clip retaining the female end to the bracket.

16   Attach the steel line fittings to the female fittings. Again, be careful not to bend the bracket or steel line. Tighten the fittings securely.

17   Make sure the hose installation didn't loosen the frame bracket. Tighten the bracket if necessary.

18   Fill the master cylinder reservoir and bleed the system (refer to Section 10).

Metal brake lines

19   When replacing brake lines, be sure to buy the correct replacement parts. Don't use copper or any other tubing for brake lines.

20   Prefabricated brake line, with the tube ends already flared and fittings installed, is available at auto parts stores and dealer parts departments. These lines are also sometimes bent to the proper shapes. If it is necessary to bend a line, use a tubing bender to prevent kinking the line.

21   When installing the new line, make sure it's securely supported in the brackets with plenty of clearance between moving or hot components.

22   After installation, check the master cylinder fluid level and add fluid as necessary. Bleed the brake system as outlined in Section 10 and test the brakes carefully before driving the vehicle in traffic.

10   Brake hydraulic system - bleeding

Refer to illustration 10.9

Note: *Bleeding the brakes is necessary to remove air that manages to find its way into the system when its been opened during removal and installation of a hose, line, caliper or master cylinder.*

Warning: *Wear eye protection when bleeding the brake system. If you get fluid in your eyes, rinse them immediately with water and seek medical attention.*

1   It'll probably be necessary to bleed the system at all four brakes if air has entered the system due to low fluid level, or if the brake lines have been disconnected at the master cylinder.

2   If a brake line was disconnected at only one wheel, then only that caliper or wheel cylinder must be bled.

3   If a brake line is disconnected at a fitting located between the master cylinder and any of the brakes, that part of the system served by the disconnected line must be bled.

4   Remove any residual vacuum from the power brake booster by applying the brake several times with the engine off.

5   Remove the master cylinder reservoir cover and fill the reservoir with brake fluid. Reinstall the cover. **Note:** *Check the fluid level often during the bleeding procedure and add fluid as necessary to prevent the level from falling low enough to allow air bubbles*

9

into the master cylinder.

6   Have an assistant on hand, as well as a supply of new (DOT 3) brake fluid, an empty, clear plastic container, a length of plastic, rubber or vinyl tubing to fit over the bleeder valve and a wrench to open and close the bleeder valve.

7   If the master cylinder has been re-placed, or a caliper or section of brake line has been replaced, it may be helpful to grav-ity-bleed the system as a first step. Open the two rear-wheel bleeder screws (with clear tubing on the bleeders leading into two jars) and allow gravity to flow fluid until it appears clear at each rear wheel. Close those bleed-ers and perform the same procedure at the front wheels. Then proceed with normal bleeding process as follows, with an assis-tant in the vehicle.

8   Beginning at the right rear wheel, loosen the bleeder valve slightly, then tighten it to a point where it's snug but can still be loos-ened quickly and easily.

9   Place one end of the tubing over the bleeder valve and submerge the other end in brake fluid in the container **(see illustration)**.

10   Have your assistant depress the brake pedal slowly and hold the pedal down firmly.

11   While the pedal is held down, open the bleeder valve just enough to allow fluid to flow out of the valve. Watch for air bubbles to exit the submerged end of the tube. When the fluid slows after a couple of seconds, close the valve and have your assistant release the pedal.

12   Repeat Steps 10 and 11 until no more air is seen leaving the tube, then tighten the bleeder valve and proceed to the left front wheel, the left rear wheel and the right front wheel, in that order, and perform the same procedure. Be sure to check the fluid in the master cylinder reservoir frequently.

13   Never use old brake fluid. It contains moisture which can boil, rendering the brakes useless.

14   Fill the master cylinder with fluid at the end of the operation.

15   Check the operation of the brakes. The pedal should feel firm when depressed. If necessary, repeat the procedure.

ABS-equipped models

16   Bleed the system starting with the left front wheel first, then the right front, followed by the left rear and right rear.

17   At each wheel, leave the bleeder screw open during at least 25 cycles of the brake pedal (fully down and fully up) or until air no longer appears. Do not pump the pedal fast during this operation, or air could be drawn back into the caliper.

18   Close that bleed screw and check the master cylinder fluid level before proceeding to the next wheel.

19   **Note:** *A 100-percent bleeding of the ABS valves may not occur in the above pro-cedure. If the pedal is still slightly spongy when you are done, you may have to take the vehicle to a dealer service department or other qualified repair shop, where they can*

10.9 When bleeding the brakes, a hose is connected to the bleeder valve (arrow) at the caliper and then submerged in brake fluid - air will be seen as bubbles in the container and in the tube (all air must be expelled before continuing to the next wheel)

perform a final bleeding procedure with a special scan tool. **Warning:** *Do not drive the vehicle if the brakes are not satisfactory!*

11 Power brake booster - check, removal and installation

Refer to illustrations 11.11 and 11.12

1   The power brake booster unit requires no special maintenance apart from periodic inspection of the vacuum hose and the case.

2   Dismantling of the brake booster requires special tools and is not ordinarily done by the home mechanic. If a problem develops, install a new or factory rebuilt unit.

Operating check

3   Depress the brake pedal several times with the engine off and make sure that there is no change in the pedal reserve distance.

4   Depress the pedal and start the engine. If the pedal goes down slightly, operation is normal.

Airtightness check

5   Start the engine and turn it off after one or two minutes. Depress the brake pedal sev-eral times slowly. If the pedal goes down far-ther the first time but gradually rises after the second or third depression, the booster is air-tight.

6   Depress the brake pedal while the engine is running, then stop the engine with the pedal depressed. If there is no change in the pedal reserve travel after holding the pedal for 30 seconds, the booster is airtight.

Removal

7   Disconnect the negative battery cable.

8   Refer to Chapter 11 and remove the cowl top vent panel.

9   Unbolt the master cylinder and pull it forward, without disconnecting the brake fluid lines (see Section 8). **Warning:** *Support*

11.11 Pull the pin (A) and disconnect the brake light switch (B) and booster pushrod (C) from the brake pedal arm - don't lose the spacers

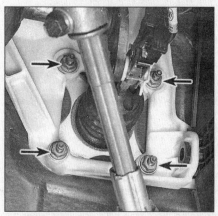

11.12 Remove the four nuts (arrows) holding the booster to the firewall

the master cylinder to prevent damage to the brake lines.

10   Disconnect the vacuum hose where it attaches to the power brake booster.

11   On 1999 and later 3.8L models, discon-nect the speed control and accelerator cables. Remove the bolts and wiring harness, then move the harness and bracket aside. Remove the speed control module bolts and move the module aside.

12   Working in the passenger compartment under the steering column, pull the hairpin clip from the brake pedal assembly and pull off the brake light switch, booster pushrod and spacers **(see illustration)**.

13   Remove the four nuts attaching the brake booster to the firewall **(see illustra-tion)**.

14   Carefully detach the booster from the firewall and lift it out of the engine compart-ment.

Installation

15   Place the booster into position on the firewall and tighten the mounting nuts to the torque listed in this Chapter's Specifications. Connect the pushrod and brake light switch to the brake pedal. Install the retaining clip in the brake pedal pin.

13.4a Wash the parking brake assembly with brake system cleaner; DO NOT blow it out with compressed air

13.4b Remove the hold-down clip from each shoe

16   Install the master cylinder and connect the vacuum hose to the booster.

17   Carefully check the operation of the brakes before driving the vehicle in traffic.

Adjustment

18   Some boosters have an adjustable pushrod. Pushrod lengths are usually set at the factory and should not require adjustment, but should be checked as a matter of precaution.

19   Some common symptoms caused by a misadjusted pushrod include dragging brakes (if the pushrod is too long) or excessive brake pedal travel accompanied by a groaning sound from the brake booster (if the pushrod is too short).

20   To check the pushrod length, unbolt the master cylinder from the booster and position it to one side. It isn't necessary to disconnect the hydraulic lines, but be careful not to bend them.

21   Block the front wheels, apply the parking brake and place the transaxle in Park or Neutral.

22   Start the engine and measure the distance that the pushrod protrudes from the master cylinder mounting surface, while exerting a force of approximately five pounds

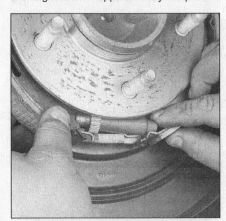

13.4c Remove the parking brake shoe adjuster, noting which end points forward . . .

to seat the pushrod in the booster. The rod measurement should be 63/64-inch on long-pushrod boosters, and 15/64-inch on short-pushrod boosters. **Note:** *The difference between a short-pushrod booster and a long-pushrod booster will be obvious. It isn't possible to accidentally adjust either one to the other's dimension. When replacing a booster, make sure the new unit is the same type as the old one. If the pushrod length isn't as specified, adjust it by holding the knurled portion of the pushrod with a pair of pliers and turning the end with a wrench.*

23   When the adjustment is complete, reinstall the master cylinder and check for proper brake operation before driving the vehicle in traffic.

12   Parking brake - adjustment

Note: *There is no parking brake adjustment procedure for models with rear drum brakes. The rear drum brakes are self-adjusting, as is the mechanism at the floor lever. The following procedure is only for models with rear disc brakes, which have a small internal drum brake assembly inside each rear disc.*

1   Raise the rear of the vehicle and support it safely on jackstands so that the rear tires are off the ground. **Warning:** *On models with air suspension, electrical power to the air suspension system must be turned off before raising the vehicle (see Chapter 10). Failure to do so can result in a sudden inflation or deflation of the air springs, causing instability of the vehicle while it's off the ground.*

2   Remove the rubber plugs from the backside of the rear disc brake backing plates on each side of the vehicle.

3   Insert a brake adjusting tool or screwdriver into the hole and turn the toothed adjuster while rotating the tire and wheel by hand to check for drag. When drag is noticed, back off the adjuster (turn it the opposite direction you had been turning it) until the brakes don't drag.

4   Perform the same adjustment for the opposite wheel, then reinstall the two rubber

plugs.

5   Lower the vehicle and test the parking brake lever operation. There should be resistance after five clicks of the ratchet mechanism.

13   Parking brake shoes (models with rear disc brakes) - replacement

Refer to illustrations 13.4a through 13.4j

1   If the vehicle is equipped with air suspension, turn off the air suspension system (see Chapter 10). **Warning:** *On models with air suspension, electrical power to the air suspension system must be turned off before raising the vehicle (see Chapter 10). Failure to do so can result in a sudden inflation or deflation of the air springs, causing instability of the vehicle while it's off the ground.*

2   Release the parking brake. Loosen the rear wheel lug nuts. Raise the vehicle and place it securely on jackstands. Remove the wheels.

3   Remove the brake calipers (see Section 4) and the brake discs (see Section 5).

4   Clean the parking brake assembly with brake system cleaner before beginning work. Follow **illustrations 13.4a through 13.4j** for the inspection and replacement of the park-

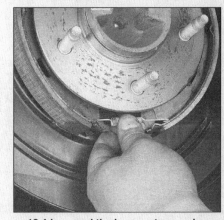

13.4d . . . and the lower return spring

9

13.4e Spread the parking brake shoes apart as shown and lift off the parking brake shoe assembly; if you're replacing the old shoes, remove the upper return spring (arrow) and transfer it to the new shoes

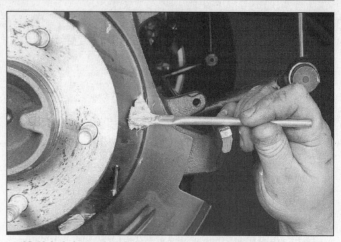

13.4f Lubricate the shoe contact points on the backing plate with high-temperature brake grease

13.4g With the upper return spring installed on the shoes as shown, spread the lower ends of the shoes apart and install the parking brake shoe assembly

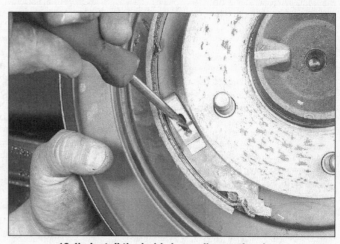

13.4h Install the hold-down clips on the shoes

ing brake shoes. Be sure to stay in order and read the caption under each illustration.

5   Clean the brake disc/parking brake drum and check it for score marks, deep grooves, hard spots (which will appear as small discolored areas) and cracks. If the disc/drum is worn, scored or out-of-round, it can be resurfaced by an automotive machine shop. If it's worn excessively, however, it will have to be replaced.

6   Repeat this procedure for the other rear brake assembly.
7   Install the brake discs (see Section 5) and the brake calipers (see Section 4).
8   Install the rear wheels, install the lug

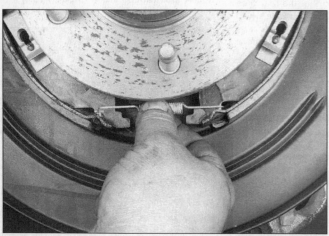

13.4i Install the lower return spring . . .

13.4j . . . and the adjuster

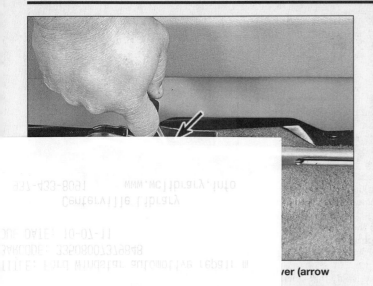

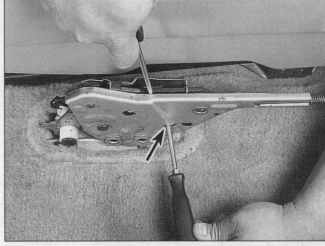

14.2b ... until a screwdriver can be inserted from the right side into the hole (arrow)

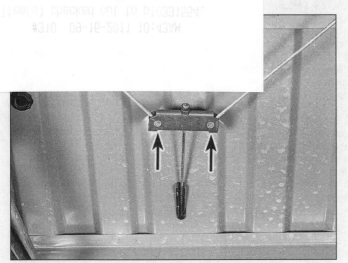

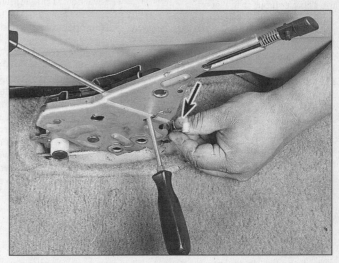

14.4 Detach the front ends of the two rear cables (arrows) from the equalizer

14.5 Remove the cable end (arrow) from the parking brake lever

nuts, lower the vehicle and tighten the wheel lug nuts to the torque listed in the Chapter 1 Specifications.

9   Refer to Section 12 for adjustment of the parking brake shoes, if necessary.

10   Reactivate the air suspension, if equipped.

14   Parking brake cables - replacement

Refer to illustrations 14.2a, 14.2b, 14.4, 14.5, 14.6 and 14.10

Front cable

1   The parking brake mechanism inside the vehicle has a spring-loaded ratchet to maintain tension on the cables. Before replacing parking brake cables, Pull up the boot over the handle by pulling the base toward the rear and up. Pull the boot up onto the handle to expose the mechanism.

2   Release the parking brake handle. To do this, push down on the tension arm with a screwdriver until you can insert an Allen wrench, screwdriver or pin through the hole in the bracket **(see illustrations)**.

3   Raise the vehicle and support it securely on jackstands. **Warning:** *On models with air suspension, electrical power to the air suspension system must be turned off before raising the vehicle (see Chapter 10). Failure to do so can result in a sudden inflation or deflation of the air springs, causing instability of the vehicle while it's off the ground.*

4   Separate the two rear cables from the equalizer under the vehicle **(see illustration)**.

5   Detach the front cable from the parking brake lever assembly **(see illustration)**.

6   Pull up the carpeting behind the parking brake assembly, enough to expose the rubber floor grommet on the cable **(see illustration)**.

7   Pull the grommet from the floor and pull the cable and equalizer up through the floor.

8   Installation is the reverse of removal.

After the cable is installed, push down on the parking brake tension arm and remove the pin. Release the tension arm and resecure the carpeting and boot.

14.6 Front parking brake cable grommet (arrow) in floor

9

Rear cable, left or right

9   Depress and secure the parking brake tension arm as in Steps 1 and 2. Raise the vehicle and support it securely on jackstands. **Warning:** *On models with air suspension, electrical power to the air suspension system must be turned off before raising the vehicle (see Chapter 10). Failure to do so can result in a sudden inflation or deflation of the air springs, causing instability of the vehicle while it's off the ground.*

10   Refer to Section 6 and remove the drums on rear-drum-brake models. Use pliers to push back the spring on the cable and pull the cable end from the lever, then squeeze the retaining clip fingers, where the cable enters the backing plate, until the cable can be pulled through the backing plate **(see illustration)**.

11   On models with rear disc brakes, unbolt the cable bracket from the rear axle and pull the cable end out of the lever.

12   On all models, slip the front ends of the rear cables from the equalizer **(see illustration 14.4)**.

13   Installation is the reverse of removal. When the new cables are in place, release the tension arm on the parking brake mechanism as in Step 2.

---

15   Brake light switch - check and replacement

Note: *The brake light switch not only controls the operation of the brake lights when the*

pedal is depressed, it also sends signals to the speed control (if equipped), the ABS microprocessor, the PCM, and the shift-lock actuator.

Check

1   The brake light switch **(see illustration 11.11)** is located on the brake pedal bracket, just to the right of the steering column mount). You'll need to remove the left side under-dash panel to get to the switch and connector (see Chapter 11).

2   With the brake pedal in the fully released position, the switch plunger is pressed into the switch housing. When the brake pedal is depressed, the plunger protrudes from the switch, which closes the circuit and sends current to the brake lights.

3   If the brake lights are inoperative, check the fuse (see Chapter 12).

4   If the fuse is okay, verify that voltage is available at one terminal of the switch.

5   If there's no voltage at the switch, find the open circuit condition between the fuse panel and the switch. If there is voltage to the switch, close the switch (depress the brake pedal) and verify that there's voltage on the other terminal of the switch.

6   If there's no voltage on the other side of the switch with the brake pedal depressed, replace the switch (see Step 7). If voltage is available, check for voltage at the brake lights. If no power is present, look for an open circuit condition between the switch and the brake lights. If power is present, check the brake light bulbs, even though it isn't likely that both of them would fail at the same time.

14.10 Squeeze the clip fingers to detach the cable casing from the backing plate

Replacement

7   Remove the left under-dash panel (see Chapter 11).

8   Lift the locking tab and unplug the electrical connector from the switch **(see illustration 11.11)**.

9   Remove the switch from the brake pedal arm (refer to Section 11, Step 11).

10   When installing the new switch, the U-shaped side goes over the pin. Push the switch up or down to lock it on the pin, then push it forward and insert the retainer clip. Make sure the bushings and brake booster rod are installed just as before.

Chapter 10
Suspension and steering systems

Contents

Specifications

Torque specifications

Ft-lbs (unless otherwise indicated)

Front suspension

Damper shaft nut	40 to 46
Lower arm-to-frame pivot bolt(s) nut(s)	98 to 113
Steering knuckle-to-balljoint pinch bolt nut	40 to 55
Strut upper mounting nuts	26 to 29
Strut-to-steering knuckle pinch bolt	85 to 97
Tension strut-to-control arm nut	85 to 97
Tension strut-to-frame nut	85 to 97
Stabilizer bracket bolts	35 to 41
Subframe mounting bolts	83 to 112

Rear suspension

Shock absorber-to-frame bolt	
1995 through 1998 models	50 to 67
1999 and later models	76
Shock absorber-to-axle bolt	50 to 67
Spindle-to-axle bolts	48 to 59
Track bar-to-frame bolt	50 to 67
Track bar-to-axle bolt	50 to 67
Trailing arm-to-frame bolt	83 to 113
Hub and bearing assembly nut	
1995 through 1997 models	
Step 1	18 to 23
Step 2	Loosen 1 to 2 turns
Step 3	18 in-lbs
1998 and later models	189 to 254

Steering system

Air bag module-to-steering wheel	106 in-lbs
Steering wheel-to-steering shaft bolt (1995 through 1998 models)	27 to 33
Steering wheel pinion shaft (1999-on models)	13 ft-lbs
Intermediate shaft-to-steering column shaft bolts	31 to 40
Intermediate shaft-to-steering gear input shaft pinch bolt	31 to 40
Steering gear mounting bolt	85 to 99
Steering gear mounting nuts	83 to 112
Tie-rod end-to-steering knuckle nut*	35 to 46

10

Torque specifications

Steering system (continued)

Tie-rod end jam nut	35 to 46
Power steering pump-to-mounting bracket bolts	30 to 40
Power steering pump bracket to engine bolts	30 to 40
Wheel lug nuts	See Chapter 1

Tighten to the minimum specified torque, then align the next castellation in the nut with the cotter pin hole by further tightening.

---

1   General information

Refer to illustrations 1.1 and 1.2

The front suspension is a MacPherson strut design **(see illustration)**. The steering knuckle is located by a lower control arm with a tension strut controlling fore-and-aft movement of the control arm. Body side roll is controlled by a stabilizer bar attached to the frame and the lower control arms.

The rear suspension on all models utilizes a beam axle with integral trailing arms **(see illustration)**. Lateral movement is controlled by a track bar between the axle and the frame, and the axle is suspended with conventional coil springs and separate shock absorbers.

An electronic, load-leveling rear suspension system is optional. This system uses air (bag) springs inflated by an on-board compressor, controlled by its own computer, and compensates for varying loads.

The rack-and-pinion steering gear is located behind the engine/transaxle assembly on the subframe and actuates the steering arms, which are integral with the steering knuckles. All models covered by this manual are equipped with power steering. The steering column is connected to the steering gear through an articulated intermediate shaft. The steering column is designed to collapse in the event of an accident.

Note: *These vehicles use a combination of standard and metric fasteners on the various suspension and steering components, so it would be a good idea to have both types of tools available when beginning work.*

Warning 1: *Do not work or place any part of your body under the vehicle when it is supported only by a jack. Jack failure could result in severe injury or death.*

Warning 2: *Whenever any of the suspension or steering fasteners are loosened or removed they must be inspected and if necessary, replaced with new ones of the same part number or of original equipment quality and design. Torque specifications must be followed for proper assembly and component retention. Never attempt to heat, straighten or weld any suspension or steering component. Instead, replace any bent or damaged part with a new one.*

Warning 3: *On models equipped with electronic rear suspension, the system's switch must be turned OFF before the vehicle is towed or raised on a jack or hoist (see Section 12). Failure to do so could result in a sudden inflation or deflation of the air springs, causing instability of the vehicle while it's off the ground.*

---

2   Stabilizer bar and bushings (front) - removal and installation

Note: *This procedure requires two floor jacks.*

Removal

Refer to illustrations 2.2 and 2.4

1   Loosen the front wheel lug nuts from both wheels. Raise the front of the vehicle and support it securely on jackstands placed behind the subframe. **Warning:** *On models*

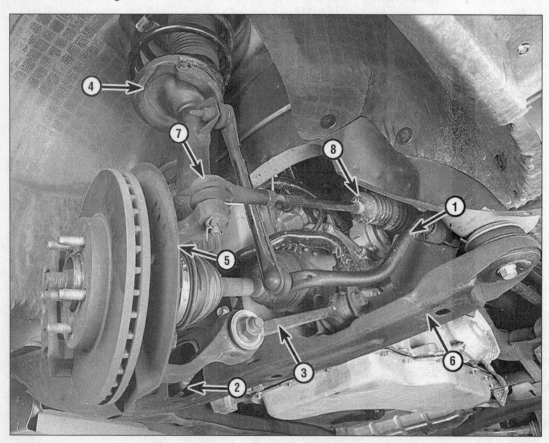

1.1 Front suspension and steering components

1   Stabilizer bar
2   Tension strut
3   Control arm
4   Strut assembly
5   Steering knuckle and hub assembly
6   Subframe
7   Tie-rod end
8   Steering gear boot

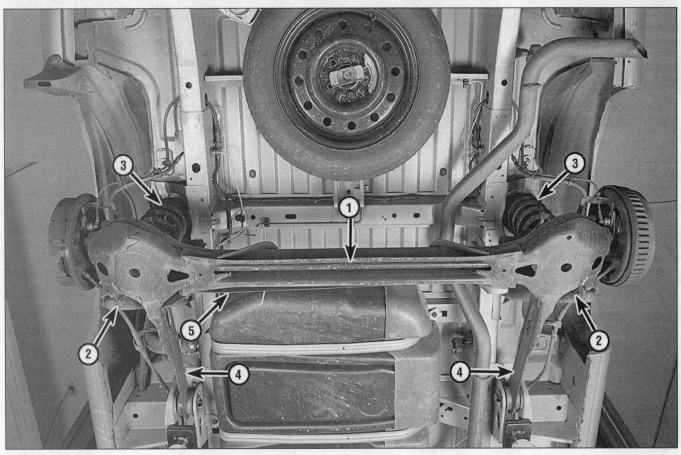

1.2 Rear suspension components (conventional)

1 Axle beam	2 Shock absorber	3 Coil spring	4 Trailing arms	5 Track bar

with air suspension, electrical power to the air suspension system must be turned off before raising the vehicle. Failure to do so can result in a sudden inflation or deflation of the air springs, causing instability of the vehicle while it's off the ground (see Section 12). Remove the front wheels.

2   Disconnect the stabilizer bar link from the strut and the stabilizer bar **(see illustration)**. If only the stabilizer bar is to be removed, or if the link isn't damaged, it isn't necessary to disconnect it from the strut bracket.

3   Remove the steering gear-to-subframe nuts (and one bolt) and push the steering gear up off of the subframe (see Section 18).

Place a jackstand under the front of the subframe.

4   Place a floor jack under each side of the rear of the subframe. Remove the two rear subframe bolts and slowly lower the jacks a little at a time until the stabilizer bar bracket bolts are accessible **(see illustration)**.

5   Remove the stabilizer bar bracket bolts

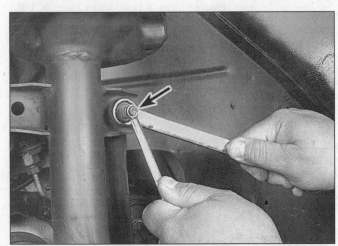

2.2 To loosen the stabilizer bar link from either the strut bracket or the bar itself, hold the stud to prevent it from turning while unscrewing the nut

2.4 The stabilizer bar bracket bolts (A) can be removed after the subframe bolts (B indicates one) have been unscrewed and the subframe lowered for clearance

10

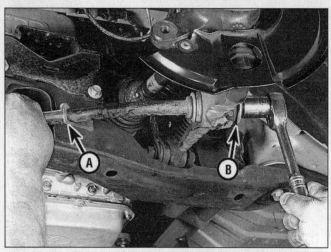

4.2 Place a wrench on the flats of the tension strut (A) to prevent it from turning while loosening the nut (B) on the control arm (1995 through 1998 models)

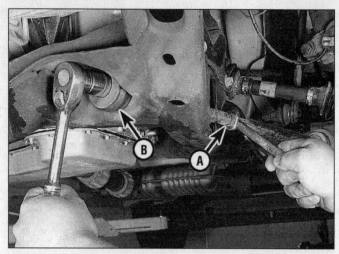

4.3 Place a wrench on the flats of the tension strut (A) to prevent it from turning while loosening the nut (B) at the front bushing in the subframe (1995 through 1998 models)

and pry the U-brackets off the bushings. The bushings may now be removed from the stabilizer bar, if desired, without removing the bar from the vehicle. If it is necessary to remove the bar, carefully guide it out from between the subframe and the body.

Installation

6   Clean the stabilizer bar in the area where the bushings ride. Position the bar on the subframe and fit the bushings over the bar in their approximate locations.

7   Push the U-brackets over the bushings and install the bolts, tightening them securely.

8   Raise the subframe until it contacts the floorpan, then install the subframe bolts, tightening them to the torque listed in this Chapter's Specifications.

9   Attach the stabilizer bar link to the strut bracket (if it was completely removed) and

5.3 The control arm is attached to the subframe by a large pivot bolt/nut (arrows) - before tightening this bolt, make sure the control arm is at a normal ride height angle (1995 through 1998 model shown; 1999 and later models similar)

the bar. Tighten the nuts securely.

10   Install the wheels and lug nuts. Lower the vehicle and tighten the nuts to the torque listed in this Chapter's Specifications.

3   Balljoints - check and replacement

The balljoints on this vehicle are not replaceable separately. The entire control arm must be replaced if the balljoints are worn out. Refer to the *Suspension and steering check* in Chapter 1 for the checking procedure. Refer to Section 5 in this Chapter for control arm removal and installation.

4   Front tension strut and bushings (1995 through 1998 models) - removal and installation

Removal

Refer to illustrations 4.2 and 4.3

1   Loosen the wheel lug nuts on the side that is to be dismantled. Raise the vehicle and support it securely on jackstands. **Warning:** *On models with air suspension, electrical power to the air suspension system must be turned off before raising the vehicle. Failure to do so can result in a sudden inflation or deflation of the air springs, causing instability of the vehicle while it's off the ground (see Section 12).* Remove the wheel.

2   Remove the tension strut-to-control arm nut and dished washer **(see illustration)**. Notice how the washer is installed - it must be reinstalled with the dished portion curling away from the bushing.

3   Remove the tension strut-to-subframe nut **(see illustration)**. Hold the tension strut from turning with a wrench on the flats.

4   Remove the control arm to subframe nut and bolt (see Section 5).

5   Pull outward on the strut/shock absorber assembly to separate the control arm from the subframe, then move the assembly to the rear just far enough to slip the control arm off the tension strut. **Caution:** *Be careful not to pull the strut assembly too far out or to the rear, as damage to the driveaxle inner CV joint may occur.*

6   Pull the tension strut out of the subframe, noting how the washers and bushings are installed. Inspect the bushings for hardness and cracking, replacing them if necessary.

7   If the tension strut bushings in the control arm are in need of replacement, refer to Section 5 for the control arm removal procedure, then take the control arm to a dealer service department or an automotive machine shop to have the new bushings installed.

Installation

8   Installation is the reverse of the removal procedure. Be sure to tighten all of the fasteners to the torque values listed in this Chapter's Specifications.

5   Control arm - removal, inspection and installation

Removal

Refer to illustrations 5.3, 5.4 and 5.5

1   Loosen the wheel lug nuts on the side to be dismantled, raise the front of the vehicle, support it securely on jackstands and remove the wheel. Warning: On models with air suspension, electrical power to the air suspension system must be turned off before raising the vehicle. Failure to do so can result in a sudden inflation or deflation of the air springs, causing instability of the vehicle while it's off the ground (see Section 12). **Note:** *The steering wheel should be unlocked during this procedure.*

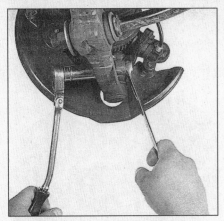

5.4 Remove the balljoint pinch bolt from the steering knuckle - a punch may be used to drive the bolt out (1995 through 1998 model shown; 1999 and later models similar)

2   On 1995 through 1998 models, remove the tension strut-to-control arm nut and dished washer **(see illustration 4.2)**.
3   On 1995 through 1998 models, remove the bolt and nut from the inner control arm pivot **(see illustration)**. On 1999 and later models, there are two inner control arm pivot bolts/nuts; remove both of them.
4   Remove and discard the balljoint pinch bolt and nut from the steering knuckle **(see illustration)**. Spread the joint slightly with a screwdriver or prybar.
5   Pry the control arm down to separate it from the steering knuckle **(see illustration)**.
6   On 1995 through 1998 models, pull the control arm off the tension strut and remove it from the vehicle. On all models, remove the arm from the vehicle.

Inspection

7   Check the control arm for distortion and the bushings for wear, damage and deterioration. Replace a damaged or bent control arm with a new one. If the inner pivot bushing

or tension strut bushings are worn, take the control arm to a dealer service department or other repair shop, as special tools are required to replace them. If the balljoint is worn or damaged, the control arm must be replaced.

Installation

8   Place the control arm balljoint stud into the steering knuckle. Note that the notch in the balljoint stud must be aligned with the hole in the knuckle before the pinch bolt is inserted. Using a new pinch bolt and nut, insert the bolt from the front of the steering knuckle and tighten the new nut to the torque listed in this Chapter's Specifications.
9   On 1995 through 1998 models, swing the control arm into position over the tension strut end.
10   Install the control arm pivot bolt(s) and nut(s) but, on 1995 through 1998 models, don't tighten the nut yet. On 1999 and later models, you can tighten the nuts now.
11   On 1995 through 1998 models, install the tension strut-to-control arm washer and nut (with the dished portion of the washer facing away from the rubber bushing). Tighten the nut to the torque listed in this Chapter's Specifications.
12   On 1995 through 1998 models, using a floor jack placed under the outer end of the control arm, raise the suspension to simulate normal ride height. Tighten the control arm pivot bolt and nut to the torque listed in this Chapter's Specifications.
13   Install the wheel and lug nuts, lower the vehicle and tighten the lug nuts to the torque listed in this Chapter's Specifications.

6   Strut and spring assembly (front) - removal and installation

Removal

Refer to illustration 6.2
1   Loosen the wheel lug nuts. Raise the vehicle and support it securely on jackstands,

then remove the front wheel. **Warning:** *On models with air suspension, electrical power to the air suspension system must be turned off before raising the vehicle. Failure to do so can result in a sudden inflation or deflation of the air springs, causing instability of the vehicle while it's off the ground (see Section 12).*
2   Loosen, but do not remove the three strut upper mounting nuts from the strut tower **(see illustration)**.
3   Remove the steering knuckle following the procedure described in Section 8.
4   Disconnect the stabilizer bar link from the bracket on the strut **(see illustration 2.2)**.
5   Remove the three upper strut mounting nuts from the shock tower while supporting the strut/spring assembly so it doesn't fall.
6   Carefully guide the strut and spring assembly out of the wheel well. If the strut/shock absorber or coil spring is to be replaced, go on to Section 7 and remove the coil spring.

Installation

7   To install the strut, place it in position with the studs extending up through the strut tower. Install the nuts and tighten them finger tight.
8   Install the steering knuckle, tightening all fasteners to the torque values listed in this Chapter's Specifications.
9   Connect the stabilizer bar link to the strut assembly, install the nut and tighten it securely.
10   Tighten the three upper strut-to-shock tower mounting nuts to the torque listed in this Chapter's Specifications.
11   Install the wheel and lower the vehicle. Tighten the lug nuts to the torque listed in this Chapter's Specifications.

7   Front strut/shock absorber or coil spring - replacement

Refer to illustrations 7.3, 7.4, 7.5 and 7.6
1   If the struts exhibit the telltale signs of

10

5.5 Pry the balljoint stud out of the steering knuckle using a long prybar (arrow) (1995 through 1998 model shown)

6.2 The strut upper mounting nuts (arrows) are on top of the strut tower in the engine compartment - DON'T remove the center nut

wear (leaking fluid, loss of dampening capability) explore all options before beginning any work. The strut/shock absorber assemblies are not serviceable and must be replaced if a problem develops. However, strut assemblies complete with springs may be available on an exchange basis, which eliminates much time and work. Whichever route you choose to take, check on the cost and availability of parts before disassembling the vehicle. **Warning:** *Disassembling a strut is dangerous - be very careful and follow all instructions or serious injury could result. Use only a high-quality spring compressor and carefully follow the manufacturer's instructions furnished with the tool.* After removing the coil spring from the strut assembly, set it aside in a safe, isolated area.

2   Remove the strut and spring assembly following the procedure described in Section 6. Mount the strut assembly in a vise with the jaws of the vise clamping onto the stabilizer bar link bracket.

3   Following the tool manufacturer's instructions, install the spring compressor (which can be obtained at most auto parts stores or equipment yards on a daily rental basis) on the spring and compress it sufficiently to relieve all pressure from the spring seat **(see illustration)**. This can be verified by wiggling the spring.

4   Loosen the damper shaft nut while using a socket wrench on the shaft hex to prevent it from turning **(see illustration)**. Remove the nut and upper concave washer.

5   Lift the spring seat assembly and upper mount off of the damper shaft **(see illustration)**. Inspect the bearing in the spring seat for smooth operation and replace it if necessary.

6   Carefully remove the compressed spring and set it in a safe place **(see illustration)**. **Warning:** *Never place your head near the end of the spring!*

7   Slide the dust boot, washer and the rubber jounce bumper off the damper shaft.

8   Assemble the strut beginning with the jounce bumper, dust boot, washer and spring, then the spring seat and bearing cap

7.3 Install a spring compressor and compress the spring until there is no pressure being exerted on the spring seat

top mount assembly. Note that the larger concave washer is installed below the mount.

9   Install the damper shaft nut and tighten it to the torque listed in this Chapter's Specifications.

10   Install the strut and spring assembly on the vehicle as outlined in Section 6.

8   Steering knuckle - removal and installation

Warning: *Dust created by the brake system contains asbestos, which is harmful to your health. Never blow it out with compressed air and don't inhale any of it. Do not, under any circumstances, use petroleum-based solvents to clean brake parts. Use brake system cleaner only.*

Removal

Refer to illustrations 8.3, 8.8, 8.9 and 8.10

1   Loosen the wheel lug nuts, raise the vehicle and support it securely on jackstands. Remove the wheel. **Warning:** *On models with air suspension, electrical power to the air suspension system must be turned off before*

7.4 After the spring has been compressed, remove the damper shaft nut - use a socket and a ratchet or breaker bar to keep the damper shaft from turning

raising the vehicle. Failure to do so can result in a sudden inflation or deflation of the air springs, causing instability of the vehicle while it's off the ground (see Section 12).

2   Remove the brake caliper and support it with a piece of wire as described in Chapter 9. Separate the brake disc from the hub.

3   Loosen, but do not remove the strut-to-steering knuckle pinch bolt **(see illustration)**.

4   Separate the tie-rod end from the steering knuckle arm as outlined in Section 17.

5   Remove and discard the balljoint pinch bolt and nut from the steering knuckle **(see illustration 5.4)**. Using a large prybar, pry the balljoint stud out of the steering knuckle **(see illustration 5.5)**.

6   Loosen but do not remove the upper strut mounting nuts.

7   Remove and discard the driveaxle hub nut. Press the driveaxle out of the hub as described in Chapter 8. Support the driveaxle with a piece of wire.

8   Mark the relationship of the strut to the steering knuckle **(see illustration)**. This will simplify reassembly. Refer to Chapter 9 and remove the ABS wheel sensor and bracket. If

7.5 Remove the spring seat/upper mount assembly

7.6 Remove the compressed spring assembly - use extreme care when handling the spring

8.3 Loosen the strut-to-knuckle pinch bolt (arrow)

8.8 Mark the relationship of the strut to the steering knuckle

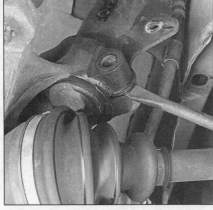

8.9 Spread the pinch joint slightly with a screwdriver

the knuckle is to be replaced, unbolt the splash shield from the back of the knuckle to transfer to the new knuckle.

9   Remove and discard the strut-to-steering knuckle pinch bolt. Apply penetrating oil to the strut-to-knuckle joint. Spread the pinch joint slightly with a screwdriver or prybar **(see illustration)**.

10   Wiggle the knuckle and hub assembly off the strut. If it is stuck, gently tap the assembly off the strut with a brass hammer, supporting it with your other hand to prevent it from falling when it comes off the strut **(see illustration)**.

Installation

11   Position the knuckle and hub assembly on the end of the strut, aligning the blade on the strut with the pinch joint in the knuckle. The previously applied alignment marks can be used to accomplish this.

12   Install a new strut-to-steering knuckle pinch bolt. Don't tighten it at this time.

13   Install the driveaxle in the hub (refer to

Chapter 8). Install a NEW driveaxle hub nut and tighten it securely.

14   Pull down on the control arm and insert the balljoint stud into the steering knuckle. Note that the notch in the balljoint stud must be aligned with the hole in the knuckle before the pinch bolt is inserted. Install the pinch bolt from the front and tighten the nut to the torque listed in this Chapter's Specifications.

15   Tighten the strut-to-knuckle pinch bolt to the torque listed in this Chapter's Specifications.

16   Tighten the upper strut mounting nuts to the torque listed in this Chapter's Specifications.

17   Attach the tie-rod end to the steering knuckle arm as described in Section 17.

18   Place the brake disc on the hub and install the caliper as outlined in Chapter 9.

19   Install the wheel and lug nuts.

20   Lower the vehicle and tighten the lug nuts to the torque listed in the Chapter 1 Specifications. Tighten the driveaxle hub nut to the torque listed in this Chapter's Specifications.

9   Front hub and wheel bearing assembly - removal and installation

Due to the special tools and expertise required to press the hub and bearing from the steering knuckle, this job should be left to a professional mechanic. However, the steering knuckle and hub may be removed and the assembly taken to a local dealer service department or other qualified repair shop. Refer to Section 8 for steering knuckle and hub removal.

10   Rear shock absorbers - removal and installation

Removal

Refer to illustration 10.3

1   If both shock absorbers are to be replaced, complete one replacement procedure before starting the other to avoid damage to the brake lines and hoses (the shock absorbers limit the downward travel of the suspension.

2   Loosen the rear wheel lug nuts, raise the vehicle and support it securely on jackstands. **Warning:** *On models with air suspension, electrical power to the air suspension system must be turned off before raising the vehicle. Failure to do so can result in a sudden inflation or deflation of the air springs, causing instability of the vehicle while it's off the ground (see Section 12).* Support the rear axle with a jack, making sure the jack pad contacts both edges of the "beam" of the axle. Remove the wheel.

3   Remove the lower shock absorber attaching bolt **(see illustration)**.

4   Lower the axle enough to ease removal of the upper mounting bolt, then remove the shock absorber **(see illustration 10.3)**.

8.10 If the steering knuckle won't easily slide off the strut, tap on it with a brass hammer - be careful not to hit the strut

10.3 Rear shock absorber mounting bolts (arrows)

10

11.7 Guide the spring/insulator into it's upper socket (arrow), while an assistant does the same on the other side

13.2 Track bar mounting bolt/nut (arrows) at the rear axle

Installation

5   Place the shock absorber in position on the chassis and install the upper bolt finger tight. Make sure the bushings and washers are installed in the correct order.

6   Raise the axle sufficiently to insert the lower bolt through the shock absorber and axle bracket.

7   Tighten the shock absorber bolts to the torque listed in this Chapter's Specifications.

8   Reinstall the wheel and lug nuts, lower the vehicle and tighten the lug nuts to the torque listed in this Chapter's Specifications.

11   Coil springs and insulators (rear) - removal and installation

Note: *If your vehicle has the optional electronic load-leveling rear suspension, refer to Section 12.*

Removal

1   Loosen the wheel lug nuts. Raise the vehicle and support it securely on jackstands placed under the frame (not the axle beam). **Warning:** *On models with air suspension, electrical power to the air suspension system must be turned off before raising the vehicle. Failure to do so can result in a sudden inflation or deflation of the air springs, causing instability of the vehicle while it's off the ground (see Section 12).*

2   Support the rear axle securely on two floor jacks. Make sure the jack pads contact both edges of the "beam" of the axle.

3   Remove the rear wheels, and raise the axle to normal ride height.

4   Remove the shock absorber lower attaching bolts (see Section 10).

5   Lower the rear axle with the jacks until the spring and insulator assemblies can be removed. Keep the springs separate so they are reinstalled in their original locations.

Installation

Refer to illustration 11.7

6   Seat the lower spring insulators firmly into their seats on the rear axle spring pads.

7   With the help of an assistant, place the springs and upper insulators in position **(see illustration)** and slowly raise the axle evenly with the jacks, guiding the springs into place.

8   Install the shock absorber bolts and tighten them to the torque listed in this Chapter's Specifications.

9   Install the wheels and lower the vehicle. Tighten the lug nuts to the torque listed in this Chapter's Specifications.

12   Rear air suspension - general information

1   The optional load-leveling electronic rear suspension uses two air (bag) springs in place of the conventional coil springs. A ride-height sensor at the rear axle determines when there is an extra load at the rear of the vehicle, and an on-board air compressor supplies air to raise the ride height to compensate for the load. An electronic module maintains normal ride height, regardless of the vehicle load, and the ride quality is smoother than would be obtainable with conventional overload coil springs. The system is operational whenever the ignition key is in the Run position, but it will continue to operate for about an hour after shutting off the ignition. This allows for maintenance of normal ride height, even while loading the vehicle with passengers or cargo.

2   Located on the right side of the instrument panel is a Warning light for the air suspension. The light comes on if a system malfunction is detected, or if the system has been manually turned Off by the service switch.

3   The compressor is located low at the left front of the vehicle, with air lines that distribute the compressed/dried air to the rear air springs. When the load is lessened and ride height must be lowered, the system air pressure is vented through a drier/vent in the compressor.

4   The air springs are attached to the axle and chassis much like the conventional coil springs, sitting in insulated pockets at the top and bottom.

5   Operation of the springs is controlled by solenoid valves on top of each air spring.

6   The ride height is determined by a sensor mounted between the chassis and the rear axle track bar.

Troubleshooting

7   If the system fails to compensate for extra loads, listen for operation of the compressor. If it doesn't operate when loads change, check the relay that supplies power to the compressor (see Chapter 12).

8   If the compressor operates but the ride height doesn't change, or if the vehicle is sagging at the rear, inspect the air springs for signs of damage or air leaks. Also inspect the solenoid valves and the air lines from the compressor back to the rear springs.

9   Further troubleshooting and diagnosis of the electronic load-leveling rear suspension should be left to a dealer service department or other qualified repair shop with the necessary scan tool to retrieve the trouble codes relating to the system.

Turning off the system

8   The service switch for the load-leveling rear suspension is located in the jack stowage area at the rear of the vehicle. **Warning:** *The system's switch must be turned OFF before the vehicle is towed or raised on a jack or hoist. Failure to do so could result in a sudden inflation or deflation of the air springs, causing instability of the vehicle while it's off the ground.*

13   Rear track bar - removal and installation

Removal

Refer to illustrations 13.2 and 13.3

1   Raise and suitably support the rear of the vehicle with jackstands, then place a floor jack under the center of the rear axle. **Warning:** *On models with electronic load-leveling rear suspension, disconnect the vehicle height sensor from the track bar and turn Off*

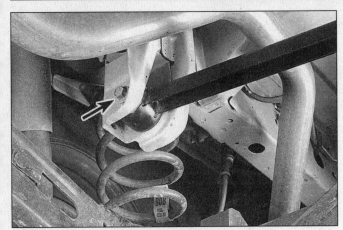

13.3 Track bar mounting bolt (arrow) at the body

14.5 Remove the cotter pin (A) and the retainer (B)

the air suspension switch (see Section 12) before beginning this procedure.

2 Raise the axle just enough to support the weight of the axle so that the rear springs aren't fully extended (but not compressed all the way to ride height position), then remove the bolt holding the left end of the track bar to the rear axle (see illustration).

3 Remove the bolt at the right end of the track bar and remove the bar (see illustration).

Installation

4 Installation is the reverse of the removal procedure. On air-suspension models, reattach the ride height sensor to the track bar, but do not turn the system switch ON until the vehicle is lowered to the ground (see Section 12). **Note:** *Make sure the track bar's "captive-nut" clips are in position at the body and axle. If they are damaged, replace them with new clips.*

---

14 Rear hub/bearing assembly and spindle - removal, inspection and installation

Warning: *On models equipped with electronic rear suspension, the system's switch*

must be turned OFF before the vehicle is towed or raised on a jack or hoist (see Section 13). Failure to do so could result in a sudden inflation or deflation of the air springs, causing instability of the vehicle while it's off the ground.

Rear hub/bearing assembly

1 1995 through 1997 models have conventional rear hubs with serviceable inner and outer bearings and races. 1998 models are equipped with sealed rear wheel bearings which do not require maintenance. They should be replaced when they become noisy or develop excessive play (see *Suspension and steering check* in Chapter 1 for more information).

2 Loosen the wheel lug nuts, raise the vehicle and support it securely on jackstands. **Warning:** *On models with air suspension, electrical power to the air suspension system must be turned off before raising the vehicle. Failure to do so can result in a sudden inflation or deflation of the air springs, causing instability of the vehicle while it's off the ground (see Section 12).* Remove the rear wheel(s).

3 Remove the rear brake drum or disc (see Chapter 9).

4 Remove the grease cap from the center of the hub.

1995 through 1997 models

Refer to illustrations 14.5, 14.7, 14.9 and 14.12

5 Remove and discard the cotter pin and remove the retainer ring (see illustration).

6 Wiggle the hub on the spindle until the outer wheel bearing and large washer come out in your hand.

7 Pull the hub from the spindle. Use a screwdriver or seal remover to remove the inner grease seal (see illustration).

8 Remove the inner bearing. Clean the inner and outer bearings and the bearing races in the hub with solvent or lacquer thinner. Inspect the bearings and races for signs or pitting, discoloration or abnormal wear. If the bearings are worn, the races must be replaced along with the bearings. Take the hubs to an automotive machine shop to have the new races pressed into the hub.

9 Use lithium-based, high-temperature wheel bearing grease to pack the bearings. Work the grease completely into the bearings, forcing it between the rollers, cone and cage from the back side (see illustration).

10 Apply a thin coat of grease to the bear-

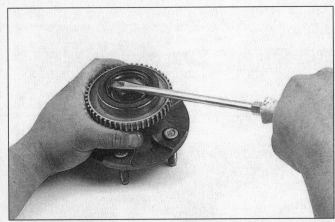

14.7 Remove the inner grease seal from the hub with a screwdriver

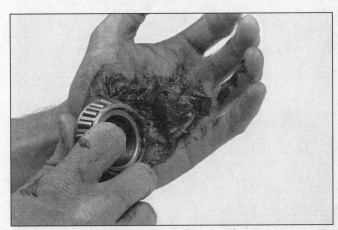

14.9 Use the palm of your hand as a cup to hold the grease while you scrape the bearing across to pack the rollers with grease

10

14.12 Tap the new grease seal evenly into the hub until it's flush

14.23 Remove the four spindle bolt nuts (arrows)

ing surfaces of the spindle.

11   Put a small quantity of grease inboard of each bearing race inside the hub. Using your finger form a dam at these points to provide extra grease availability and to keep thinned grease from flowing out of the bearing.

12   Place the grease-packed inner bearing into the rear of the hub, place a new grease seal over the bearing and tap it evenly into the hub with a hammer and large socket until it is flush with the hub **(see illustration)**.

13   Carefully place the hub onto the spindle and push the grease-packed outer bearing into position.

14   Install the large washer and hand-tighten the spindle nut. Now tighten the spindle nut to 18 to 23 ft-lbs while rotating the hub.

15   Back off the nut until loose and retighten it to the final torque listed in this Chapter's Specifications.

16   Install the cotter pin retainer so one set of notches lines up with the holes in the end of the spindle. Insert a new cotter pin and bend the ends until they are flat against the nut.

17   The remainder of the installation is the reverse of the removal procedure.

1998 models

18   Remove the hub retaining nut and remove the hub and bearing assembly from the spindle.

19   Check with a dealer service department or automotive machine shop to see if a new bearing can be pressed into the old hub assembly. Otherwise, the entire hub assembly will have to be replaced.

20   Installation of the hub and bearing assembly is basically the reverse of removal. Tighten a *new* hub retaining nut to the torque listed in this Chapter's Specifications and install a new grease cap. **Note:** *You may have to rent a torque wrench with sufficient capacity to achieve the high torque to which the nut must be tightened.*

Rear spindle

Refer to illustrations 14.23 and 14.24

21   Remove the hub/bearing assembly as described.

22   Refer to Chapter 9 and remove the speed sensor, disconnect the brake line from the wheel cylinder (on drum-brake models) and disconnect the parking brake cable.

23   Remove the four nuts holding the spindle bolts to the axle **(see illustration)**. **Note:** *On disc brake models these bolts hold the caliper adapter as well as the spindle, and on drum brake models they hold the backing plate and spindle.*

24   Serrations on the shank of these bolts hold the bolts to the backing plate or adapter - you'll have to drive the bolts out with a hammer and punch **(see illustration)**. The spindle can now be separated from the backing plate or adapter.

25   To install, reverse the disassembly process. Put the backing plate or adapter and spindle against the axle and tap in the four bolts. Tighten the nuts to the torque listed in this Chapter's Specifications.

26   Install the hub/bearing assembly as described.

27   Install the wheel and lug nuts. Lower the vehicle and tighten the lug nuts to the torque listed in this Chapter's Specifications.

14.24 Use a punch and hammer to drive the spindle bolts out

15.6 Remove the front trailing arm bolts

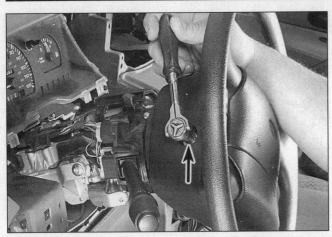

16.2 The airbag module is attached to the steering wheel by two screws (arrow indicates left-side screw location)

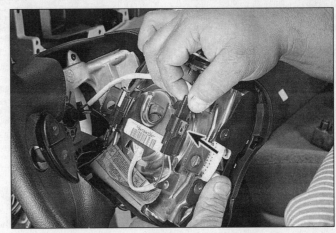

16.3 Pull the airbag module away from the steering wheel enough to disconnect the airbag connector (arrow)

15   Rear axle beam - removal and installation

Removal

Refer to illustration 15.6

1   The rear axle and its trailing arms are a single unit. If one part of the assembly is damaged, the entire assembly must be replaced. The manufacturer does not recommend repairing or straightening either the trailing arms or the axle.

2   Raise the rear of the vehicle and support it securely on jackstands placed under the frame (not under the axle beam) and support each side of the rear axle with a floor jack. **Warning:** *On models with air suspension, electrical power to the air suspension system must be turned off before raising the vehicle. Failure to do so can result in a sudden inflation or deflation of the air springs, causing instability of the vehicle while it's off the ground (see Section 12).*

3   Refer to Chapter 9 for removal of the brakes, and Section 14 for removal of the rear hubs and spindles.

4   Detach the lower ends of the rear shock absorbers from the rear axle (see Section 10). Refer to Section 11 and remove the rear coil springs.

5   Refer to Section 13 for removal of the rear track bar.

6   Remove the front trailing arm bolts and lower the axle from the vehicle **(see illustration)**. **Note:** *The nuts for these bolts are captive in the chassis.*

Installation

7   Installation is the reverse of the removal procedure. If the original axle assembly is being installed and the vehicle has a lot of mileage, inspect the front trailing arm bushings. If they show signs or wear or cracking, have new ones pressed in at an automotive machine shop before reinstalling the rear axle assembly. Tighten all fasteners to the torque listed in this Chapter's Specifications.

16   Steering wheel - removal and installation

Warning: *These models are equipped with airbags. Always turn the steering wheel to the straight ahead position, place the ignition switch in the Lock position and disable the airbag system (see Chapter 12) before working in the vicinity of the impact sensors, steering column or instrument panel to avoid the possibility of accidental deployment of the airbag, which could cause personal injury.*
Note: *A new steering wheel retaining bolt must be used. Make sure you have one before beginning the procedure.*

1995 through 1998 models

Removal

Refer to illustrations 16.2, 16.3, 16.4, 16.5a and 16.5b

1   Disconnect the negative battery cable, then the positive battery cable and wait two minutes before proceeding. **Note:** *Make sure the wheels of the vehicle are in the straight-ahead position, place the ignition switch in*

the Lock position and remove the key.

2   Using a ratchet and socket, remove the two screws that secure the airbag module to the steering wheel **(see illustration)**. **Note:** *There are plastic covers over the screw locations. Pry the covers out with a small screwdriver.*

3   Lift the airbag module carefully away from the steering wheel and disconnect the yellow and gray airbag electrical connectors **(see illustration)**. On models so equipped, disconnect the electrical connector for the cruise control. Remove the airbag module. **Warning:** *When carrying the airbag module, keep the trim side of it facing away from your body, and when you set it down, make sure the trim side is facing up.*

4   Use a ratchet and deep socket to remove the steering wheel bolt **(see illustration)**.

5   Make two alignment marks to indicate the exact orientation of the steering wheel to the shaft, and use a steering wheel puller (available at most auto parts stores) to remove the wheel **(see illustrations)**.

6   Route the wires through the wheel as it is removed.

16.4 Remove the steering wheel mounting bolt

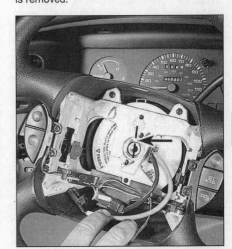

16.5a Make alignment marks (arrow) on the shaft and wheel before removing the wheel

16.5b Use a steering wheel puller to remove the wheel

16.7 Make sure the airbag coil assembly is centered (arrow) - the arrow on the coil must point to the notch at the top

16.10 Remove the trim panel from the steering wheel with a small screwdriver, and then carefully disconnect the two electrical connectors

Installation

Refer to illustration 16.7

7   Installation is the reverse of removal. Before installing the steering wheel, make sure the airbag coil assembly is centered **(see illustration)**. Connect the airbag connectors to the back of the airbag module just as it was before steering wheel removal. Use a NEW steering wheel bolt and tighten it to the torque listed in this Chapter's Specifications.

8   Refer to Chapter 12 for the procedure to enable the airbag system.

1999 and later models

Removal

Refer to illustrations 16.10 and 16.11

9   Disconnect the negative battery cable, then the positive battery cable and wait two minutes before proceeding.

10   From the straight-ahead position, turn the steering wheel counterclockwise a 1/4-turn so that the access panel is facing to the right (or, if you're left-handed, turn the wheel to the left instead). Using a thin-bladed screwdriver, carefully remove the trim panel from the steering wheel **(see illustration)**.and then unplug the two electrical connectors.

11   Remove the steering wheel-to-steering shaft retaining bolt. Unlike the steering wheel bolt on a conventional steering wheel design, which can be removed with a socket wrench after removing the horn pad, the steering wheel bolt on 1999 and later models cannot be accessed from the front side of the steering wheel, which has no removable horn pad. Instead, loosen the bolt by turning the pinion shaft **(see illustration)** located in the same recess as the electrical connectors. **Note:** *You may have to turn the pinion shaft as much as 120 to 130 turns to remove the steering wheel bolt from the steering shaft.*

12   Turn the steering wheel back to the straight-ahead position and then carefully pull it from the steering shaft, being extremely careful not to damage the clockspring harness and connector, which must be threaded through the hole in the steering wheel as the wheel is removed. **Warning:** *When carrying*

the steering wheel and airbag module assembly, keep the air bag trim side (the side that faces toward you when the steering wheel is installed) away from your body. When setting the steering wheel down, make sure that the trim side is facing up.

13   No further disassembly of the steering wheel/airbag assembly is recommended. Unlike earlier steering wheels, on which the airbag module is a separate, removable unit, the steering wheel/airbag module is an integral, one-piece assembly on these models. If the airbag has deployed, you must replace the steering wheel and airbag as a single assembly.

Installation

14   Installation is the reverse of removal. Make sure that the wheels are still in the straight-ahead position. Center the steering wheel, install it on the steering shaft and then pull the clockspring and airbag module connectors through the small opening in the back of the steering wheel. Then push the steering wheel down onto the steering shaft until it seats.

15   Turn the pinion shaft 120 to 130 times and tighten it securely.

16   Plug in the electrical connectors and install the access panel.

17   Tie-rod ends - removal and installation

Removal

Refer to illustrations 17.2a, 17.2b and 17.4

1   Loosen the wheel lug nuts. Raise the front of the vehicle, support it securely, block the rear wheels and set the parking brake. **Warning:** *On models with air suspension, electrical power to the air suspension system must be turned off before raising the vehicle. Failure to do so can result in a sudden inflation or deflation of the air springs, causing instability of the vehicle while it's off the ground (see Section 12).* Remove the front wheel.

16.11 To unscrew the steering wheel retaining bolt, unscrew the pinion shaft (lower arrow) until the steering wheel is free from the steering shaft (this may take over 100 turns); when removing the steering wheel, carefully thread the clockspring connector (upper arrow) through the small opening in the back of the steering wheel

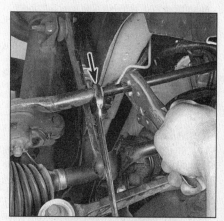

17.2a Hold the tie-rod with a pair of locking pliers (clamped onto the serrated portion of the tie-rod) and loosen the jam nut (arrow)

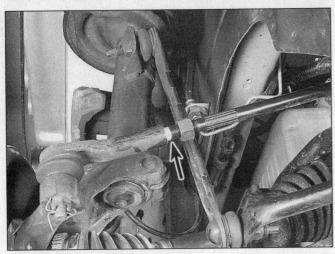

17.2b Mark the position of the tie-rod end on the tie-rod (arrow)

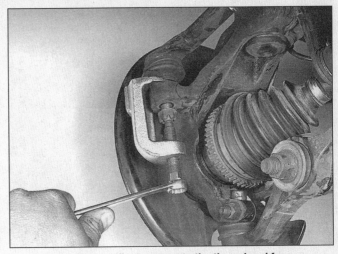

17.4 Use a puller to separate the tie-rod end from the steering knuckle

2   Hold the tie-rod with a pair of locking pliers and loosen the jam nut enough to mark the position of the tie-rod end in relation to the threads **(see illustrations)**.
3   Remove the cotter pin and loosen the nut on the tie-rod end stud.
4   Disconnect the tie-rod end from the steering knuckle arm with a puller **(see illustration)**. Remove the nut and separate the tie-rod end from the steering knuckle.
5   Unscrew the tie-rod end from the tie-rod.

Installation

6   Thread the tie-rod end on to the marked position and insert the tie-rod stud into the steering knuckle arm. Tighten the jam nut securely.
7   Install the nut on the stud and tighten it to the torque listed in this Chapter's Specifications (see note in Specifications table at the beginning of this Chapter). Install a new cotter pin.
8   Install the wheel and lug nuts. Lower the vehicle and tighten the lug nuts to the torque listed in the Chapter 1 Specifications.

9   Have the alignment checked by a dealer service department or an alignment shop.

18 Steering gear boots - replacement

Refer to illustrations 18.3a and 18.3b
1   Loosen the lug nuts, raise the vehicle and support it securely on jackstands. **Warning:** *On models with air suspension, electrical power to the air suspension system must be turned off before raising the vehicle. Failure to do so can result in a sudden inflation or deflation of the air springs, causing instability of the vehicle while it's off the ground (see Section 12).* Remove the wheel.
2   Referring to Section 17, remove the tie-rod end.
3   Remove the steering gear boot clamps and slide the boot off the tie-rod **(see illustrations)**.
4   Before installing the new boot, wrap the threads and serrations on the end of the steering rod with a layer of tape so the small

end of the new boot isn't damaged.
5   Slide the new boot into position on the steering gear until it seats in the groove in the steering rod and install new clamps.
6   Remove the tape and install the tie-rod end (see Section 17).
7   Install the wheel and lug nuts. Lower the vehicle and tighten the lug nuts to the torque listed in this Chapter's Specifications.

19 Steering gear - removal and installation

Removal

Refer to illustrations 19.4 and 19.6
Warning: *Don't allow the steering shaft to turn with the steering wheel removed. If the shaft turns, the airbag coil assembly (the mechanism which protects the airbag wiring when the steering wheel is turned) will become uncentered, which will cause the airbag harness to break when the vehicle is returned to service. To prevent the shaft from turning, turn the ignition key to the Lock posi-*

18.3a The outer boot clamp (arrow) can be squeezed with a pair of pliers and slid off the boot

18.3b The inner boot clamp (arrow) must be cut off

10

19.4 Pull back the boot (A) and remove the steering gear pinch bolt (B)

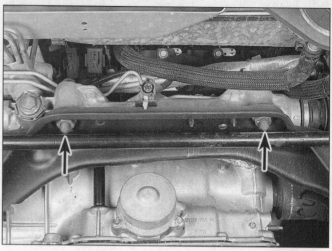

19.6 Remove the steering gear nuts and bolts (arrows)

tion before beginning work or run the seat belt through the steering wheel and clip the seat belt into place.

Note: *The following is a difficult procedure for the home mechanic. It is best performed with a vehicle hoist and specialized (tall and adjustable) jackstands or jacks, because the subframe must be lowered four inches to remove the steering gear.*

1   Disconnect the cable from the negative battery terminal.

2   Raise the vehicle and support it securely on jackstands. **Warning:** *On models with air suspension, electrical power to the air suspension system must be turned off before raising the vehicle. Failure to do so can result in a sudden inflation or deflation of the air springs, causing instability of the vehicle while it's off the ground (see Section 12).*

3   Refer to Section 17 to remove the tie-rod ends. Refer to Section 2 to remove the front stabilizer bar.

4   Raise the rubber boot over the steering column intermediate shaft-to-steering gear joint and remove the pinch bolt **(see illustra-**

tion).

5   Refer to Chapter 4 and remove the exhaust system flex pipe, being careful not to bend it.

6   Remove the bolts and nuts retaining the steering gear to the subframe **(see illustration).**

7   Support each side of the rear of the subframe with a floor jack and remove the rear subframe mounting bolts.

8   Lower the rear of the subframe four inches and remove the heat shield from the subframe. Disconnect the pressure and return lines from the steering gear and allow the fluid to drain into a container.

9   Rotate the steering gear to dislocate the mounting bolts from the holes, then carefully guide the steering gear out the left side wheel well.

Installation

10   Installation is the reverse of the removal procedure. Be sure to tighten all fasteners to the torque listed in this Chapter's Specifications. Bleed the power steering system follow-

ing the procedure outlined in Section 21.
Note: *Use new Teflon O-rings on the fluid line fittings and a new pinch bolt when connecting the steering gear to the intermediate shaft.*

20 Power steering pump - removal and installation

Removal

Refer to illustrations 20.5a, 20.5b, 20.7 and 20.8

1   Disconnect the cable from the negative terminal of the battery.

2   Place a drain pan under the power steering pump. Remove the drivebelt (see Chapter 1).

3   Refer to Chapter 5 and remove the alternator.

4   On 3.0L models, refer to Chapter 2 and remove the drivebelt tensioner.

5   Remove the pressure and return hoses from the back side of the pump and allow the fluid to drain **(see illustrations)**. Plug the

20.5a Disconnect the return hose (arrow) and allow the fluid to drain into a container

20.5b Use a flare-nut wrench to disconnect the pressure hose (arrow) from the pump

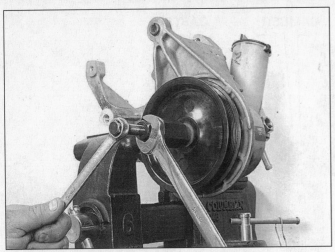

20.7 Use a special tool to remove the power steering pump pulley

20.8 Remove the mounting bolts (arrows)

hoses to prevent contaminants from entering.

6   Remove the power steering bracket bolts from the engine and remove the pump and bracket assembly, taking care not to spill fluid on the painted surfaces.

7   With the pump and bracket mounted in a vise, use a special power steering pump pulley remover to remove the pulley from the pump **(see illustration)**. **Note:** *The pulley removal and installer are usually sold at auto parts stores as a set.*

8   Remove the pump mounting bolts **(see illustration)** and lift the pump from the bracket.

Installation

9   Position the pump in the mounting bracket and install the bolts. Tighten the bolts securely.

10   Press the pulley onto the pump shaft using a special pulley-installer tool. Push the pulley onto the shaft until the front of the hub is flush with the end of the shaft, but no further.

11   Install the pump and bracket assembly, tightening the fasteners securely.

12   Connect the hoses to the pump. Tighten the fittings securely.

13   Install the drivebelt.

14   Fill the power steering reservoir with the recommended fluid and bleed the system following the procedure described in Section 21.

21 Power steering system - bleeding

1   Following any operation in which the power steering fluid lines have been disconnected, the power steering system must be bled to remove all air and obtain proper steering performance.

2   With the front wheels in the straight ahead position, check the power steering fluid level and, if low, add fluid until it reaches the Cold mark on the dipstick.

3   Start the engine and allow it to run at

fast idle. Recheck the fluid level and add more if necessary to reach the Cold mark on the dipstick.

4   Bleed the system by turning the wheels from side-to-side, without hitting the stops. This will work the air out of the system. Keep the reservoir full of fluid as this is done.

5   When the air is worked out of the system, return the wheels to the straight ahead position and leave the vehicle running for several more minutes before shutting it off.

6   Road test the vehicle to be sure the steering system is functioning normally and noise free.

7   Recheck the fluid level to be sure it is up to the Hot mark on the dipstick while the engine is at normal operating temperature. Add fluid if necessary (see Chapter 1).

22 Wheels and tires - general information

Refer to illustration 22.1

All vehicles covered by this manual are equipped with metric-size fiberglass or steel belted radial tires **(see illustration)**. The use of other size or type tires may affect the ride

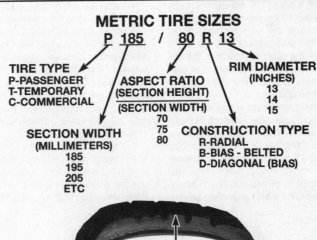

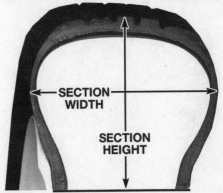

22.1 Metric tire size code

10

and handling of the vehicle. Don't mix different types of tires, such as radials and bias belted, on the same vehicle, since handling may be seriously affected. Tires should be replaced in pairs on the same axle, but if only one tire is being replaced, be sure it's the same size, structure and tread design as the other.

Because tire pressure affects handling and wear, the tire pressures should be checked at least once a month or before any extended trips (see Chapter 1).

Wheels must be replaced if they're bent, dented, leak air, have elongated bolt holes, are heavily rusted, out of vertical symmetry or if the lug nuts won't stay tight. Wheel repairs by welding or peening aren't recommended.

Tire and wheel balance is important to the overall handling, braking and performance of the vehicle. Unbalanced wheels can adversely affect handling and ride characteristics as well as tire life. Whenever a tire is installed on a wheel, the tire and wheel should be balanced by a shop with the proper equipment.

23　Wheel alignment - general information

Refer to illustration 23.1

A wheel alignment refers to the adjustments made to the wheels so they're in proper angular relationship to the suspension and the ground. Wheels that are out of proper alignment not only affect steering control, but also increase tire wear. The adjustment most commonly required on the front end is the toe-in adjustment, but camber and caster adjustments are also possible with a slight modification of the alignment plate at the top of the strut tower **(see illustration)**. The rear wheel alignment angles are preset at the factory and can't be adjusted.

Getting the proper wheel alignment is a very exacting process, one in which complicated and expensive machines are necessary to perform the job properly. Because of this, you should have a technician with the proper equipment perform these tasks. We will, however, attempt to give you a basic idea of what's involved with wheel alignment so you

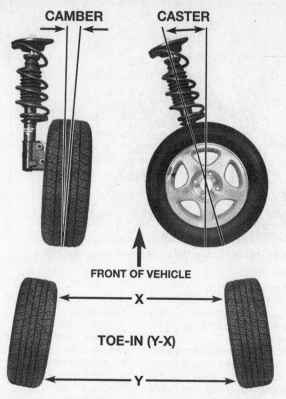

23.1　Front end alignment details

can better understand the process and deal intelligently with the shop that does the work.

Toe-in is the turning in of the wheels. The purpose of a toe specification is to ensure parallel rolling of the wheels. In a vehicle with zero toe-in, the distance between the front edges of the wheels will be the same as the distance between the rear edges of the wheels. The actual amount of toe-in is normally only a fraction of an inch. Incorrect toe-in will cause the tires to wear improperly by making them scrub against the road surface. Toe adjustment is controlled by the tie-rod end position on the tie-rod.

Camber is the tilting of the wheels from the vertical when viewed from the front or rear of the vehicle. When the wheels tilt out at the top, the camber is said to be positive (+). When the wheels tilt in at the top the camber

is negative (-). The amount of tilt is measured in degrees from the vertical and this measurement is called the camber angle. This angle affects the amount of tire tread which contacts the road and compensates for changes in the suspension geometry when the vehicle is cornering or traveling over an undulating surface.

Caster and camber adjustments on the front suspension can only be done by an alignment shop, while the vehicle is on the alignment machine. Some corrections can be made by drilling out the spot-welds at the top of the strut plate, realigning, and riveting the plate in the new position. If the caster/camber can't be corrected within this range of adjustment, the suspension components must be checked for damage.

Chapter 11 Body

Contents

1 General information

These models feature a "unibody" construction, using a floor pan with left and right frame side rails which support the body components, front and rear suspension systems and other mechanical components. Certain components are particularly vulnerable to accident damage and can be unbolted and repaired or replaced. Among these parts are the body moldings, bumpers, hood and trunk lids and all glass.

Only general body maintenance practices and body panel repair procedures within the scope of the do-it-yourselfer are included in this Chapter.

2 Body - maintenance

1 The condition of your vehicle's body is very important, because the resale value depends a great deal on it. It's much more difficult to repair a neglected or damaged body than it is to repair mechanical components. The hidden areas of the body, such as the wheel wells, the frame and the engine compartment, are equally important, al-

though they don't require as frequent attention as the rest of the body.

2 Once a year, or every 12,000 miles, it's a good idea to have the underside of the body steam cleaned. All traces of dirt and oil will be removed and the area can then be inspected carefully for rust, damaged brake lines, frayed electrical wires, damaged cables and other problems. The front suspension components should be greased after completion of this job.

3 At the same time, clean the engine and the engine compartment with a steam cleaner or water soluble degreaser.

4 The wheel wells should be given close

11

attention, since undercoating can peel away and stones and dirt thrown up by the tires can cause the paint to chip and flake, allowing rust to set in. If rust is found, clean down to the bare metal and apply an anti-rust paint.

5   The body should be washed about once a week. Wet the vehicle thoroughly to soften the dirt, then wash it down with a soft sponge and plenty of clean soapy water. If the surplus dirt is not washed off very carefully, it can wear down the paint.

6   Spots of tar or asphalt thrown up from the road should be removed with a cloth soaked in solvent.

7   Once every six months, wax the body and chrome trim. If a chrome cleaner is used to remove rust from any of the vehicle's plated parts, remember that the cleaner also removes part of the chrome, so use it sparingly.

3   Vinyl trim - maintenance

Don't clean vinyl trim with detergents, caustic soap or petroleum-based cleaners. Plain soap and water works just fine, with a soft brush to clean dirt that may be ingrained. Wash the vinyl as frequently as the rest of the vehicle.

After cleaning, application of a high quality rubber and vinyl protectant will help prevent oxidation and cracks. The protectant can also be applied to weatherstripping, vacuum lines and rubber hoses (which often fail as a result of chemical degradation) and to the tires.

4   Upholstery and carpets - maintenance

1   Every three months remove the carpets or mats and clean the interior of the vehicle (more frequently if necessary). Vacuum the upholstery and carpets to remove loose dirt and dust.

2   Leather upholstery requires special care. Stains should be removed with warm water and a very mild soap solution. Use a clean, damp cloth to remove the soap, then wipe again with a dry cloth. Never use alcohol, gasoline, nail polish remover or thinner to clean leather upholstery.

3   After cleaning, regularly treat leather upholstery with a leather wax. Never use car wax on leather upholstery.

4   In areas where the interior of the vehicle is subject to bright sunlight, cover leather seats with a sheet if the vehicle is to be left out for any length of time.

5   Body repair - minor damage

Plastic body panels

The following repair procedures are for minor scratches and gouges. Repair of more

serious damage should be left to a dealer service department or qualified auto body shop. Below is a list of the equipment and materials necessary to perform the following repair procedures on plastic body panels. Although a specific brand of material may be mentioned, it should be noted that equivalent products from other manufacturers may be used instead.

> *Wax, grease and silicone removing solvent*
> *Cloth-backed body tape*
> *Sanding discs*
> *Drill motor with three-inch disc holder*
> *Hand sanding block*
> *Rubber squeegees*
> *Sandpaper*
> *Non-porous mixing palette*
> *Wood paddle or putty knife*
> *Curved tooth body file*
> *Flexible parts repair material*

Flexible panels (front and rear bumper fascia)

1   Remove the damaged panel, if necessary or desirable. In most cases, repairs can be carried out with the panel installed.

2   Clean the area(s) to be repaired with a wax, grease and silicone removing solvent applied with a water-dampened cloth.

3   If the damage is structural, that is, if it extends through the panel, clean the backside of the panel area to be repaired as well. Wipe dry.

4   Sand the rear surface about 1-1/2 inches beyond the break.

5   Cut two pieces of fiberglass cloth large enough to overlap the break by about 1-1/2 inches. Cut only to the required length.

6   Mix the adhesive from the repair kit according to the instructions included with the kit, and apply a layer of the mixture approximately 1/8-inch thick on the backside of the panel. Overlap the break by at least 1-1/2 inches.

7   Apply one piece of fiberglass cloth to the adhesive and cover the cloth with additional adhesive. Apply a second piece of fiberglass cloth to the adhesive and immediately cover the cloth with additional adhesive insufficient quantity to fill the weave.

8   Allow the repair to cure for 20 to 30 minutes at 60-degrees to 80-degrees F.

9   If necessary, trim the excess repair material at the edge.

10   Remove all of the paint film over and around the area(s) to be repaired. The repair material should not overlap the painted surface.

11   With a drill motor and a sanding disc (or a rotary file), cut a "V" along the break line approximately 1/2-inch wide. Remove all dust and loose particles from the repair area.

12   Mix and apply the repair material. Apply a light coat first over the damaged area; then continue applying material until it reaches a level slightly higher than the surrounding finish.

13   Cure the mixture for 20 to 30 minutes at 60-degrees to 80-degrees F.

14   Roughly establish the contour of the area being repaired with a body file. If low areas or pits remain, mix and apply additional adhesive.

15   Block sand the damaged area with sandpaper to establish the actual contour of the surrounding surface.

16   If desired, the repaired area can be temporarily protected with several light coats of primer. Because of the special paints and techniques required for flexible body panels, it is recommended that the vehicle be taken to a paint shop for completion of the body repair.

Steel body panels

See photo sequence

Repair of minor scratches

17   If the scratch is superficial and does not penetrate to the metal of the body, repair is very simple. Lightly rub the scratched area with a fine rubbing compound to remove loose paint and built-up wax. Rinse the area with clean water.

18   Apply touch-up paint to the scratch, using a small brush. Continue to apply thin layers of paint until the surface of the paint in the scratch is level with the surrounding paint. Allow the new paint at least two weeks to harden, then blend it into the surrounding paint by rubbing with a very fine rubbing compound. Finally, apply a coat of wax to the scratch area.

19   If the scratch has penetrated the paint and exposed the metal of the body, causing the metal to rust, a different repair technique is required. Remove all loose rust from the bottom of the scratch with a pocket knife, then apply rust inhibiting paint to prevent the formation of rust in the future. Using a rubber or nylon applicator, coat the scratched area with glaze-type filler. If required, the filler can be mixed with thinner to provide a very thin paste, which is ideal for filling narrow scratches. Before the glaze filler in the scratch hardens, wrap a piece of smooth cotton cloth around the tip of a finger. Dip the cloth in thinner and then quickly wipe it along the surface of the scratch. This will ensure that the surface of the filler is slightly hollow. The scratch can now be painted over as described earlier in this section.

Repair of dents

20   When repairing dents, the first job is to pull the dent out until the affected area is as close as possible to its original shape. There is no point in trying to restore the original shape completely as the metal in the damaged area will have stretched on impact and cannot be restored to its original contours. It is better to bring the level of the dent up to a point which is about 1/8-inch below the level of the surrounding metal. In cases where the dent is very shallow, it is not worth trying to pull it out at all.

21   If the back side of the dent is accessible, it can be hammered out gently from behind using a soft-face hammer. While doing this,

hold a block of wood firmly against the opposite side of the metal to absorb the hammer blows and prevent the metal from being stretched.

22 If the dent is in a section of the body which has double layers, or some other factor makes it inaccessible from behind, a different technique is required. Drill several small holes through the metal inside the damaged area, particularly in the deeper sections. Screw long, self-tapping screws into the holes just enough for them to get a good grip in the metal. Now the dent can be pulled out by pulling on the protruding heads of the screws with locking pliers.

23 The next stage of repair is the removal of paint from the damaged area and from an inch or so of the surrounding metal. This is done with a wire brush or sanding disk in a drill motor, although it can be done just as effectively by hand with sandpaper. To complete the preparation for filling, score the surface of the bare metal with a screwdriver or the tang of a file, or drill small holes in the affected area. This will provide a good grip for the filler material. To complete the repair, see the subsection on filling and painting later in this Section.

Repair of rust holes or gashes

24 Remove all paint from the affected area and from an inch or so of the surrounding metal using a sanding disk or wire brush mounted in a drill motor. If these are not available, a few sheets of sandpaper will do the job just as effectively.

25 With the paint removed, you will be able to determine the severity of the corrosion and decide whether to replace the whole panel, if possible, or repair the affected area. New body panels are not as expensive as most people think and it is often quicker to install a new panel than to repair large areas of rust.

26 Remove all trim pieces from the affected area except those which will act as a guide to the original shape of the damaged body, such as headlight shells, etc. Using metal snips or a hacksaw blade, remove all loose metal and any other metal that is badly affected by rust. Hammer the edges of the hole in to create a slight depression for the filler material.

27 Wire brush the affected area to remove the powdery rust from the surface of the metal. If the back of the rusted area is accessible, treat it with rust inhibiting paint.

28 Before filling is done, block the hole in some way. This can be done with sheet metal riveted or screwed into place, or by stuffing the hole with wire mesh.

29 Once the hole is blocked off, the affected area can be filled and painted. See the following subsection on filling and painting.

Filling and painting

30 Many types of body fillers are available, but generally speaking, body repair kits which contain filler paste and a tube of resin hardener are best for this type of repair work.

A wide, flexible plastic or nylon applicator will be necessary for imparting a smooth and contoured finish to the surface of the filler material. Mix up a small amount of filler on a clean piece of wood or cardboard (use the hardener sparingly). Follow the manufacturer's instructions on the package, otherwise the filler will set incorrectly.

31 Using the applicator, apply the filler paste to the prepared area. Draw the applicator across the surface of the filler to achieve the desired contour and to level the filler surface. As soon as a contour that approximates the original one is achieved, stop working the paste. If you continue, the paste will begin to stick to the applicator. Continue to add thin layers of paste at 20-minute intervals until the level of the filler is just above the surrounding metal.

32 Once the filler has hardened, the excess can be removed with a body file. From then on, progressively finer grades of sandpaper should be used, starting with a 180-grit paper and finishing with 600-grit wet-or-dry paper. Always wrap the sandpaper around a flat rubber or wooden block, otherwise the surface of the filler will not be completely flat. During the sanding of the filler surface, the wet-or-dry paper should be periodically rinsed in water. This will ensure that a very smooth finish is produced in the final stage.

33 At this point, the repair area should be surrounded by a ring of bare metal, which in turn should be encircled by the finely feathered edge of good paint. Rinse the repair area with clean water until all of the dust produced by the sanding operation is gone.

34 Spray the entire area with a light coat of primer. This will reveal any imperfections in the surface of the filler. Repair the imperfections with fresh filler paste or glaze filler and once more smooth the surface with sandpaper. Repeat this spray-and-repair procedure until you are satisfied that the surface of the filler and the feathered edge of the paint are perfect. Rinse the area with clean water and allow it to dry completely.

35 The repair area is now ready for painting. Spray painting must be carried out in a warm, dry, windless and dust free atmosphere. These conditions can be created if you have access to a large indoor work area, but if you are forced to work in the open, you will have to pick the day very carefully. If you are working indoors, dousing the floor in the work area with water will help settle the dust which would otherwise be in the air. If the repair area is confined to one body panel, mask off the surrounding panels. This will help minimize the effects of a slight mismatch in paint color. Trim pieces such as chrome strips, door handles, etc., will also need to be masked off or removed. Use masking tape and several thickness of newspaper for the masking operations.

36 Before spraying, shake the paint can thoroughly, then spray a test area until the spray painting technique is mastered. Cover the repair area with a thick coat of primer. The thickness should be built up using several thin layers of primer rather than one thick one. Using 600-grit wet-or-dry sandpaper, rub down the surface of the primer until it is very smooth. While doing this, the work area should be thoroughly rinsed with water and the wet-or-dry sandpaper periodically rinsed as well. Allow the primer to dry before spraying additional coats.

37 Spray on the top coat, again building up the thickness by using several thin layers of paint. Begin spraying in the center of the repair area and then, using a circular motion, work out until the whole repair area and about two inches of the surrounding original paint is covered. Remove all masking material 10 to 15 minutes after spraying on the final coat of paint. Allow the new paint at least two weeks to harden, then use a very fine rubbing compound to blend the edges of the new paint into the existing paint. Finally, apply a coat of wax.

6   Body repair - major damage

1 Major damage must be repaired by an auto body shop specifically equipped to perform unibody repairs. These shops have the specialized equipment required to do the job properly.

2 If the damage is extensive, the body must be checked for proper alignment or the vehicle's handling characteristics may be adversely affected and other components may wear at an accelerated rate.

3 Due to the fact that all of the major body components (hood, fenders, etc.) are separate and replaceable units, any seriously damaged components should be replaced rather than repaired. Sometimes the components can be found in a wrecking yard that specializes in used vehicle components, often at considerable savings over the cost of new parts.

7   Hinges and locks - maintenance

Once every 3000 miles, or every three months, the hinges and latch assemblies on the doors, hood and trunk should be given a few drops of light oil or lock lubricant. The door latch strikers should also be lubricated with a thin coat of grease to reduce wear and ensure free movement. Lubricate the door and trunk locks with spray-on graphite lubricant.

8   Windshield and fixed glass - replacement

11

Replacement of the windshield and fixed glass requires the use of special fast-setting adhesive/caulk materials and some specialized tools. It is recommended that these operations be left to a dealer or a shop specializing in glass work.

These photos illustrate a method of repairing simple dents. They are intended to supplement *Body repair - minor damage* in this Chapter and should not be used as the sole instructions for body repair on these vehicles.

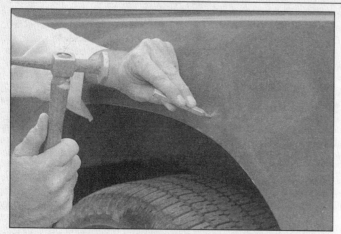

1   If you can't access the backside of the body panel to hammer out the dent, pull it out with a slide-hammer-type dent puller. In the deepest portion of the dent or along the crease line, drill or punch hole(s) at least one inch apart . . .

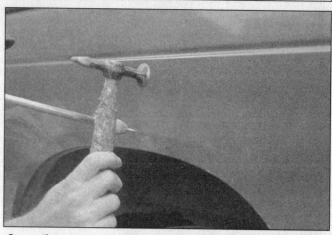

2   . . . then screw the slide-hammer into the hole and operate it. Tap with a hammer near the edge of the dent to help 'pop' the metal back to its original shape. When you're finished, the dent area should be close to its original contour and about 1/8-inch below the surface of the surrounding metal

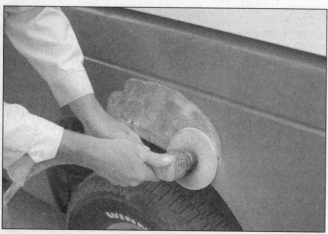

3   Using coarse-grit sandpaper, remove the paint down to the bare metal. Hand sanding works fine, but the disc sander shown here makes the job faster. Use finer (about 320-grit) sandpaper to feather-edge the paint at least one inch around the dent area

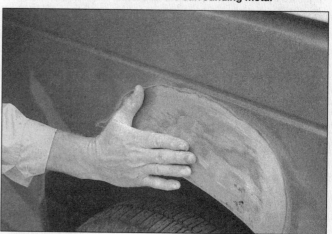

4   When the paint is removed, touch will probably be more helpful than sight for telling if the metal is straight. Hammer down the high spots or raise the low spots as necessary. Clean the repair area with wax/silicone remover

5   Following label instructions, mix up a batch of plastic filler and hardener. The ratio of filler to hardener is critical, and, if you mix it incorrectly, it will either not cure properly or cure too quickly (you won't have time to file and sand it into shape)

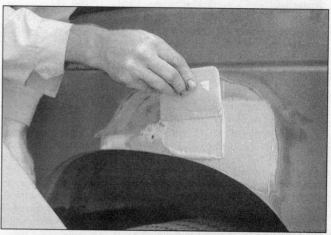

6   Working quickly so the filler doesn't harden, use a plastic applicator to press the body filler firmly into the metal, assuring it bonds completely. Work the filler until it matches the original contour and is slightly above the surrounding metal

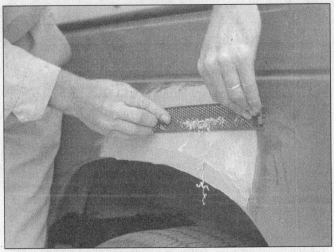

7   Let the filler harden until you can just dent it with your fingernail. Use a body file or Surform tool (shown here) to rough-shape the filler

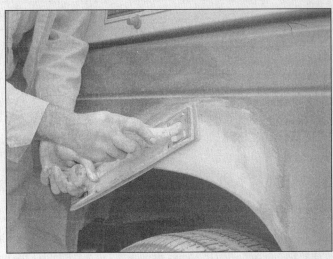

8   Use coarse-grit sandpaper and a sanding board or block to work the filler down until it's smooth and even. Work down to finer grits of sandpaper - always using a board or block - ending up with 360 or 400 grit

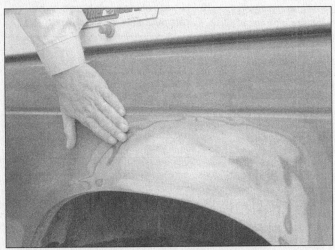

9   You shouldn't be able to feel any ridge at the transition from the filler to the bare metal or from the bare metal to the old paint. As soon as the repair is flat and uniform, remove the dust and mask off the adjacent panels or trim pieces

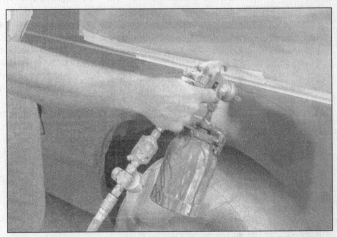

10   Apply several layers of primer to the area. Don't spray the primer on too heavy, so it sags or runs, and make sure each coat is dry before you spray on the next one. A professional-type spray gun is being used here, but aerosol spray primer is available inexpensively from auto parts stores

11   The primer will help reveal imperfections or scratches. Fill these with glazing compound. Follow the label instructions and sand it with 360 or 400-grit sandpaper until it's smooth. Repeat the glazing, sanding and respraying until the primer reveals a perfectly smooth surface

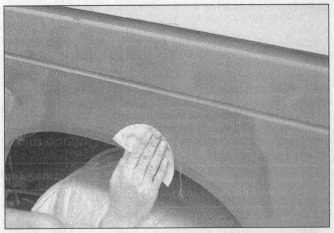

12   Finish sand the primer with very fine sandpaper (400 or 600-grit) to remove the primer overspray. Clean the area with water and allow it to dry. Use a tack rag to remove any dust, then apply the finish coat. Don't attempt to rub out or wax the repair area until the paint has dried completely (at least two weeks)

9.2 Before removing the hood, draw a mark around the hinge plate

9.4 Remove the hinge-to-hood retaining bolts and lift off the hood with the help of an assistant

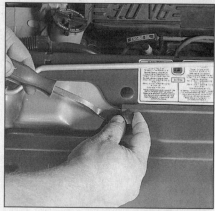

9.10a Use a trim removal tool or a small screwdriver to pry out the plastic clips securing the radiator grille opening cover

9   Hood - removal, installation and adjustment

Note: *The hood is heavy and somewhat awkward to remove and install - at least two people should perform this procedure.*

Removal and installation

Refer to illustrations 9.2 and 9.4

1   Use blankets or pads to cover the cowl area of the body and fenders. This will protect the body and paint as the hood is lifted off.

2   Make marks or scribe a line around the hood hinge to ensure proper alignment during installation **(see illustration)**.

3   Disconnect any cables or wires that will interfere with removal.

4   Have an assistant support one side of the the hood. Take turns removing the hinge-to-hood retaining bolts **(see illustration)**.

5   Lift off the hood.

6   Installation is the reverse of removal.

Adjustment

Refer to illustrations 9.10a, 9.10b and 9.11

7   Fore-and-aft and side-to-side adjustment of the hood is done by moving the hinge plate

slot after loosening the bolts or nuts.

8   Scribe a line around the entire hinge plate so you can determine the amount of movement **(see illustration 9.2)**.

9   Loosen the bolts or nuts and move the hood into correct alignment. Move it only a little at a time. Tighten the hinge bolts and carefully lower the hood to check the position.

10   If necessary after installation, the entire hood latch assembly can be adjusted up-and-down as well as from side-to-side on the radiator support so the hood closes securely and flush with the fenders. To make the adjustment, first remove the radiator grille cover, then scribe a line or mark around the hood latch mounting bolts to provide a reference point, then loosen them and reposition the latch assembly, as necessary **(see illustrations)**. Following adjustment, retighten the mounting bolts.

11   Finally, adjust the hood bumpers on the radiator support so the hood, when closed, is flush with the fenders **(see illustration)**.

12   The hood latch assembly, as well as the hinges, should be periodically lubricated with white, lithium-base grease to prevent binding and wear.

10   Hood release latch and cable - removal and installation

Latch

Refer to illustration 10.2

1   Remove the radiator grille opening cover, then scribe a line around the latch to aid alignment when installing, detach the latch retaining bolts from the radiator support **(see illustrations 9.10a and 9.10b)** and remove the latch.

2   Disconnect the hood release cable by disengaging the cable from the latch assembly **(see illustration)**.

3   Installation is reverse of the removal.

Note: *Adjust the latch so the hood engages securely when closed and the hood bumpers are slightly compressed.*

Cable

Refer to illustration 10.5

4   Disconnect the hood release cable from the latch assembly as described above.

5   Working in the passenger's compartment, remove the driver side kick panel. Then remove the two release lever mounting bolts

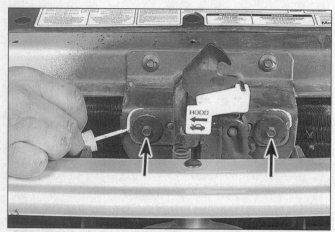

9.10b Make a line around the latch to use as a reference point - to adjust the hood latch, loosen the retaining bolts (arrows), move the latch and retighten bolts, then close the hood to check the fit

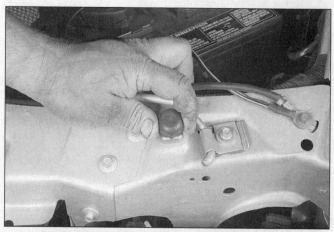

9.11 Adjust the hood closing height by turning the hood bumpers in or out

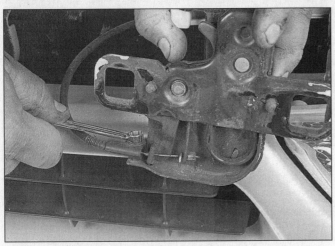

10.2 Unscrew the cable retaining bolt from the backside of the hood latch assembly, then disengage the cable

10.5 Remove the hood release lever retaining screws (arrows), detach the cable and pull it into the engine compartment

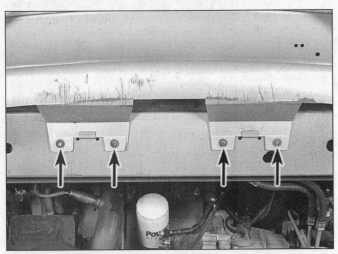

11.3a Remove the screws (arrows) at the center . . .

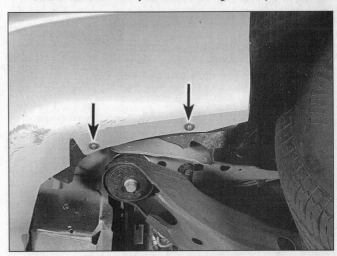

11.3b . . . and in front of each wheel housing that secure the lower edge of the bumper cover

and detach the hood release lever **(see illustration)**. Detach the cable from the lever.

6   Attach a piece of stiff wire to the latch end of the cable, then detach all the cable

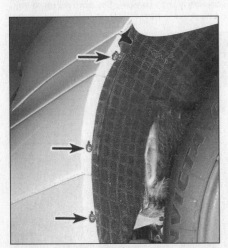

11.4 Detach the retaining screws (arrows) securing the bumper cover to the inner fenderwell splash shields

retaining clips.

7   Push the grommet through the firewall and pull the cable into the engine compartment. Ensure that the new cable has a grommet attached, then remove the old cable from the wire and replace it with the new cable.

8   Pull the wire back through the firewall.

9   Installation is the reverse of the removal
Note: *Push on the grommet to seat it in the firewall completely.*

11   Bumpers - removal and installation

Warning 1: *The models covered by this manual are equipped with Supplemental Restraint systems (SRS), more commonly known as airbags. Always disconnect the negative battery cable, then the positive battery cable and wait two minutes before working in the vicinity of the impact sensors, steering column or instrument panel to avoid the possibility of accidental deployment of the airbag, which could cause personal injury (see Chapter 12). Do not use electrical test equipment on any of*

the airbag system wiring or tamper with them in any way.
Warning 2: *Some models covered by this manual are equipped with air suspension systems. Always disconnect electrical power to the suspension system before lifting or towing the vehicle (see Chapter 10). Failure to perform this procedure may result in unexpected shifting or movement of the vehicle which could cause personal injury.*

Front bumper

Refer to illustrations 11.3a, 11.3b, 11.4, 11.5, 11.6 and 11.7

1   Apply the parking brake, raise the vehicle and support it securely on jackstands.

2   Disconnect the negative battery cable, then the positive battery cable and wait two minutes before proceeding any further.

3   Working under the vehicle, detach the screws securing the lower edges of the bumper cover **(see illustrations)**.

4   Working in the front wheel opening, detach the retaining screws securing the bumper cover to inner fenderwell splash shields **(see illustration)**.

11

5   Pry out the lower edge of the splash shield, then reach up behind the bumper cover and remove the bumper cover-to-fender retaining nuts **(see illustration)**.

6   On 1995-1998 models, remove the side marker lights (see Chapter 12) and detach the screws located in the side marker light opening **(see illustration)**. On 1999 and later models, remove the headlight assembly (see Chapter 12) and release the three tabs at the upper front corner located in the headlight opening.

7   On 1995-1998 models, drill out the retaining rivets securing the upper portion of the bumper cover **(see illustration)**, depress the retaining tabs below each headlight housing and pull the bumper cover out and away from the vehicle. **Note:** *New retaining rivets must be purchased at a dealer service department. On 1999 and later models, remove the pin type retainers used in place of the rivets.*

8   Disconnect any electrical connections which would interfere with removal.

9   To remove the bumper, drill out the rivets securing the foam isolator, then remove the bumper retaining bolts and pull the bumper assembly out and away from the vehicle.

10   Installation is the reverse of removal.

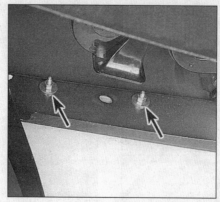

11.5 Peel back the splash shield and remove the bumper cover-to-fender retaining nuts (arrows)

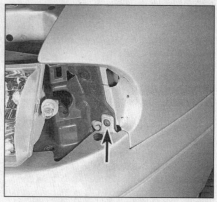

11.6 Remove the side marker lights and detach the screw (arrow) located in the side marker light opening

Rear bumper

Refer to illustrations 11.12, 11.13 and 11.14

11   Apply the parking brake, raise the vehicle and support it securely on jackstands.

12   Working under the vehicle, detach the plastic clips and screws securing the lower edge of the bumper cover **(see illustration)**.

13   Remove the screws securing the bumper cover in the rear wheel openings **(see illustration)**.

14   Open the rear liftgate and remove the screws securing the upper edge of the bumper cover **(see illustration)**. Pull the bumper cover assembly out and away from the vehicle.

15   To remove the bumper, drill out the rivets securing the foam isolator, then remove the bumper retaining bolts and pull the bumper assembly out and away from the vehicle.

16   Installation is the reverse of removal.

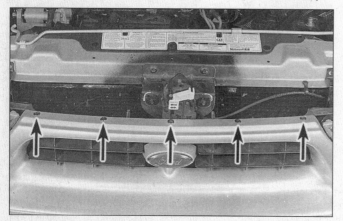

11.7 Drill out the rivets (arrows) securing the upper half of the bumper cover

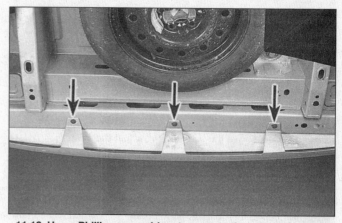

11.12 Use a Phillips screwdriver to remove the screws (arrows) securing the lower edge of the bumper cover

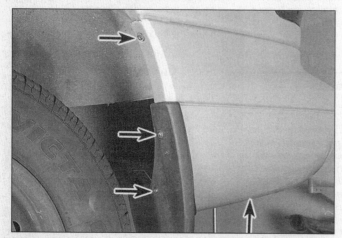

11.13 Remove the screws (arrows) securing the bumper cover to the wheel opening

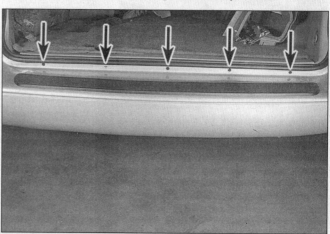

11.14 Remove the screws (arrows) securing the upper edge in the liftgate opening

12.3 Detach inner fenderwell screws and clips (arrows) and remove it from the vehicle

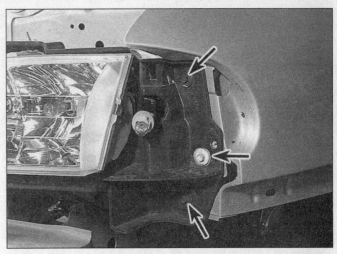

12.4 Detach the bolts (arrows) securing the front edge of the fender to the radiator support

12 Front fender - removal and installation

Refer to illustrations 12.3, 12.4, 12.5a, 12.5b, 12.5c, 12.5d and 12.5e

1   Raise the vehicle, support it securely on jackstands and remove the front wheel.

2   Remove the side marker lights and the front bumper cover (see Section 11).

3   Detach the inner fenderwell screws and clips, then remove the inner fenderwell and mud shield **(see illustration)**.

4   Detach the retaining bolts securing the headlight housing panel and the front edge of the fender to the radiator support **(see illustration)**. Remove the cowl top cover (see

Chapter 11).

5   Remove the remaining fender mounting bolts **(see illustrations)**.

6   Detach the fender. It's a good idea to have an assistant support the fender while it's being moved away from the vehicle to prevent damage to the surrounding body panels.

7   Installation is the reverse of removal.

12.5a Remove the fender-to-rocker panel retaining nut (arrow)

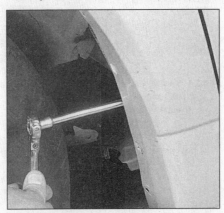

12.5b Remove the fender-to-door pillar lower bolt through the wheel opening

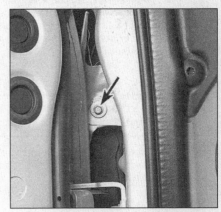

12.5c Open the door to access and remove the upper fender-to-door pillar bolt (arrow)

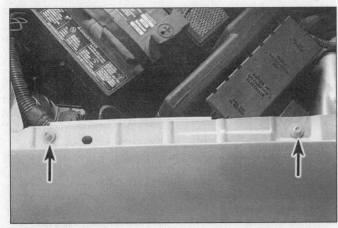

12.5d Detach the remaining bolts located in the hood opening . . .

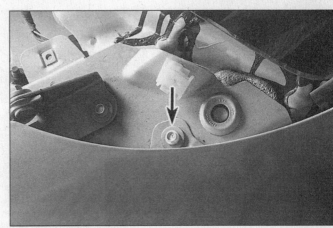

12.5e . . . and the cowl area

11

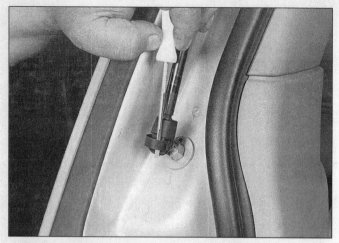

13.2 Use a small screwdriver to pry the clips out of its locking groove, then detach both ends of the strut from the locating studs

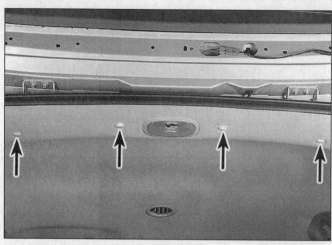

14.2 Detach the upper trim moulding retaining clips (arrows) and disconnect all wiring harness connectors leading to the liftgate.

13 Liftgate support struts - removal and installation

Refer to illustration 13.2

Note: *The rear liftgate is heavy and somewhat awkward to hold - at least two people should perform this procedure.*

1   Open the liftgate and support it securely.
2   Use a small screwdriver to detach the retaining clips at both ends of the support strut. Then pry or pull sharply to detach it from the vehicle **(see illustration)**.
3   Installation is the reverse of removal.

14 Liftgate - removal, installation and adjustment

Note: *The liftgate is heavy and somewhat awkward to hold - at least two people should perform this procedure.*

Removal and installation

Refer to illustrations 14.2 and 14.4

1   Open the liftgate and support it securely.
2   Remove the upper trim moulding from the lift gate opening **(see illustration)** and disconnect all wiring harness connectors leading to the liftgate.
3   While an assistant supports the liftgate, detach both ends of the support struts. Then pry or pull sharply to remove them from the vehicle.
4   Detach the hinge-to-liftgate bolts **(see illustration)** and remove the liftgate from the vehicle.
5   Installation is the reverse of removal.

Adjustment

Refer to illustration 14.7

6   Adjustments are made by loosening the hinge-to-liftgate bolts and moving the liftgate. Proper alignment is achieved when the edges

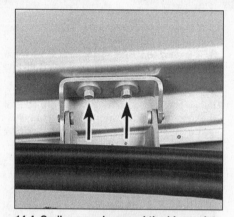

14.4 Scribe a mark around the hinge plate for realignment purposes - then remove the retaining bolts (arrows) on each side of the liftgate

of the liftgate are parallel with the rear quarter panels and the roof panel.
7   Finally, adjust the latch striker assembly as necessary (up and down) to provide positive engagement with the latch mechanism **(see illustration)**. **Note:** *The tail light housing must be removed to access the liftgate latch striker retaining bolts (see Chapter 12).*

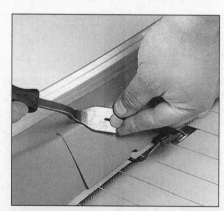

15.1a Using a trim removal tool or a small screwdriver, detach the clips and remove the upper window moulding

14.7 If the liftgate does not close properly, it will be necessary to remove the tail light housing and loosen the nuts (arrows) to adjust the striker plate

15 Liftgate latch, handle and lock cylinder - removal and installation

Refer to illustrations 15.1a, 15.1b, 15.1c, 15.2, 15.6, 15.7 and 15.10

1   Open the liftgate and remove the liftgate trim panel retaining screws and clips **(see illustrations)**.

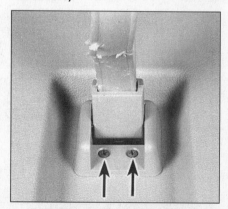

15.1b Remove the inside pull handle retaining screws (arrows)

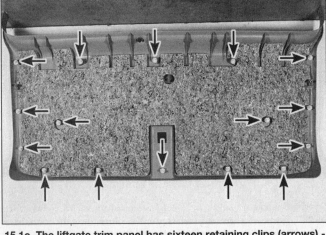

15.1c The liftgate trim panel has sixteen retaining clips (arrows) - pry only at the clip locations, being careful not to distort the panel

15.2 Remove the liftgate latch retaining screws (arrows), then detach the actuating cables and rods

15.6 Remove the liftgate handle cover retaining bolts (arrows)

15.7 Use a center punch to knock out the rivet center pin, then use a 5/16-inch drill bit to drill out the rivet head - drill only until the rivet head comes off, DO NOT drill all the way through the sheet metal

Latch

2   Remove the latch mounting screws (**see illustration**) from the liftgate.
3   Disconnect the actuating rods and cables from the latch and any electrical connections, then remove the latch from the liftgate.
4   Installation is the reverse of removal.

Handle

5   Working through the access hole in the liftgate, disconnect the actuating rod from the back of the handle.
6   Close the liftgate and remove the handle cover from the outside of the liftgate (**see illustration**).
7   Drill out the handle retaining rivets and remove the handle from the liftgate (**see illustration**).
8   Installation is the reverse of removal.

Lock cylinder

9   Detach the plastic clip securing the lock actuating rod.
10   Remove the lock retaining clip securing the lock to the liftgate (**see illustration**).
11   Pull the lock cylinder out to remove it from the door.
12   Installation is the reverse of removal.

---

16   Door trim panel - removal and installation

1   Disconnect the cable from the negative terminal of the battery.

Front door

Refer to illustrations 16.3a, 16.3b, 16.4, 16.5 and 16.6

2   On manual window equipped models, remove the window crank by prying open the

11

15.10 Detach the actuating rod, then pry off the retaining clip (arrow) and remove the lock cylinder

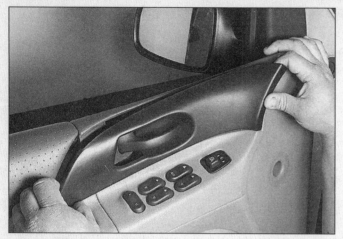

16.3a Using a small screwdriver, pry off the inside handle trim cover . . .

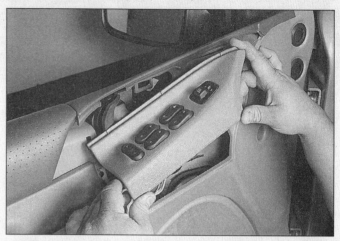

16.3b . . . and the armrest switch control plate, then disconnect the electrical connectors from the backside of the switches

cover from the base of the handle and removing the handle retaining screw. With the retaining screw removed, pull off the handle.

3   On power window equipped models, pry out the armrest switch control plate and disconnect the electrical connections (see illustrations).

4   Detach the retaining screw located behind the armrest switch control plate (see illustration).

5   Detach the side view mirror trim cover by pulling straight out (see illustration).

6   Remove the screws securing the outer edge of the door trim panel (see illustration).

7   Once all of the clips and screws are disengaged, pull the lower edge of the trim panel away from the door, disconnect any electrical connectors and remove the trim panel from the vehicle by gently pulling it up and out.

8   For access to the inner door, peel back the watershield, taking care not to tear it. To install the trim panel, first press the watershield back into place. If necessary, add more sealant to hold it in place.

9   The remainder of the installation is the reverse of removal.

16.4 Remove the retaining screw located behind the armrest switch control plate

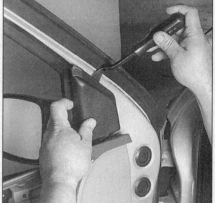

16.5 Pull straight out to remove the side view mirror trim cover

Sliding door

Refer to illustrations 16.10 and 16.11

10   Unscrew the inside door lock knob if equipped. Remove the inside door handle by working a cloth back-and-forth behind the handle to dislodge the retaining clip (see illustration). A special tool is available for

this purpose, but it is not essential. With the retaining clip removed, pull off the handle.

11   Detach the clips and remove the upper window moulding. Insert a wide putty knife or a special trim panel removal tool between the door trim panel and the head of the retaining clip to disengage the door panel retaining

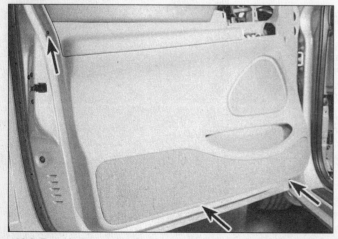

16.6 Detach the screws (arrows) securing the outer edge of the door trim panel

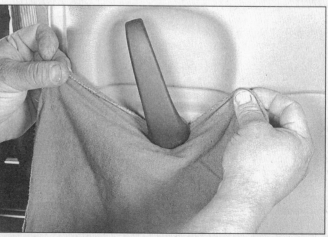

16.10 Work a cloth up behind the inside door handle, then move it back-and-forth until the handle retaining clip releases itself from the shaft

16.11 Use a trim panel removal tool to detach the trim panel retaining clips, then pull the sliding door trim panel up and out to remove it

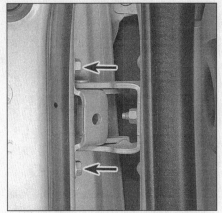

17.8 Before loosening the door retaining bolts (arrows), draw a line around the hinge plates for a reinstallation reference

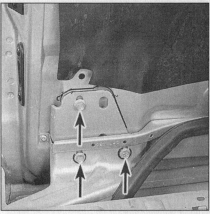

17.12a Remove the bolts (arrows) from the lower guide roller bracket

clips **(see illustration)**. **Note:** *Door trim panel retaining clips are approximately six to ten inches apart. Pry at the clip location only. Prying in between clips will result in distorted or damaged door trim panels.*

12   Once all of the clips are disengaged, detach the trim panel and remove the trim panel from the vehicle by gently pulling it up and out.

13   For access to the inner door, peel back the watershield, taking care not to tear it. To install the trim panel, first press the watershield back into place. If necessary, add more sealant to hold it in place.

14   Installation is the reverse of removal.

---

17   Door - removal, installation and adjustment

Note: *The doors are heavy and somewhat awkward to remove and install - at least two people should perform this procedure.*

Removal and installation
Front door
Refer to illustration 17.8

1   Raise the window completely in the door and then disconnect the cable from negative terminal of the battery.

2   Open the door all the way and support it on jacks or blocks covered with rags to prevent damaging the paint.

3   Remove the door trim panel and water deflector as described in (Section 16).

4   Remove the door speaker (see Chapter 12).

5   Unplug all electrical connections, ground wires and harness retaining clips from the door. **Note:** *It is a good idea to label all connections to aid the reassembly process.*

6   Working through the door speaker hole and the door opening, detach the rubber conduit between the body and the door. Then pull wiring harness through conduit hole and remove from door.

7   Mark around the door hinges with a pen or a scribe to facilitate realignment during

reassembly.

8   Have an assistant hold the door, remove the hinge-to-door bolts **(see illustration)** from the upper and lower hinge and lift the door off.

9   Installation is the reverse of the removal.

Sliding door
Refer to illustrations 17.12a, 17.12b and 17.12c

Note: *On power sliding door models, disconnect the negative battery cable.*

10   Open the sliding door several inches and support it on jacks or blocks covered with rags to prevent damaging the paint.

11   Remove the door trim panel and water deflector as described in (Section 16).

12   Mark around the sliding door guide roller brackets with a pen or a scribe to facilitate realignment during reassembly. Remove the bolts securing the sliding door to the upper, center and lower guide rollers brackets **(see illustrations)**.

13   With the help of an assistant, lift the sliding door upward three to four inches and remove it from the vehicle.

14   Installation is the reverse of the removal.

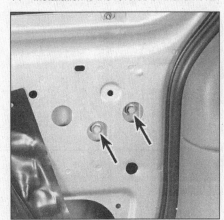

17.12b Working on the inside of the vehicle, remove the bolts (arrows) from the center guide roller bracket, then slide the center guide roller bracket out of the body

Adjustment
Front door
Refer to illustration 17.18

15   Having proper door-to-body alignment is a critical part of a well-functioning door assembly. First check the door hinge pins for excessive play. Fully open the door and lift up and down on the door without lifting the body. If a door has 1/16-inch or more excessive play, the hinges should be replaced.

16   Door to body alignment adjustments are made by loosening the hinge-to-body or hinge to door bolts and moving the door. Proper body alignment is achieved when the top of door is aligned parallel with the roof panel and the bottom of the door is aligned parallel with the lower rocker panel. If these goals can't be reached by adjusting the hinge-to-body or hinge-to-door bolts, body alignment shims may have to be purchased and inserted behind the hinges to achieve correct alignment.

17   To adjust the door closed position, first check that the door latch is contacting the center of the latch striker. If not, remove

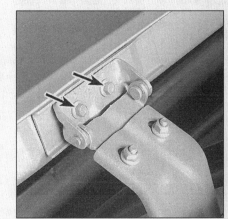

17.12c Remove the bolts (arrows) from the upper guide roller bracket, move the top of the sliding door outward slightly and remove the upper guide roller assembly

11

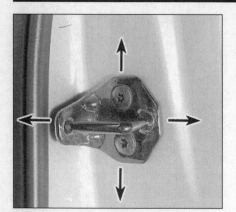

17.18 Adjust the door latch striker by loosening the mounting screws and gently tapping the striker in the desired direction (arrows)

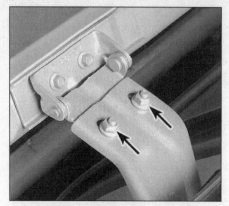

17.20a Upper guide roller adjustment nut locations

17.20b Lower guide roller adjustment bolt location

striker and add or subtract shims to achieve correct alignment.

18   Finally, adjust the latch striker as necessary (up and down or sideways) to provide positive engagement with the latch mechanism **(see illustration)** and so the door panel is flush with the center pillar.

Sliding door

Refer to illustrations 17.20a, 17.20b and 17.22

19   First adjust the up-and-down position by loosening the lower guide bracket retaining bolts and moving the door as necessary **(see illustration 17.12a)**. Proper door alignment is achieved when the top of door is aligned parallel with the roof panel and the bottom of the door is aligned parallel with the lower rocker panel. **Note:** *The door trim panel must first be removed for this step (see Section 16).*

20   Next, adjust the in-and-out position of the door by loosening the upper guide roller adjustment nuts and the lower guide roller adjustment bolt **(see illustrations)**. Proper door alignment is achieved when the top of door is flush with the roof panel and the cen-

ter pillar, and the bottom of the door is flush with the lower rocker panel and the center pillar with the door closed.

21   To adjust the door closed position, first check that the door latch is contacting the center of the rear latch striker. If not, remove striker and add or subtract shims to achieve correct alignment.

22   Finally adjust latch striker as necessary (sideways) to provide positive engagement with the latch mechanism **(see illustration)** and the door panel is flush with the rear quarter panel.

---

18   Door latch, lock cylinder and handles - removal and installation

---

Note: *The procedures described below apply to both front doors and the sliding door except where noted.*

Door latch

Refer to illustrations 18.2 and 18.5

1   Remove the door trim panel and watershield as described in Section 16.

2   Remove the screws securing the latch

to the door **(see illustration)**.

3   Working through the large access hole, position the latch as necessary to disengage the outside door handle and outside lock cylinder to latch rods and the inside handle to latch cable.

4   All door locking rods are attached by plastic clips. The plastic clips can be removed by unsnapping the portion engaging the connecting rod and then by pulling the rod out of its locating hole.

5   Position the latch as necessary to disengage the door lock solenoid hook (if equipped). Then remove the latch assembly from the door. **Note:** *On power door lock equipped models, it may be necessary to drill out the door lock solenoid retaining rivet from the end of the door to allow the solenoid hook to disengage from the latch* **(see illustration)**. *New retaining rivets must be purchased at a dealer service department.*

6   Installation is the reverse of removal.

Door lock cylinder and outside handle

Refer to illustrations 18.9, 18.11a and 18.11b

7   To remove the lock cylinder, raise the window and remove the door trim panel and watershield as described in Section 16.

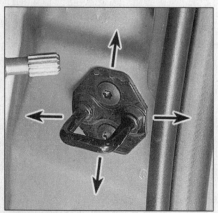

17.22 Adjust the sliding door lock striker by loosening the mounting screws and gently tapping the striker in the desired direction (arrows)

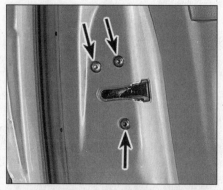

18.2 Remove the latch retaining screws (arrows) from the end of the door, then detach the locking rods and cable and pull the latch assembly through the access hole (front door shown, sliding door similar)

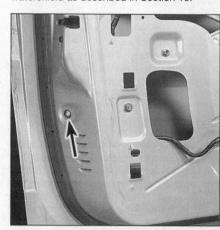

18.5 Power door lock solenoid retaining rivet location (arrow)

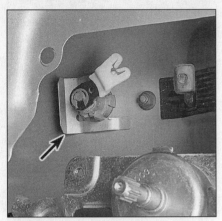

18.9 To remove the lock cylinder, detach the plastic clip securing the lock rod, then pry off the lock cylinder retaining clip (arrow)

8   Working through the large access hole, disengage the plastic clip that secures the lock cylinder to latch rod.

9   Using a screwdriver, slide the lock cylinder retaining clip out of engagement and remove the lock cylinder from the door **(see illustration)**.

10   To remove the outside handle, work through the access hole and disengage the plastic clip that secures the outside handle-to-latch rod.

11   On front doors, drill out the handle retaining rivets and remove the handle from the door **(see illustration)**. On sliding doors, remove the outside handle retaining nuts **(see illustration)** and pull the handle from the door.

12   Installation is the reverse of removal.

Inside handle

Refer to illustrations 18.14 and 18.15

13   Remove the door trim panel as described in Section 16 and peel away the watershield.

14   On front doors, drill out the rivets securing the inside handle **(see illustration)**. Pull rearward on the handle to disengage it from the inner door panel. Detach the latch cable from the backside of the handle and remove it from the vehicle.

15   On sliding doors, remove the screws securing the inside handle control mechanism to the door **(see illustration)**. Working through the large access hole, position the inside handle control mechanism as necessary to disengage the outside door handle and outside lock cylinder to latch rods and the inside handle to latch cable, then remove the handle control mechanism from the door.
Note: *All door locking rods are attached by plastic clips. The plastic clips can be removed by unsnapping the portion engaging the connecting rod and then by pulling the rod out of its locating hole.*

16   Installation is the reverse of removal.

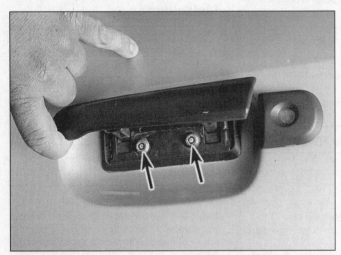

18.11a On front doors, use a center punch to knock out the rivet center pin, then use a 5/16-inch drill bit to drill out the rivet head - drill only until the rivet head comes off, DO NOT drill all the way through the sheet metal

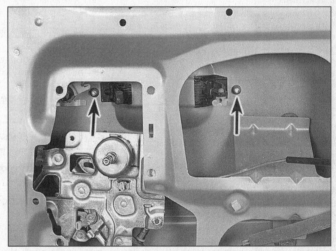

18.11b On sliding doors, the outside handle retaining nuts (arrows) can be reached through the access hole in the door frame

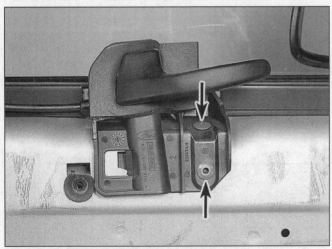

18.14 On front doors, drill out the handle retaining rivets, then rotate the handle out and detach the latch cable from the backside

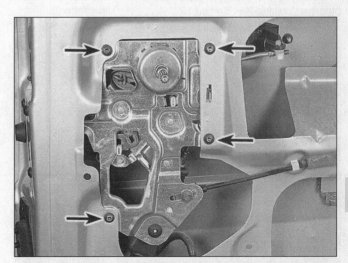

18.15 On sliding doors, remove the handle control mechanism retaining screws (arrows), then position the control mechanism as necessary to disengage the actuating rods and cables from the backside

11

19   Door window glass - removal and installation

Refer to illustrations 19.4 and 19.5
Note: *The procedure described below applies to front doors only.*
1   Remove the door trim panel and the plastic watershield (see Section 16).
2   Lower the window glass all the way down into the door.
3   Carefully pry the inner and outer weatherstripping out of the door window opening.
4   Raise the window glass all the way up in the door frame. Remove the window glass run channel retaining bolt **(see illustration)**, then pull the run channel down and out through the access hole in the door frame to remove it.
5   Lower the window just enough to access both of the window retaining rivets through the holes in the door frame **(see illustration)**.
6   Place a rag over the glass to help prevent scratching the glass, then drill out the two glass mounting rivets.
7   Remove the glass by pulling it up and out.
8   Installation is the reverse of removal.

20   Door window glass regulator - removal and installation

Refer to illustrations 20.4a and 20.4b
Warning: *The regulator arms are under extreme pressure and can cause serious injury if the motor or counter-balance spring is removed without locking the sector gear. This can be done by inserting a bolt and nut through the holes in the backing plate and sector gear to lock them together.*
Note: *The procedure described below applies to the front doors only.*
1   Remove the door trim panel and the plastic watershield (see Section 16).
2   Remove the window glass assembly

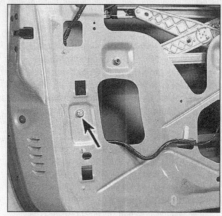

19.4 Remove the window run channel bolts (arrow) from the end of the door

(see Section 19).
3   On power operated windows, disconnect the electrical connector from the window regulator motor.
4   Remove the equalizer arm bracket and the regulator mounting rivets **(see illustrations)**.
5   Pull the equalizer arm and regulator assembly through the service hole in the door frame to remove it.
6   Installation is the reverse of removal.

21   Mirrors - removal and installation

Outside mirrors

Refer to illustration 21.4
1   Remove the door trim panel and the plastic watershield (see Section 16).
2   Pry off the mirror trim cover **(see illustration 16.5)**.
3   Disconnect the electrical connector from the mirror (if equipped).
4   Remove the three mirror retaining nuts and detach the mirror from the vehicle **(see illustration)**.
5   Installation is the reverse of removal.

19.5 Raise the window just enough to access the glass retaining rivets (arrow) through the holes in the door frame - drill out the rivets securing the glass to the equalizer arm

Inside mirror

6   Disconnect the electrical connector from the mirror (if equipped).
7   Insert a small screwdriver into the slot located at the base of the mirror and gently pry rearward until the mirror snaps off its retaining bracket.
8   To install the mirror, simply snap it back into place on the mirror retaining bracket.

22   Rear quarter window - replacement

Refer to illustrations 22.1 and 22.3
Note: *The rear quarter window glass is fragile and somewhat awkward to remove and install - at least two people should perform this procedure.*
1   Open the window latch assembly. On power window equipped models, remove the retaining screw securing the latch to the glass **(see illustration)**. On manual window equipped models, pry out the retaining pin on

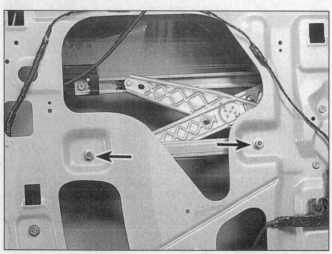

20.4a Detach the window equalizer arm bracket retaining nuts (arrows) . . .

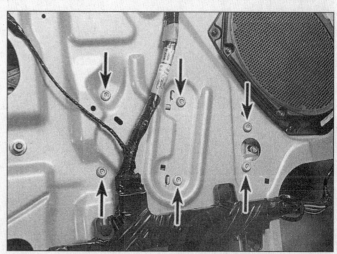

20.4b . . . then drill out the window regulator rivets (arrows) (power window equipped model shown)

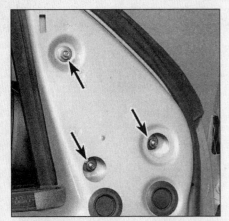

21.4 Pry out the body plugs, then detach the mirror retaining nuts (arrows) and remove the mirror from the vehicle

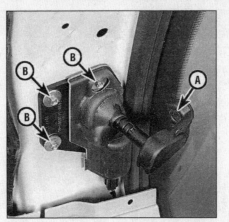

22.1 Rear quarter window latch to glass retaining screw (A) and latch to body bolts (B)

the bottom of the latch that secures the latch to the glass.

2   Remove the upper rear quarter trim panel located in front of the rear quarter window (see Section 27).

3   With an assistant holding the window

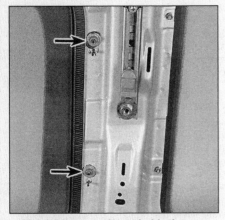

22.3 While an assistant holds the rear quarter window glass remove the hinge retaining nuts (arrows)

glass, remove the retaining screws securing the hinges to the body **(see illustration)**.

4   Remove the rear quarter window glass from the vehicle.

5   To remove the rear quarter window latch, first remove the upper rear quarter trim panel located between the rear quarter window and the liftgate, disconnect the electrical connector (If equipped) and remove the latch retaining bolts **(see illustrations 22.1)**.

6   Installation is the reverse of removal.

23 Center console - removal and installation

Refer to illustration 23.2

Warning: *The models covered by this manual are equipped with Supplemental Restraint systems (SRS), more commonly known as airbags. Always disconnect the negative battery cable, then the positive battery cable and wait two minutes before working in the vicinity of the impact sensors, steering column or instrument panel to avoid the possibility of accidental deployment of the airbag, which*

could cause personal injury (see Chapter 12). Do not use electrical test equipment on any of the airbag system wiring or tamper with them in any way.

1   Disconnect the negative battery cable. then the positive battery cable and wait two minutes before proceeding any further.

2   Lift up the spare change/cup holder mats from the front and rear of the console and remove the retaining screws beneath **(see illustration)**.

3   Disconnect any electrical connections and remove the console from the vehicle.

4   Installation is the reverse of removal.

24 Instrument cluster bezel - removal and installation

Refer to illustration 24.3

Warning: *The models covered by this manual are equipped with Supplemental Restraint systems (SRS), more commonly known as airbags. Always disconnect the negative battery cable, then the positive battery cable and wait two minutes before working in the vicinity of the impact sensors, steering column or instrument panel to avoid the possibility of accidental deployment of the airbag, which could cause personal injury (see Chapter 12). Do not use electrical test equipment on any of the airbag system wiring or tamper with them in any way.*

1   Disconnect the negative battery cable. then the positive battery cable and wait two minutes before proceeding any further.

2   Remove the knee bolster, radio trim bezel, headlight switch bezel and the center trim panel from the instrument panel (see Section 25).

3   Remove the bezel retaining screws **(see illustration)**.

4   Tilt the steering wheel down to the lowest position. Pull the top of the instrument cluster bezel outward while disengaging the lower clips on the bezel. Remove the bezel from the vehicle.

5   Installation is the reverse of removal.

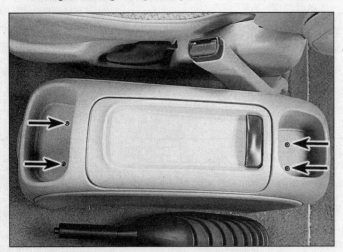

23.2 Remove the screws (arrows) located under the cup holder mats

24.3 Detach the bezel retaining screws (arrow) then pull the bezel outward to disengage the retaining clips

11

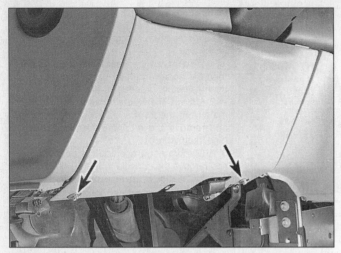

25.2 Remove the screws (arrows) at the lower edge of the knee bolster - pull the bottom edge outward slightly, then downward to remove it

25.4 The knee bolster reinforcement panel can be removed after the retaining bolts (arrows) are unscrewed

25 Dashboard trim panels - removal and installation

Warning: *The models covered by this manual are equipped with Supplemental Restraint systems (SRS), more commonly known as airbags. Always disconnect the negative battery cable, then the positive battery cable and wait two minutes before working in the vicinity of the impact sensors, steering column or instrument panel to avoid the possibility of accidental deployment of the airbag, which could cause personal injury (see Chapter 12). Do not use electrical test equipment on any of the airbag system wiring or tamper with them in any way.*

1   Disconnect the negative battery cable. then the positive battery cable and wait two minutes before proceeding any further.

Knee bolster

Refer to illustrations 25.2 and 25.4

2   Working in the drivers side passenger compartment, remove the retaining screws along the lower edge of the knee bolster **(see illustration)**.

3   Pull outward on the lower edge of the knee bolster and detach it from the vehicle.

4   Remove the retaining bolts securing the knee bolster reinforcement panel **(see illustration)**. Pull outward on the lower edge of the knee bolster reinforcement panel and detach it from the vehicle.

5   Installation is the reverse of removal.

Headlight switch bezel

Refer to illustrations 25.6 and 25.7

6   Detach the headlight switch knob from the headlight switch **(see illustration)**.

7   Using two hands, pull headlight switch bezel straight out to release the bezel retaining clips **(see illustration)**.

8   Disconnect the electrical connector from the headlight switch illumination bulb and remove the bezel from the instrument panel.

9   Installation is the reverse of removal.

Radio trim bezel

Refer to illustrations 25.10 and 25.11

10   Remove the ash tray **(see illustration)**.

11   Remove the two retaining screws located along the lower edge of the bezel, then grasp the bezel securely and detach it from the instrument panel by pulling it straight back **(see illustration)**.

12   Disconnect any electrical connections and remove the bezel from the instrument panel.

13   Installation is the reverse of the removal.

Center trim panel

Refer to illustration 25.15

14   Remove the knee bolster, headlight switch bezel and the radio trim bezel as described above.

15   Detach the retaining screws from the lower half of the trim panel **(see illustration)**.

16   Pull the bottom of the trim panel outward while disengaging the upper clips on the trim panel and remove it from the instrument panel.

17   Installation is the reverse of removal.

25.6 Simply pull the headlight switch control knob straight off to remove it

25.7 Pull the headlight switch bezel straight off to disengage it from the instrument panel, then disconnect the electrical connector

25.10 Pull the ashtray outward and depress the retaining tab to remove it

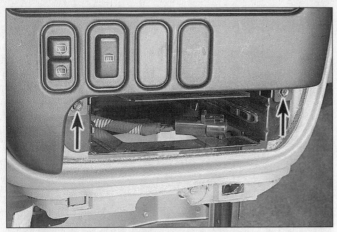

25.11 Remove the screws (arrows) securing the lower edge of the
radio trim bezel, then pull it straight back to release
the bezel retaining clips

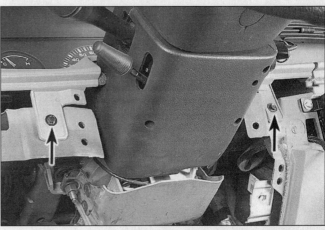

25.15 Remove the screws (arrows) securing the lower edge of the
center trim panel, then carefully pry it off the instrument panel

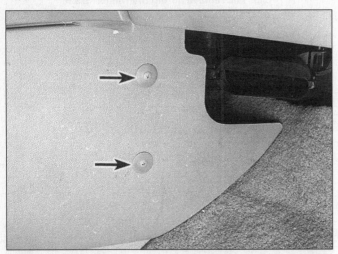

25.18 Remove the two screws (arrows) from each side of the
utility compartment and carefully pry it off the instrument panel

25.21 Detach the glove box door hinge retaining screws (arrows)

Utility compartment

Refer to illustration 25.18

18   Remove the retaining screws located on
each side of the utility compartment, then pull
the compartment straight out to disengage
the clips securing the front and remove it
from the instrument panel **(see illustration)**.
19   Installation is the reverse of removal.

Glove box

Refer to illustration 25.21

20   Open the glove box door. Release the
door stops by pressing in on the sides, then
let the glove box hang down to access the
hinge retaining screws.
21   Detach the door hinge retaining screws
along the lower edge of the glove box and
remove it from the instrument panel **(see
illustration)**.
22   Installation is the reverse of removal.

26   Steering column cover - removal
and installation

Refer to illustrations 26.2a and 26.2b

1   Remove the knee bolster, headlight
switch bezel, radio trim bezel and the center
trim panel (see Section 25).
2   Detach the steering column tilt lever and
remove the screws from the lower steering
column cover **(see illustrations)**.
3   Remove the key lock cylinder (see
Chapter 12).
4   Separate the cover halves and detach
them from the steering column.
5   Installation is the reverse of removal.

11

26.2a Use a small wrench to unscrew the
steering column tilt lever

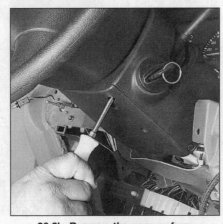

26.2b Remove the screws from
the lower cover

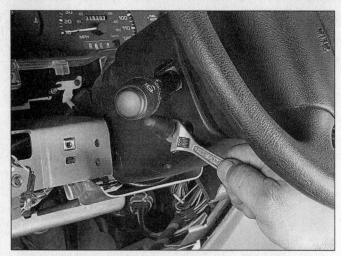

27.2a Pry open the trim cover and remove the
seat belt anchor bolt . . .

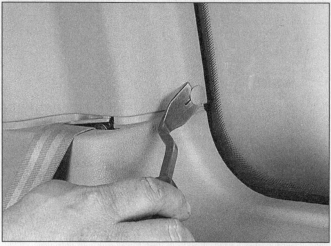

27.2b . . . then pry out the upper rear quarter trim panel retaining
clips and remove it from the window pillar

27.3 Detach the clips, disconnect the electrical connector and
remove the rear air conditioning service cover

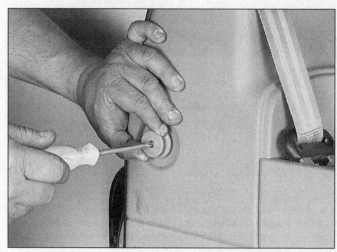

27.4 Remove the cargo net retainers if equipped

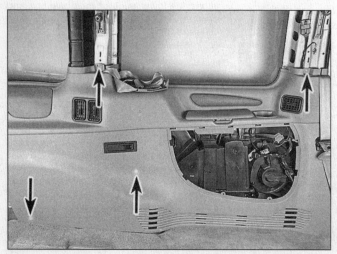

27.5 Detach the trim panel retaining clips (arrows)
(right hand panel shown, left side similar)

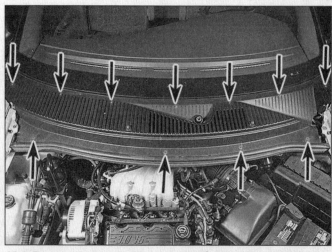

28.2 Remove the screws (arrows) securing the top cowl cover . . .

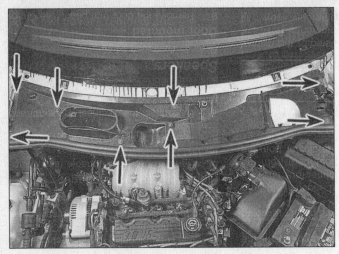

28.3 . . . then remove the screws (arrows) from the lower cowl cover

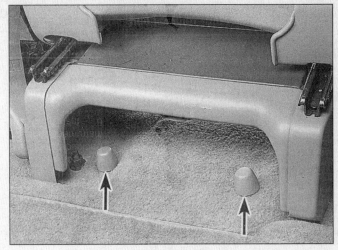

29.2 Detach the trim covers to access the front seat retaining bolts

27 Rear quarter trim panels - removal and installation

Refer to illustrations 27.2a, 27.2b, 27.3, 27.4 and 27.5

1   Remove the rear seat (see Section 29).
2   Remove the upper rear quarter trim panels surrounding the rear quarter window glass **(see illustrations)**.
3   When removing right side trim panels on vehicles equipped with rear air conditioning, detach the air conditioning service cover and disconnect the blower switch electrical connector **(see illustration)**.
4   Remove the screws from the rear cargo net if equipped **(see illustration)**.
5   Using a trim removal tool or a small screwdriver, detach the clips and remove the rear quarter trim panel **(see illustration)**.
6   Installation is the reverse of removal.

28 Cowl cover - removal and installation

Refer to illustrations 28.2 and 28.3

1   Remove the windshield wiper arms and the antenna mast (see Chapter 12).
2   Remove the retaining screws securing the cowl top cover **(see illustration)**.
3   Remove the retaining screws securing the lower cowl cover **(see illustration)**.
4   Detach the lower cover and remove it from the vehicle.
5   Installation is the reverse of removal.

29 Seats - removal and installation

Front seat

Refer to illustration 29.2

1   Position the seat all the way forward or all the way to the rear to access the front seat retaining bolts.
2   Detach any bolt trim covers and remove the retaining bolts **(see illustration)**.
3   Tilt the seat upward to access the underneath, then disconnect any electrical connectors and lift the seat from the vehicle.
4   Installation is the reverse of removal.

Rear seats

Refer to illustration 29.6

5   Fold the rear seat back to the down position.

29.6 Lift up release handle (arrow) and disengage the rear seat from the hooks in the floor

6   Lift up the rear seat latch handle **(see illustration)**, then disengage the rear seat from the hooks in the floor and remove it from the vehicle.
7   Installation is the reverse of removal.

11

Notes

Chapter 12
Chassis electrical system

Contents

1   General information

The electrical system is a 12-volt, negative ground type. Power for the lights and all electrical accessories is supplied by a lead/acid-type battery which is charged by the alternator.

This Chapter covers repair and service procedures for the various electrical components not associated with the engine.

Information on the battery, alternator, distributor and starter motor can be found in Chapter 5.

It should be noted that when portions of the electrical system are serviced, the negative battery cable should be disconnected from the battery to prevent electrical shorts and/or fires.

2   Electrical troubleshooting - general information

Refer to illustration 2.15

A typical electrical circuit consists of an electrical component, any switches, relays, motors, fuses, fusible links or circuit breakers related to that component and the wiring and connectors that link the component to both the battery and the chassis. To help you pinpoint an electrical circuit problem, wiring diagrams are included at the end of this Chapter.

Before tackling any troublesome electrical circuit, first study the appropriate wiring diagrams to get a complete understanding of what makes up that individual circuit. Trouble

spots, for instance, can often be narrowed down by noting if other components related to the circuit are operating properly. If several components or circuits fail at one time, chances are the problem is in a fuse or ground connection, because several circuits are often routed through the same fuse and ground connections.

Electrical problems usually stem from simple causes, such as loose or corroded connections, a blown fuse, a melted fusible link or a failed relay. Visually inspect the condition of all fuses, wires and connections in a problem circuit before troubleshooting the circuit.

If test equipment and instruments are going to be utilized, use the diagrams to plan ahead of time where you will make the necessary connections in order to accurately pinpoint the trouble spot.

The basic tools needed for electrical troubleshooting include a circuit tester or voltmeter (a 12-volt bulb with a set of test leads can also be used), a continuity tester, which includes a bulb, battery and set of test leads, and a jumper wire, preferably with a circuit breaker incorporated, which can be used to bypass electrical components. Before attempting to locate a problem with test instruments, use the wiring diagram(s) to decide where to make the connections.

Voltage checks

Voltage checks should be performed if a circuit is not functioning properly. Connect one lead of a circuit tester to either the negative battery terminal or a known good ground. Connect the other lead to a connector in the circuit being tested, preferably nearest to the battery or fuse. If the bulb of the tester lights, voltage is present, which means that the part of the circuit between the connector and the battery is problem free. Continue checking the rest of the circuit in the same fashion. When you reach a point at which no voltage is present, the problem lies between that point and the last test point with voltage. Most of the time the problem can be traced to a loose connection. **Note:** *Keep in mind that some circuits receive voltage only when the ignition key is in the Accessory or Run position.*

Finding a short

One method of finding shorts in a circuit is to remove the fuse and connect a test light or voltmeter in place of the fuse terminals. There should be no voltage present in the circuit. Move the wiring harness from side-to-side while watching the test light. If the bulb goes on, there is a short to ground somewhere in that area, probably where the insulation has rubbed through. The same test can be performed on each component in the circuit, even a switch.

Ground check

Perform a ground test to check whether a component is properly grounded. Disconnect the battery and connect one lead of a self-powered test light, known as a continuity tester, to a known good ground. Connect the other lead to the wire or ground connection being tested. If the bulb goes on, the ground is good. If the bulb does not go on, the ground is not good.

Continuity check

A continuity check is done to determine if there are any breaks in a circuit - if it is passing electricity properly. With the circuit off (no power in the circuit), a self-powered continuity tester can be used to check the circuit. Connect the test leads to both ends of the circuit (or to the "power" end and a good ground), and if the test light comes on the circuit is passing current properly. If the light doesn't come on, there is a break somewhere

2.15 To backprobe a connector, insert a small, sharp probe (such as a straight-pin) into the back of the connector along side the desired wire until it contacts the metal terminal inside; connect your meter leads to the probes - this allows you to test a functioning circuit

in the circuit. The same procedure can be used to test a switch, by connecting the continuity tester to the switch terminals. With the switch turned On, the test light should come on.

Finding an open circuit

When diagnosing for possible open circuits, it is often difficult to locate them by sight because oxidation or terminal misalignment are hidden by the connectors. Merely wiggling a connector on a sensor or in the wiring harness may correct the open circuit condition. Remember this when an open circuit is indicated when troubleshooting a circuit. Intermittent problems may also be caused by oxidized or loose connections.

Electrical troubleshooting is simple if you keep in mind that all electrical circuits are basically electricity running from the battery, through the wires, switches, relays, fuses and fusible links to each electrical component (light bulb, motor, etc.) and to ground, from which it is passed back to the battery. Any electrical problem is an interruption in the flow of electricity to and from the battery.

Connectors

Most electrical connections on these vehicles are made with multiwire plastic connectors. The mating halves of many connectors are secured with locking clips molded into the plastic connector shells. The mating halves of large connectors, such as some of those under the instrument panel, are held together by a bolt through the center of the connector.

To separate a connector with locking clips, use a small screwdriver to pry the clips apart carefully, then separate the connector halves. Pull only on the shell, never pull on the wiring harness as you may damage the individual wires and terminals inside the connectors. Look at the connector closely before

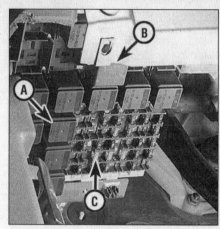

3.1a The passenger compartment fuse block is located under the driver's side of the instrument panel - It contains the flasher unit (A), relays (B) and miniaturized fuses (C)

trying to separate the halves. Often the locking clips are engaged in a way that is not immediately clear. Additionally, many connectors have more than one set of clips.

Each pair of connector terminals has a male half and a female half. When you look at the end view of a connector in a diagram, be sure to understand whether the view shows the harness side or the component side of the connector. Connector halves are mirror images of each other, and a terminal shown on the right side end view of one half will be on the left side end view of the other half.

It is often necessary to take circuit voltage measurements with a connector connected. Whenever possible, carefully insert the test probes of your meter into the rear of the connector shell to contact the terminal inside. This kind of connection is called "backprobing" **(see illustration)**. When inserting a test probe into a male terminal, be careful not to distort the terminal opening. Doing so can lead to a poor connection and corrosion at that terminal later.

3   Fuses - general information

Refer to illustrations 3.1a, 3.1b and 3.2

The electrical circuits of the vehicle are protected by a combination of fuses, circuit breakers and cartridge type fusible links. All models covered in this manual have two fuse blocks, one located under the instrument panel on the left side of the dashboard and one in the engine compartment adjacent to the battery called the power distribution box **(see illustrations)**. Always disconnect the cable to the negative battery terminal before replacing high current fuses.

Miniaturized fuses are employed in the passenger compartment fuse block and the power distribution box. These compact fuses, with blade terminal design, allow fingertip removal and replacement. If an electri-

3.1b The power distribution box is located in the engine compartment adjacent to the battery - it contains cartridge type fusible links (A), miniaturized fuses (B), and relays (C)

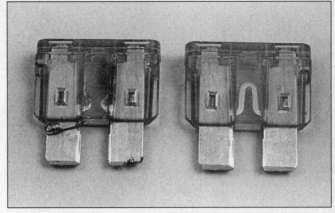

3.2 The fuses can easily be checked visually to see if they are blown

cal component fails, always check the fuse first. The best way to check the fuses is with a test light. Check for power at the exposed terminal tips of each fuse. If power is present at one side of the fuse but not the other, the fuse is blown. A blown fuse can also be identified by visually inspecting it **(see illustration)**.

Be sure to replace blown fuses with the correct type. Fuses of different ratings are physically interchangeable, but only fuses of the proper rating should be used. Replacing a fuse with one of a higher or lower value than specified is not recommended. Each electrical circuit needs a specific amount of protection. The amperage value of each fuse is molded into the fuse body.

If the replacement fuse immediately fails, don't replace it again until the cause of the problem is isolated and corrected. In most cases, the cause will be a short circuit in the wiring caused by a broken or deteriorated wire.

4 Fusible links - general information

Refer to illustration 4.2

Some circuits are protected by fusible links. The links are used in circuits which are not ordinarily fused, such as the ignition circuit.

In addition to the conventional type of fusible link (described below) which is located in the wiring harness **(see illustration)**, there is also cartridge type fusible links located in the engine compartment fuse block that are similar to a large fuses **(see illustration 3.1b)**, and, after disconnecting the negative battery cable, are simply unplugged and replaced by a unit of the same amperage. Some fusible links are held in place by a screw which must be loosened before removing the link.

Conventional type fusible links cannot be repaired, a new link of the same size wire should be installed in its place. The procedure is as follows:

a) *Disconnect the cable from the negative battery terminal.*
b) *Disconnect the fusible link from the wiring harness.*
c) *Cut the damaged fusible link out of the wiring just behind the connector.*
d) *Strip the insulation back approximately 1/2-inch.*
e) *Position the connector on the new fusible link and crimp it into place.*
f) *Use rosin core solder at each end of the new link to obtain a good solder joint.*
g) *Use plenty of electrical tape around the soldered joint. No wires should be exposed.*
h) *Connect the battery ground cable. Test the circuit for proper operation.*

5 Circuit breakers - general information

Circuit breakers protect components such as power windows, power door locks and headlights.

On some models the circuit breaker resets itself automatically, so an electrical overload in a circuit breaker protected system will cause the circuit to fail momentarily, then come back on. If the circuit does not come back on, check it immediately. Once the condition is corrected, the circuit breaker will resume its normal function.

6 Relays - general information and testing

General information

1 Several electrical circuits in the vehicle, such as the electronic engine control system, the fuel pump, horns, starter, fog lamps, and others use relays to switch high current to system components. Relays use a low-current control circuit to open and close a high-current power circuit. If the relay is defective, the circuit load will not operate properly. Most of the relays are in the fuse blocks

4.2 Conventional type fusible links (arrow) can be found exiting the power distribution box in the engine compartment

described in section 4. If you suspect a faulty relay, you can remove it and test it using the procedure below. Alternatively, have the relay tested by a dealer service department or a repair shop. Defective relays must be replaced.

2 Most of the relays used in these vehicles are often called "ISO" relays, which refers to the International Standards Organization. The terminals of ISO relays are numbered to indicate their usual circuit connections and functions.

Testing

Refer to illustration 6.5

3 Refer to the wiring diagram for the circuit to determine the proper connections for the relay you're testing. If you can't determine the correct connection from the wiring diagrams, however, you may be able to determine the test connections from the information that follows.

4 Two of the terminals are the relay control circuit and connect to the relay coil. The other relay terminals are the power circuit. When the relay is energized, the coil creates a magnetic field that closes the larger con-

12

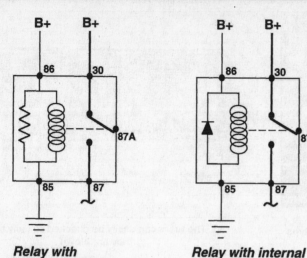

6.11a The CCRM module is mounted next to the battery

Relay with internal resistor

Relay with internal clamping diode

6.5 Typical ISO relay designs, terminal numbering and circuit connections

tacts of the power circuit to provide power to the circuit loads.

5   Terminals 85 and 86 are normally the control circuit **(see illustration)**. If the relay contains a diode, terminal 86 must be connected to battery positive (B+) voltage and terminal 85 to ground. If the relay contains a resistor, terminals 85 and 86 can be connected in either direction with respect to B+ and ground. Terminal 30 is normally connected to the battery voltage (B+) source for the circuit loads. Terminal 87 is connected to the ground side of the circuit, either directly or through a load. If the relay has several alternate terminals for load or ground connections, they usually are numbered 87A, 87B, 87C, and so on.

6   Use an ohmmeter to check continuity through the relay control coil.

a)   *Connect the meter according to the polarity shown in illustration 6.5 for one check; then reverse the ohmmeter leads and check continuity in the other direction.*

b)   *If the relay contains a resistor, resistance should be the specified value with the ohmmeter in either direction.*

c)   *If the resistor contains a diode, resistance should be the specified coil resistance value with the ohmmeter in the forward polarity direction. With the meter leads reversed, resistance should be lower.*

d)   *If the ohmmeter shows infinite resistance in both directions, replace the relay.*

7   Remove the relay from the vehicle and use the ohmmeter to check for continuity

between the relay power circuit terminals. There should be no continuity between terminal 30 and 87 with the relay deenergized.

8   Connect a fused jumper wire to terminal 86 and the positive battery terminal. Connect another jumper wire between terminal 85 and ground. When the connections are made, the relay should click.

9   With the jumper wires connected, check for continuity between the power circuit terminals. Now, there should be continuity between terminals 30 and 87.

10   If the relay fails any of the above tests, replace it.

Constant Control Relay Module

Refer to illustrations 6.11a and 6.11b

11   1995 through 1997 models are equipped with a relay assembly called the Constant Control Relay Module (CCRM). The CCRM is mounted next to the battery and

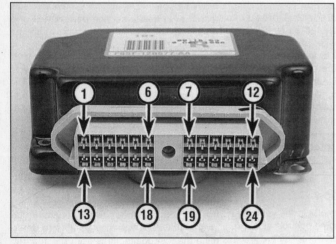

6.11b CCRM terminal identification

1   *Battery power to cooling fan control relay (black/orange)*
2   *Battery power to cooling fan control relay (black/orange)*
3   *Power to cooling fan motors (red/orange)*
4   *Power to cooling fan motors (red/orange)*

5   *Power to fuel pump (dark green/yellow)*
6   *Reduced power from cooling fan dropping resistor to low speed cooling fan control relay (brown/orange)*
7   *Reduced power from cooling fan dropping resistor to low speed cooling fan control relay (brown/orange)*
8   *Power to cooling fan motors (red/orange)*
9   *PCM power relay coil ground (black/light blue)*
10   *Power to cooling fan motors (red/orange)*
11   *Battery power to fuel pump relay (pink/black)*
12   *Battery power to cooling fan control relay (black/orange)*
13   *Battery power from PCM power relay (red)*
14   *Low speed fan control from PCM (dark blue)*
15   *Ground (black/white)*
16   *Air conditioning compressor clutch ground (green/white)*
17   *High speed fan control from PCM (light green/pink)*
18   *Fuel pump relay control from PCM (light blue/orange)*
19   *Not used*
20   *Not used*
21   *Air conditioning high pressure cutout switch signal to PCM (black/yellow)*
22   *Wide-open throttle signal to PCM (pink/yellow)*
23   *Power to air conditioning compressor clutch (pink/light blue)*
24   *Battery power to cooling fan control relay (black/orange)*

contains the fuel pump relay, the low-speed fan control relay, the high-speed fan control relay, and the air conditioning clutch control relay **(see illustrations)**.

12   The CCRM is serviced only as an assembly, the relays inside are not replaceable individually. To test the separate operation of each relay in the CCRM. Refer to **illustration 6.11b** and the wiring diagrams at the end of this chapter.

13   Use your voltmeter or ohmmeter to test operation and continuity of the power circuits as necessary. Further testing of the CCRM should be performed by a dealership or other qualified repair shop.

7   Turn signal/hazard flasher - check and replacement

Warning: *The models covered by this manual are equipped with Supplemental Restraint systems (SRS), more commonly known as airbags. Always disconnect the negative battery cable, then the positive battery cable and wait two minutes before working in the vicinity of the impact sensors, steering column or instrument panel to avoid the possibility of accidental deployment of the airbag, which could cause personal injury (see Section 27). Do not use electrical test equipment on any of the airbag system wiring or tamper with them in any way.*

1   The turn signal and hazard flashers are controlled from a single electronic flasher unit which is mounted to the side of the interior fuse block **(see illustration 3.1a)**.

2   When the flasher unit is functioning properly, an audible click can be heard during its operation. If the turn signals fail on one side or the other and the flasher unit does not make its characteristic clicking sound, a faulty turn signal bulb is indicated.

3   If both turn signals fail to blink, the prob-

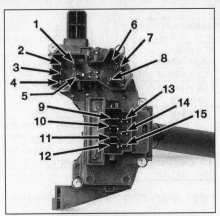

8.3a Multi- function switch terminal identification guide

lem may be due to a blown fuse, a faulty flasher unit, a broken switch or a loose or open connection. If a quick check of the fuse box indicates that the turn signal fuse has blown, check the wiring for a short before installing a new fuse.

4   To replace the flasher, simply unplug it from the instrument panel fuse block.

5   Make sure that the replacement unit is identical to the original. Compare the old one to the new one before installing it.

6   Installation is the reverse of removal.

8   Steering column switches - check and replacement

Warning: *The models covered by this manual are equipped with Supplemental Restraint systems (SRS), more commonly known as airbags. Always disconnect the negative battery cable, then the positive battery cable and wait two minutes before working in the vicinity of the impact sensors, steering column or*

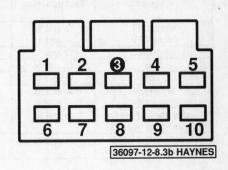

36097-12-8.3b HAYNES

8.3b Multi-function switch terminal identification guide (1999 and later)

instrument panel to avoid the possibility of accidental deployment of the airbag, which could cause personal injury (see Section 27). Do not use electrical test equipment on any of the airbag system wiring or tamper with them in any way.

Multi-function switch
Check
Refer to illustrations 8.3a, 8.3b, 8.3c, 8.3d, 8.3e, 8.3f, 8.3g and 8.3h

1   The multi-function switch is located on the left side of the steering column. It incorporates the turn signal, headlight dimmer and windshield wiper/washer functions into one switch.

2   Remove the multi-function switch (see Step 4).

3   Using an ohmmeter or self-powered test light and the accompanying diagrams, check for continuity between the indicated switch terminals with the switch in each of the indicated positions **(see illustrations)**. If the continuity isn't as specified, replace the switch.

Switch positions	Hazard warning	Continuity between
Neutral	OFF ON	1 and 6; 1 and 8 3 and 5; 3 and 8 3 and 2; 3 and 6
Left	OFF	4 and 6; 4 and 2
Right	OFF	4 and 8; 4 and 5

8.3c Turn signal and hazard switch continuity chart (1995 through 1998)

Switch positions	Hazard warning	Continuity between
Neutral	OFF ON	9 and 10 8 and 10
Left	OFF	8 and 10
Right	OFF	9 and 10

8.3d Turn signal and hazard warning switch continuity chart (1999 and later)

Switch positions	Continuity between
Flash to Pass hold lever in this position	12 and 11 9 and 10
High beam	9 and 11
Low beam	9 and 10

8.3e Headlight dimmer switch continuity chart (1995 through1998)

Switch position	Continuity between
Flash to Pass hold lever in this position	7 and 10 6 and 10
High beam	7 and 10
Low beam	6 and 10

8.3f Headlight dimmer switch continuity chart (1999 and later)

12

Switch positions	Test terminals	Ohmmeter readings (+ or -- 10%)
Wiper OFF Wash OFF	13 and 14	47.6 K-ohms
	14 and 15	103.3 K-ohms
Intermitent	13 and 14	11.3 K-ohms
Low	13 and 14	4.08 K-ohms
High	13 and 14	0-ohms
Wash ON Wiper OFF	14 and 15	0-ohms
Delay control knob @ maximum position	14 and 15	103.3 K-ohms
Delay control knob @ minimum position	14 and 15	3.3 K-ohms

8.3g Windshield wiper switch continuity chart (1995 through 1998)

Switch positions	Test terminals	Ohmmeter readings
Wiper OFF Wash ON	3 and 4	103.3 K-ohms
Intermitent Low High	3 and 5	11.33 K-ohms (+ or -- 5%) 4.0 K-ohms (+ or -- 5%) 0-ohms
Delay control knob @ maximum position	3 and 4	103.3 K-ohms (+10%)
Delay control knob @ minimum position	3 and 4	3.4 K-ohms (+10%)

8.3h Windshield wiper switch continuity chart (1999 and later)

8.6a Remove the retaining screws (arrow) . . .

8.6b . . . unplug the electrical connectors, then remove the multi-function switch

8.8 The cruise control amplifier/servo electrical connector (arrow) is located in the left front corner of the engine compartment - on 1996 and later 3.8L engines its located beneath the air filter housing

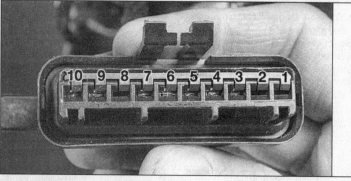

8.10a Cruise control speed amplifier/servo terminal identification guide

Cruise control switches

Check

Refer to illustrations 8.8, 8.10a and 8.10b

8   Open the hood and locate the cruise control speed amplifier/servo (**see Illustration**).

9   Disconnect the electrical connector from the cruise control speed amplifier/servo.

10   Using an ohmmeter and the accompanying diagrams, measure the resistance between the indicated terminals leading to the speed amplifier/servo with the switch in each of the indicated positions (**see illustrations**). If the resistance isn't as specified, replace the switch.

Replacement

Refer to illustrations 8.6a and 8.6b

4   Disconnect the negative battery cable, then the positive battery cable and wait two minutes before proceeding any further.

5   Remove the ignition lock cylinder (see Section 10) and the steering column covers (see Chapter 11).

6   Remove the switch retaining screws, disconnect the electrical connectors, then detach the switch from the steering column (**see illustrations**).

7   Installation is the reverse of removal.

Switch positions	Test terminals	Ohmmeter readings (+ or -- 10%)
OFF	5 and 6	less than 4 K ohms
ON with ignition in RUN position	5 and 10	Battery voltage
SET/ACCEL	5 and 6	720 and 640 K ohms
COAST	5 and 6	126 and 114 K ohms
RESUME	5 and 6	2310 and 2090 K ohms

Rotate the steering wheel through its full range of travel while performing each of the above tests.

8.10b Cruise control actuator switch continuity chart

8.13 Detach the driver side airbag module to access the cruise control actuator switch retaining screws (arrows)

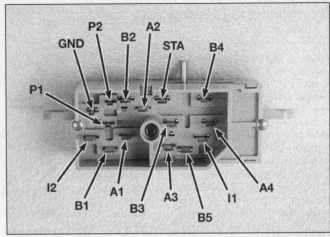

9.2a Check the ignition switch terminals for continuity in each of the indicated positions

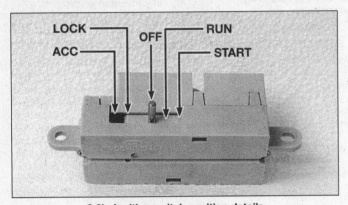

9.2b Ignition switch position details

Switch positions	Continuity between
ACC	B5 and A1
LOCK	No continuity
OFF	No continuity
RUN	B5 and I1; B1 and A1 B3 and A3; B4 and A4
START	B4 and STA P1 and GND

9.2c Ignition switch continuity chart

Replacement

Refer to illustration 8.13

11   Disconnect the negative battery cable, then the positive battery cable and wait two minutes before proceeding any further.

12   Remove the driver's side airbag module from the steering wheel (see Chapter 10).

13   Detach the switch retaining screws **(see illustration)**. Then disconnect the electrical connections and remove the switches from the steering wheel.

14   Installation is the reverse of removal.

9   Ignition switch - check and replacement

Warning: *The models covered by this manual are equipped with Supplemental Restraint systems (SRS), more commonly known as airbags. Always disconnect the negative battery cable, then the positive battery cable and wait two minutes before working in the vicinity of the impact sensors, steering column or instrument panel to avoid the possibility of accidental deployment of the airbag, which could cause personal injury (see Section 27). Do not use electrical test equipment on any of the airbag system wiring or tamper with them in any way.*

Check

Refer to illustrations 9.2a, 9.2b and 9.2c

1   The ignition switch is mounted below the steering column behind the knee bolster. To check the switch it must first be removed (see Step 3).

2   Using an ohmmeter or self-powered test light and the accompanying diagrams, check for continuity between the indicated switch terminals with the switch in each of the indicated positions **(see illustrations).** If the continuity isn't as specified, replace the switch.

Replacement

Refer to illustration 9.6

3   Disconnect the negative battery cable, then the positive battery cable and wait two minutes before proceeding any further.

4   Turn the ignition key lock cylinder to the Run position.

5   Remove the driver side knee bolster and the lower steering column cover (see Chapter 11).

6   Unplug the ignition switch electrical connector and remove the switch retaining screws **(see illustration)**.

7   Disengage the ignition switch from the actuator pin and remove the switch from the vehicle.

8   Make sure the actuator pin slot in the

new ignition switch is in the Run position **(see illustration 9.2b)**. **Note:** *A new replacement switch assembly will be set in this position.*

9   Place the new switch in position on the actuator pin and install the retaining screws. It may be necessary to move the switch back and forth to line up the screw holes.

10   The remainder of the installation is the reverse of removal. Check for proper operation of the ignition switch in the lock, start and accessory positions.

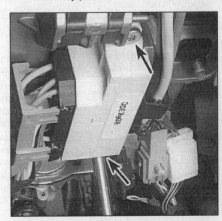

9.6 Unplug the electrical connector and remove the ignition switch retaining screws (arrows)

12

10 Ignition lock cylinder - removal and installation

Warning: *The models covered by this manual are equipped with Supplemental Restraint systems (SRS), more commonly known as airbags. Always disconnect the negative battery cable, then the positive battery cable and wait two minutes before working in the vicinity of the impact sensors, steering column or instrument panel to avoid the possibility of accidental deployment of the airbag, which could cause personal injury (see Section 27). Do not use electrical test equipment on any of the airbag system wiring or tamper with them in any way.*

Removal

Refer to illustration 10.3
1   Disconnect the negative battery cable, then the positive battery cable and wait two minutes before proceeding any further.
2   Turn the ignition key/lock cylinder to the Run position.
3   Insert an 1/8-inch diameter tool into the hole at the bottom of the steering column cover surrounding the lock cylinder. Depress the punch while pulling out on the lock cylinder to remove it from the column housing **(see illustration)**.

Installation

4   Depress the retaining pin on the side of the lock cylinder and rotate the ignition

10.3 To remove the ignition lock cylinder, place the key in the "RUN" position, push in on the release tab with the proper tool and pull the cylinder straight out

key/lock cylinder to the Run position.
5   Install the lock cylinder into the steering column housing, making sure it's fully seated and aligned in the interlocking washer.
6   Rotate the key back to the Off position. This will allow the retaining pin to extend itself back into the locating hole in the steering column housing.
7   Turn the lock to ensure that operation is correct in all positions.
8   The remainder of installation is the reverse of removal.

11.2a Headlight switch terminal identification guide

11 Instrument panel switches - check and replacement

Warning: *The models covered by this manual are equipped with Supplemental Restraint systems (SRS), more commonly known as airbags. Always disconnect the negative battery cable, then the positive battery cable and wait two minutes before working in the vicinity of the impact sensors, steering column or instrument panel to avoid the possibility of accidental deployment of the airbag, which*

Switch positions	Test terminals	Ohmmeter readings
OFF	4 and 5; 10 and 11	open circuit
PARK	4 and 5	closed circuit
ON	4 and 5; 10 and 11	closed circuit
Dimmer knob rotated to full left position	6 and 7	closed circuit
Dimmer knob rotating right from the full left position	1 and 12	varied resistance
Autolamp knob rotated from off to full delay position	10 and 15	varied resistance
Autolamp knob rotated to off position	10 and 14	varied resistance from 3 K to 200 K-ohms
Autolamp knob rotated to all other positions	10 and 14	open circuit
		closed circuit

11.2b Headlight switch continuity chart

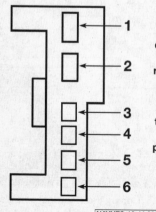

11.2c Interior light dimmer switch terminal guide - with the knob rotating UP from the full DOWN position there should be varied resistance between terminals 4 and 5 - with the knob in the full UP position there should be continuity between terminals 1 and 2

HAYNES-12-11.02C.L

11.5a Detach the headlight switch retaining screws (arrows) - unplug the electrical connector and withdraw the switch from the instrument panel

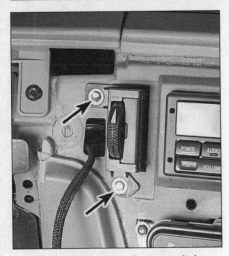

11.5b Interior light dimmer switch mounting screws (arrows)

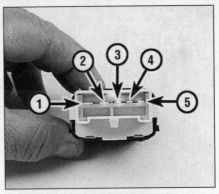

11.9a Rear defogger switch terminal guide - with the defogger switch in the ON position, check for continuity between terminals 4 and 5 - to check the illumination section of the switch, simply switch the ohmmeter to the diode function and check for continuity between terminals 1 and 5 and between terminals 2 and 5

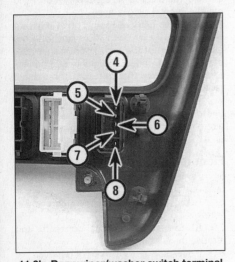

11.9b Rear wiper/washer switch terminal guide - with the rear wiper switch in the ON position, check for continuity between terminals 6 and 7 - with the rear washer switch in the ON position, check for continuity between terminals 6 and 8 - to check the illumination bulb, check for continuity between terminals 4 and 5

could cause personal injury (see Section 27). Do not use electrical test equipment on any of the airbag system wiring or tamper with them in any way.

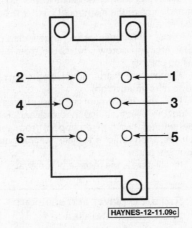

HAYNES-12-11.09c

11.9c Rear quarter window switch terminal guide (1995 through 1997)

Headlight switch and interior light dimmer switch

Check

Refer to illustrations 11.2a, 11.2b and 11.2c

1   To check the switch it must first be removed (see Step 3).

2   Using an ohmmeter or self-powered test light and the accompanying diagrams, check for continuity between the indicated switch terminals with the switch in each of the indicated positions (see illustrations). If the continuity isn't as specified, replace the switch.

Replacement

Refer to illustrations 11.5a and 11.5b

3   To remove the headlight switch, detach the knob by pulling it straight off, then remove the headlight switch trim bezel (see Chapter 11). Remove the knee bolster and the knee bolster reinforcement panel (see Chapter 11).

4   To remove the interior light dimmer switch, first remove the radio trim bezel (see Chapter 11).

5   Detach the switch retaining screws (see illustrations).

6   Unplug the electrical connector and remove the switch from the vehicle.

7   Installation is the reverse of removal.

Rear window defogger, rear wiper, rear quarter window and fog lamp switches

Check

Refer to illustrations 11.9a. 11.9b, 11.9c, 11.9d and 11.9e

8   To check the switch it must first be removed (see Step 10).

9   Using an ohmmeter or self-powered test light and the accompanying diagrams, check for continuity between the indicated switch terminals with the switch in each of the indicated positions (see illustrations). If the continuity isn't as specified, replace the switch.

Replacement

Refer to illustration 11.11

10   Remove the radio trim bezel from the instrument panel (see Chapter 11).

11   Detach the plastic locking tabs on each side of the switch and remove it from the radio trim bezel (see illustration).

12   Installation is the reverse of removal.

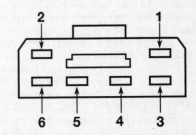

11.9e Fog lamp switch terminal guide - with the switch in the ON position, check for continuity between terminals 5 and 6 - to check the illumination bulb, check for continuity between terminals 3 and 4 - to check the indicator ON lamp, check for continuity between terminals 4 and 6

Rear quarter window switch	Switch positions	Continuity between
LEFT	UP	4 and 2; 1 and 3
	DOWN	4 and 1; 2 and 3
RIGHT	UP	4 and 5; 6 and 3
	DOWN	4 and 6; 5 and 3

11.9d Rear quarter window switch continuity chart (1995 through 1997)

12

12 Instrument panel gauges - check

Warning: *The models covered by this manual are equipped with Supplemental Restraint systems (SRS), more commonly known as airbags. Always disconnect the negative battery cable, then the positive battery cable and wait two minutes before working in the vicinity of the impact sensors, steering column or instrument panel to avoid the possibility of accidental deployment of the airbag, which could cause personal injury (see Section 27). Do not use electrical test equipment on any of the airbag system wiring or tamper with them in any way.*

Note: *This procedure applies to conventional analog type gauges (NON-digital) only.*

Fuel and temperature gauges

1   All tests below require the ignition switch to be turned to ON position when testing.

2   If the gauge pointer does not move from the empty or cold positions, check the fuse. If the fuse is OK, locate the particular sending unit for the circuit you're working on (see Chapter 4 for fuel sending unit location or Chapter 3 for the temperature sending unit location). Connect the sending unit connector to ground momentarily. If the pointer goes to the full or hot position replace the sending unit. If the pointer stays in same position, use a jumper wire to ground the sending unit terminal on the back of the gauge. If necessary, refer to the wiring diagrams at the end of this Chapter. If the pointer moves, the problem lies in the wiring between the gauge and the sending unit. If the pointer does not move with the sending unit terminal on the back of the gauge grounded, check for voltage at the other terminal of the gauge. If voltage is present, replace the gauge.

13 Instrument cluster - removal and installation

Refer to illustration 13.6

Warning: *The models covered by this manual are equipped with Supplemental Restraint systems (SRS), more commonly known as airbags. Always disconnect the negative battery cable, then the positive battery cable and wait two minutes before working in the vicinity of the impact sensors, steering column or instrument panel to avoid the possibility of accidental deployment of the airbag, which could cause personal injury (see Section 27). Do not use electrical test equipment on any of the airbag system wiring or tamper with them in any way.*

1   Disconnect the negative battery cable, then the positive battery cable and wait two minutes before proceeding any further.

2   Remove the knee bolster, headlight switch bezel, radio trim bezel and the center trim panel from the instrument panel (see Chapter 11).

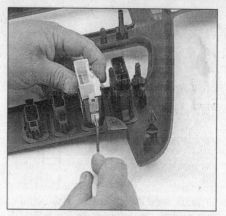

11.11 Detach the retaining clips on each side of the switches and detach them from the radio trim bezel

3   On 1999 and later models, turn the windshield wiper control stalk counterclockwise and remove it from the left side of the steering column.

4   Tilt the steering wheel to its lowest position and remove the instrument cluster bezel (see Chapter 11).

5   Detach the transmission range indicator cable retaining screw and cable from the steering column.

6   Remove the instrument cluster retaining screws **(see illustration)**.

7   Pull the instrument cluster out and unplug the electrical connectors and the speedometer cable (if equipped) from the backside, then remove the cluster from the instrument panel.

8   Installation is the reverse of removal.

14 Radio/CD player and speakers - removal and installation

Warning: *The models covered by this manual are equipped with Supplemental Restraint systems (SRS), more commonly known as airbags. Always disconnect the negative battery cable, then the positive battery cable and wait two minutes before working in the vicinity of the impact sensors, steering column or instrument panel to avoid the possibility of accidental deployment of the airbag, which could cause personal injury (see Section 27). Do not use electrical test equipment on any of the airbag system wiring or tamper with them in any way.*

1   Disconnect the negative battery cable, then the positive battery cable and wait two minutes before proceeding any further.

Radio/CD player

Refer to illustration 14.3

2   For theft protection, the radio receiver/CD player is retained in the instrument panel by special clips. Releasing these clips requires the use of two special removal tools (available at most auto parts stores) or two short lengths of coat hanger wire bent into U-shapes. Insert the tools into the holes

13.6 The instrument cluster is held in place by screws (arrows) at both sides of the housing

at the corners of the radio/CD player until you feel the internal clips release.

3   With the clips released, push outward simultaneously on both tools and pull the assembly out of the instrument panel, disconnect the antenna and electrical connectors and remove the unit from the vehicle **(see illustration)**.

4   Install by plugging in the electrical connectors, then sliding the radio/CD player along the track and into the instrument panel until the clips can be felt snapping in place.

Floor mounted CD changer

5   Remove the floor console (see Chapter 11).

6   Remove the seven CD changer retaining screws.

7   Disconnect the electrical connector and remove the CD changer.

8   Installation is the reverse of removal.

Speakers

Refer to illustration 14.10

Door mounted

9   Remove the door trim panel (see Chapter 11).

14.3 Insert the radio removal tools until they seat, then push outward simultaneously on both tools to release the clips and withdraw the radio from the dash

14.10 After removing the trim panel, the speaker retaining screws (arrows) are easily accessible

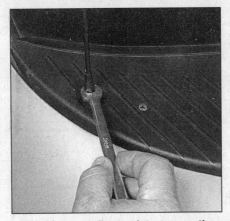

15.2 Use a small wrench to remove the antenna mast

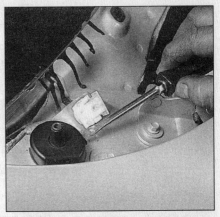

15.4 Pry off the antenna base trim cap

15.5 Remove the retaining screws (arrows) securing the antenna base to the cowl

16.2 Disengage the headlight housing retaining clips (arrows) - pull up and out to remove them

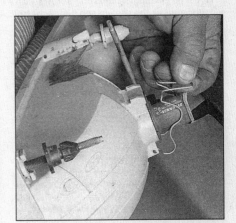

16.4 Detach the headlight bulb retaining clip and pull the bulb and socket assembly out of the housing - when installing the new bulb, don't touch the surface; clean it with rubbing alcohol if you do

10   Remove the mounting screws, withdraw the speaker, unplug the electrical connector and remove the speaker from the vehicle (see illustration).
11   Installation is the reverse of removal.

Rear quarter panel mounted

12   Remove the rear seat and the rear quarter trim panels (see Chapter 11).
13   Remove the speaker retaining screws, withdraw the speaker, unplug the electrical connector and remove the speaker from the vehicle.
14   Installation is the reverse of removal.

15   Antenna - removal and installation

Refer to illustrations 15.2, 15.4 and 15.5

1   Detach the radio and disconnect the antenna lead from the rear side of the radio (see Section 12). Remove the cable retaining clips securing the antenna lead to the underside of the instrument panel.
2   Working from the outside of the vehicle, use a small wrench and remove the mast (see illustration).

3   Remove the cowl top cover (see Chapter 11).
4   Pry off the antenna base trim cap (see illustration).
5   Remove the screws securing the antenna to the body (see illustration), then pull the antenna up and out to remove it.
6   Installation is the reverse of removal.

16   Headlights - replacement

Refer to illustrations 16.2 and 16.4
Warning: *Halogen gas filled bulbs are under pressure and may shatter if the surface is scratched or the bulb is dropped. Wear eye protection and handle the bulbs carefully, grasping only the base whenever possible. Do not touch the surface of the bulb with your fingers because the oil from your skin could cause it to overheat and fail prematurely. If you do touch the bulb surface, clean it with rubbing alcohol.*
1   Open the hood.
2   Using a pair of pliers or similar tool, grasp the top of the headlight housing retaining clips, then pull upward on the clips to release them from the headlight housing (see

illustration).
3   Pull the headlight housing out and away from the vehicle and unplug the electrical connectors from the rear of the housing.
4   Detach the bulb retaining clip (see illustration).
5   Remove the bulb and socket assembly by pulling it straight out from the headlight housing.
6   Without touching the glass with your bare fingers, insert the new bulb and socket assembly into the headlight housing, then install the bulb retaining clip.
7   Plug in electrical connectors, then reposition the headlight housing back into place and install the housing retaining clips. Test headlight operation, then close the hood.

17   Headlights - adjustment

Refer to illustrations 17.1 and 17.3
Note: *The headlights must be aimed correctly. If adjusted incorrectly they could blind the driver of an oncoming vehicle and cause a serious accident or seriously reduce your*

12

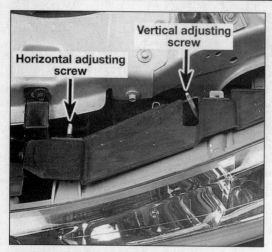

17.1 The headlight adjusting screws (arrows) are located at the rear of the headlight housing

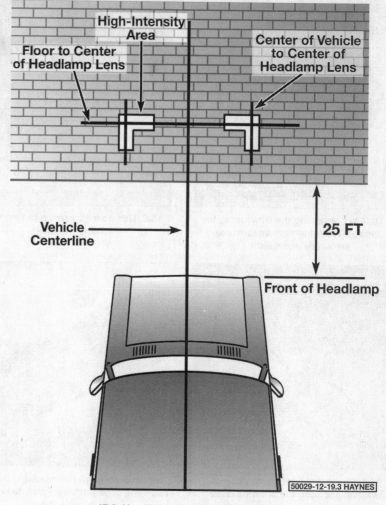

17.3 Headlight adjustment details

ability to see the road. The headlights should be checked for proper aim every 12 months and any time a new headlight is installed or front end body work is performed. It should be emphasized that the following procedure is only an interim step which will provide temporary adjustment until the headlights can be adjusted by a properly equipped shop.

1   The headlights have two adjusting screws each, one inboard and one outboard. Both adjusters are accessible from the rear of the headlight assembly and are turned using a small wrench or pliers **(see illustration)**.

2   There are several methods of adjusting the headlights. The simplest method requires masking tape, a blank wall and a level floor.

3   Position masking tape vertically on the wall in reference to the vehicle centerline and the centerlines of both headlights **(see illustration)**.

4   Position a horizontal tape line in reference to the centerline of all the headlights. **Note:** *It may be easier to position the tape on the wall with the vehicle parked only a few inches away.*

5   Adjustment should be made with the vehicle parked 25 feet from the wall, sitting level, the gas tank half-full and no unusually heavy load in the vehicle.

6   Starting with the low beam adjustment, position the high intensity zone so it is two inches below the horizontal line and two inches to the right of the headlight vertical line. Adjustment is made by turning the top adjusting screw clockwise to raise the beam and counterclockwise to lower the beam. The adjusting screw on the side should be used in the same manner to move the beam left or right.

7   With the high beams on, the high intensity zone should be vertically centered with the exact center just below the horizontal line. **Note:** *It may not be possible to position the headlight aim exactly for both high and low beams. If a compromise must be made, keep in mind that the low beams are the most used and have the greatest effect on safety.*

8   Have the headlights adjusted by a dealer service department or service station at the earliest opportunity.

18   Bulb replacement

Front turn signal bulbs

Refer to illustration 18.2

1   Detach the headlight housing (see Section 16).

2   Twist the bulb socket a quarter turn counterclockwise, then remove the bulb assembly from the housing **(see illustration)**.

3   The defective bulb can then be removed from the socket and replaced.

4   Installation of the headlight housing is the reverse of removal.

Side marker lights

Refer to illustrations 18.5 and 18.6

5   On 1995 through 1998 models, detach the side marker light retaining nut(s), then pull the side marker light housing outward away from the vehicle **(see illustration)**. On 1999 and later models, remove the headlight assembly (see Section 16).

6   Twist the bulb socket a quarter turn counterclockwise, then remove the bulb assembly from the housing **(see illustration)**.

7   The defective bulb can then be removed from the socket and replaced.

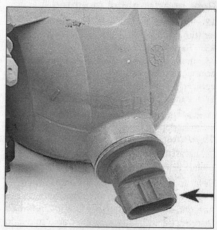

18.2 Rotate the bulb holder (arrow) counterclockwise and withdraw it from the headlight housing, then grasp the bulb and pull it straight out

18.5 Detach the side marker light housing retaining nut - pull outward on the top of the housing slightly, then lift upward to detach it from the vehicle

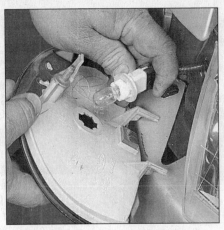

18.6 Rotate the bulb holder counterclockwise and withdraw it from the side marker light housing, then grasp the bulb and pull it straight out

18.10 Tail light housing retaining screw locations

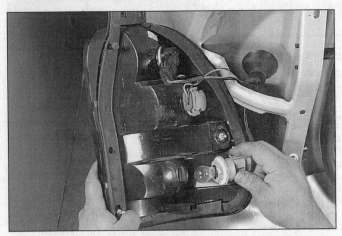

18.11 Rotate the bulb holder counterclockwise and withdraw it from the tail light housing

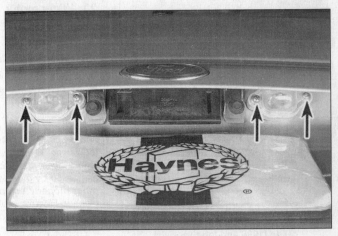

18.14a License plate lamp retaining screw locations

8   Installation of the side marker light housing is the reverse of removal.

Rear turn signal, brake, tail and back-up lights

Refer to illustrations 18.10 and 18.11

9   Open the liftgate.

10   Detach the retaining screws securing the rear tail light housing, then pull the tail light assembly outward to access the tail light bulbs **(see illustration)**.

11   Twist the bulb socket a quarter turn counterclockwise, then remove the bulb and socket assembly from the housing **(see illustration)**.

12   The defective bulb can then be removed from the socket and replaced.

13   Installation of the tail light housing is the reverse of removal.

License plate and high-mounted brake lights

Refer to illustrations 18.14a and 18.4b

14   Detach the retaining screws which secure the light housing to the liftgate **(see illustrations)**.

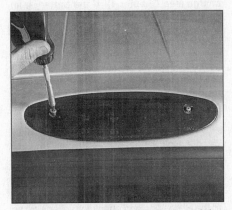

18.14b High mounted brake light retaining screw locations

15   Twist the bulb socket a quarter turn counterclockwise, then remove the bulb and socket assembly from the housing.

16   The defective bulb can then be pulled straight out of the socket and replaced.

17   Installation of the housing is the reverse of removal.

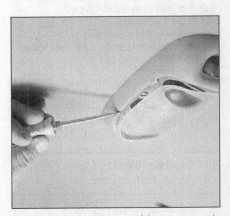

18.18 Use a small screwdriver to pry out the dome light lens

Dome light

Refer to illustration 18.18

18   Using a small screwdriver, remove the lens and replace the bulb **(see illustration)**.

Instrument cluster illumination

Refer to illustration 18.19

19   To gain access to the instrument cluster

12

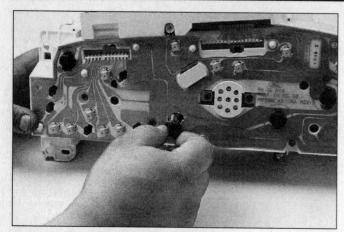

18.19 To remove an instrument cluster bulb, depress the bulb and rotate it counterclockwise

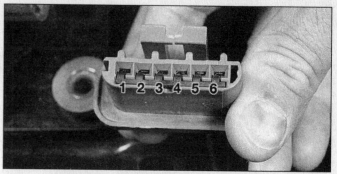

20.2a Windshield wiper motor terminal guide (1995 through 1997 models)

1   ground (black)
2   Park switch (black/pink)
3   Battery power (dark green)
4   Ground (black)
5   High speed power (white)
6   Low speed power (dark blue/orange)

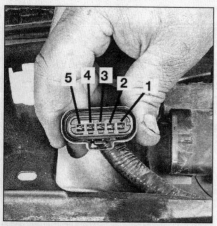

20.2b Windshield wiper motor terminal guide (1998 and later models)

1   Park switch (black/pink)
2   Battery power (dark green)
3   Ground (black)
4   Low speed power (white)
5   High speed power (dark blue/orange)

illumination lights, the instrument cluster will have to be removed (see Section 13). The bulbs can then be removed and replaced from the rear of the cluster **(see illustration).**

19   Daytime Running Lights (DRL) - general information

The Daytime Running Lights (DRL) system used on Canadian models illuminates the headlights whenever the engine is running. The only exception is with the engine running and the parking brake engaged. Once the parking brake is released, the lights will remain on as long as the ignition switch is on, even if the parking brake is later applied.

The DRL system supplies reduced power to the headlights so they won't be too bright for daytime use, while prolonging headlight life.

20   Wiper motor - check and replacement

Wiper motor circuit check

Refer to illustrations 20.2a and 20.2b

Note: *Refer to the wiring diagrams for wire colors and locations in the following checks. When checking for voltage, probe a grounded 12-volt test light to each terminal at a connector until it lights; this verifies voltage (power) at the terminal. If the following checks fail to locate the problem, have the system diagnosed by a dealer service department or other properly equipped repair facility. The Generic Electronic Module (GEM) is capable of storing trouble codes that can be retrieved with the proper equipment.*

1   If the wipers work slowly, make sure the battery is in good condition and has a strong charge (see Chapter 1). If the battery is in good condition, remove the wiper motor (see below) and operate the wiper arms by hand. Check for binding linkage and pivots. Lubricate or repair the linkage or pivots as necessary. Reinstall the wiper motor. If the wipers still operate slowly, check for loose or corroded connections, especially the ground connection. If all connections look OK, replace the motor.

2   If the wipers fail to operate when activated, check the fuse. If the fuse is OK, connect a jumper wire between the wiper motor and ground, then retest. If the motor works now, repair the ground connection. If the motor still doesn't work, turn the wiper switch to the HI position and check for voltage at the motor **(see illustrations).** If there's voltage at the motor, remove the motor and check it off the vehicle with fused jumper wires from the battery. If the motor now works, check for binding linkage (see Step 1 above). If the motor still doesn't work, replace it. If there's no voltage at the motor, check for voltage at the wiper control relays. If there's voltage at the wiper control relays and no voltage at the at the wiper motor, check the switch for continuity (see Section 8). If the switch is OK, the

20.7a Pry open the trim cover and remove the wiper arm retaining nut(s)

wiper control relays or the Generic Electronic Module (GEM) are probably bad. Have the GEM diagnosed by a dealer service department or other properly equipped repair facility.

3   If the interval (delay) function is inoperative, check the continuity of all the wiring between the switch and wiper control module. If the wiring is OK, check the resistance of the delay control knob of the multi-function switch (see Section 8). If the delay control knob is within the specified resistance, have the GEM diagnosed by a dealer service department or other properly equipped repair facility.

4   If the wipers stop at the position they're in when the switch is turned off (fail to park), check for voltage at the park feed wire of the wiper motor connector when the wiper switch is OFF but the ignition is ON. If no voltage is present, check for an open circuit between the wiper motor and the fuse panel.

5   If the wipers won't shut off unless the ignition is OFF, disconnect the wiring from the wiper control switch. If the wipers stop, replace the switch. If the wipers keep running, there's a defective limit switch in the motor; replace the motor.

6   If the wipers won't retract below the

20.7b Detach the washer hoses, then pull the wiper arm straight off the spindle shaft to remove it

20.10 Remove the wiper assembly retaining bolts (arrows)

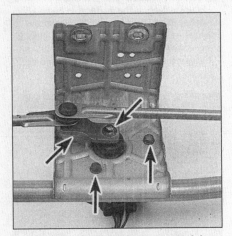

20.11 Detach the wiper motor retaining bolts (arrows)

20.14 Pull back the rear wiper arm cover, then remove the nut and pull the arm straight off its splined shaft

20.17 Remove the rear wiper motor retaining bolts (arrows)

hood line, check for mechanical obstructions in the wiper linkage or on the vehicle's body which would prevent the wipers from parking. If there are no obstructions, check the wiring between the switch and motor for continuity. If the wiring is OK, replace the wiper motor.

Wiper motor replacement

Front

Refer to illustrations 20.7a, 20.7b, 20.10 and 20.11

7   Remove the windshield wiper arms **(see illustrations).**
8   Remove the cowl upper and lower covers (see Chapter 11).
9   Disconnect the electrical connector from the wiper motor.
10   Detach the wiper linkage assembly from the lower cowl cover **(see illustration).**
11   Remove the wiper motor retaining bolts **(see illustration).**
12   Remove the wiper motor from the wiper linkage assembly.
13   Installation is the reverse of removal.

Rear

Refer to illustrations 20.14 and 20.17

14   Pull the rear wiper arm cover back to access the wiper arm retaining nut. Detach

the nut and pull the wiper arm straight off the shaft to remove it **(see illustration).**
15   Open the liftgate and remove the liftgate trim panel (see Chapter 11).
16   Disconnect the electrical connector from the wiper motor.
17   Remove the wiper motor mounting bracket bolts **(see illustration).**
18   Pull the wiper motor out from the liftgate and remove it from the vehicle.
19   Installation is the reverse of removal.

21 Horn - check and replacement

Check

Refer to illustration 21.3
Note: *Check the fuses before beginning electrical diagnosis.*
1   Disconnect the electrical connector from the horn.
2   To test the horn(s), connect battery voltage to the horn terminal with a pair of jumper wires. If the horn doesn't sound, replace it.
3   If the horn does sound, check for voltage at the horn connector when the horn switch is depressed **(see illustration).** If

there's voltage at the connector, check for a bad ground at the horn.
4   If there's no voltage at the horn, check the relay (see Section 6).
5   If the relay is OK, check for voltage to the relay power and control circuits. If either of the circuits is not receiving voltage, inspect the wiring between the relay and the fuse panel.

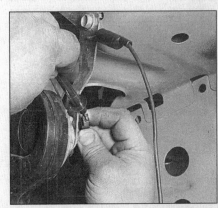

21.3 Connect a voltmeter to the horn wire and ground - test for voltage while the switch is depressed

12

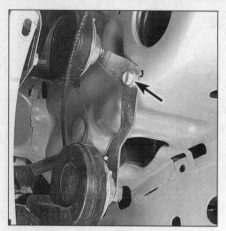

21.9 Disconnect the electrical connector, remove the bolt (arrow) and detach the horns

22.5 When measuring the voltage at the rear window defogger grid, wrap a piece of aluminum foil around the positive probe of the voltmeter and press the foil against the wire with your finger

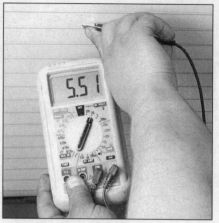

22.6 To determine if a wire has broken, check the voltage at the center of each wire. If the voltage is 5 to 6 volts, the wire is unbroken; if the voltage is 10 to 12 volts, the wire is broken between the center of the wire and the ground side; if the voltage is 0 volts, the wire is broken between the center of the wire and the power side

6   If both relay circuits are receiving voltage, depress the horn switch and check the circuit from the relay to the horn switch for continuity to ground. If there's no continuity, check the circuit for an open. If there's no open circuit, replace the horn switch.

7   If there's continuity to ground through the horn switch, check for an open or short in the circuit from the relay to the switch.

Replacement

Refer to illustration 21.9

8   To access the horns, the left front inner fenderwell must first be removed (see Chapter 11).

9   Disconnect the electrical connectors and remove the bracket bolt **(see illustration)**.

10   Installation is the reverse of removal.

22   Rear window defogger - check and repair

1   The rear window defogger consists of a number of horizontal heating elements baked onto the inside surface of the glass. Power is supplied through a 60-ampere fuse from the power distribution box in the engine compartment. The heater is controlled by the instrument panel switch and the Generic Electronic Module (GEM).

2   Small breaks in the element can be repaired without removing the rear window.

Check

Refer to illustrations 22.5, 22.6 and 22.8

3   Turn the ignition switch and defogger system switches to the ON position.

4   Using a voltmeter, place the positive probe against the defogger grid positive terminal and the negative probe against the ground terminal. If battery voltage is not indicated, check the fuse, defogger switch and related wiring. If voltage is indicated, but all or part of the defogger doesn't heat, proceed with the following tests.

22.8 To find the break, place the voltmeter negative lead against the defogger ground terminal, place the voltmeter positive lead with the foil strip against the heat wire at the positive terminal end and slide it toward the negative terminal end - the point at which the voltmeter deflects from several volts to zero volts is the point at which the wire is broken

5   When measuring voltage during the next two tests, wrap a piece of aluminum foil around the tip of the voltmeter positive probe and press the foil against the heating element with your finger **(see illustration)**. Place the negative probe on the defogger grid ground terminal.

6   Check the voltage at the center of each heating element **(see illustration)**. If the voltage is 5 to 6 volts, the element is okay (there is no break). If the voltage is 0 volts, the element is broken between the center of the element and the positive end. If the voltage is 10 to 12 volts the element is broken between the center of the element and the ground side. Check each heating element.

7   If none of the elements are broken, connect the negative probe to a good chassis ground. The voltage reading should stay the

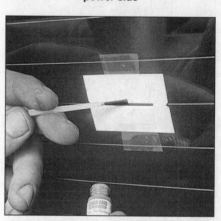

22.14 To use a defogger repair kit, apply masking to the inside of the window at the damaged area, then brush on the special conductive coating

same, if it doesn't the ground connection is bad.

8   To find the break, place the voltmeter negative probe against the defogger ground terminal. Place the voltmeter positive probe with the foil strip against the heating element at the positive side and slide it toward the negative side. The point at which the voltmeter deflects from several volts to zero is the point where the heating element is broken **(see illustration)**.

Repair

Refer to illustration 22.14

9   Repair the break in the element using a repair kit specifically for this purpose, such as Dupont paste No. 4817 (or equivalent). The kit includes plastic conductive epoxy.

10   Before repairing a break, turn off the system and allow it to cool for a few minutes.

11   Lightly buff the element area with fine

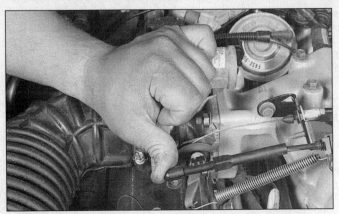

23.5 Make sure the cruise control and accelerator linkage mounted on the throttle body are not damaged and that they operate smoothly together when the throttle is opened

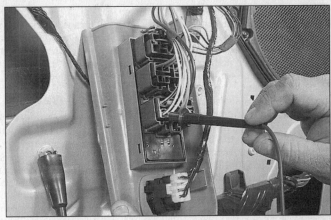

24.12 If no voltage is found at the motor with the switch depressed - check for voltage at the switch

steel wool; then clean it thoroughly with rubbing alcohol.

12   Use masking tape to mask off the area being repaired.

13   Thoroughly mix the epoxy, following the kit instructions.

14   Apply the epoxy material to the slit in the masking tape, overlapping the undamaged area about 3/4-inch on either end **(see illustration)**.

15   Allow the repair to cure for 24 hours before removing the tape and using the system.

23   Cruise control system - description and check

Refer to illustration 23.5

1   The cruise control system maintains vehicle speed with a servo motor located in the engine compartment on the driver's side fenderwell, which is connected to the throttle linkage by a cable. The system consists of the servo motor, brake switch, control switches, speed sensors and relays. Some features of the system require special testers and diagnostic procedures which are beyond the scope of this manual. Listed below are some general procedures that may be used to locate common problems.

2   Locate and check the fuse (see Section 3).

3   The brake pedal position (BPP) switch (or stop lamp switch) deactivates the cruise control system. Have an assistant press the brake pedal while you check the stop lamp operation.

4   If the brake lights do not operate properly, correct the problem and retest the cruise control.

5   Check the control cable between the cruise control servo/amplifier and the throttle linkage and replace as necessary **(see illustration)**.

6   The cruise control system uses a speed sensing device. The speed sensor is located in the transmission. To test the speed sensor, see Chapter 6, Section 4.

7   Test drive the vehicle to determine if the cruise control is now working. If it isn't, take it to a dealer service department or an automotive electrical specialist for further diagnosis.

24   Power window system - description and check

Refer to illustration 24.12

1   The power window system operates electric motors, mounted in the doors, which lower and raise the windows. The system consists of the control switches, the motors, regulators, glass mechanisms and associated wiring.

2   The power windows can be lowered and raised from the master control switch by the driver or by remote switches located at the individual windows. Each window has a separate motor which is reversible. The position of the control switch determines the polarity and therefore the direction of operation.

3   The circuit is protected by a fuse and a circuit breaker. Each motor is also equipped with an internal circuit breaker, this prevents one stuck window from disabling the whole system.

4   The power window system will only operate when the ignition switch is ON. In addition, many models have a window lockout switch at the master control switch which, when activated, disables the switches at the rear windows and, sometimes, the switch at the passenger's window also. Always check these items before troubleshooting a window problem.

5   These procedures are general in nature, so if you can't find the problem using them, take the vehicle to a dealer service department or other properly equipped repair facility.

6   If the power windows won't operate, always check the fuse and circuit breaker first.

7   If only the rear windows are inoperative, or if the windows only operate from the master control switch, check the rear window lockout switch for continuity in the unlocked

position. Replace it if it doesn't have continuity.

8   Check the wiring between the switches and fuse panel for continuity. Repair the wiring, if necessary.

9   If only one window is inoperative from the master control switch, try the other control switch at the window. **Note:** *This doesn't apply to the drivers door window.*

10   If the same window works from one switch, but not the other, check the switch for continuity.

11   If the switch tests OK, check for a short or open in the circuit between the affected switch and the window motor.

12   If one window is inoperative from both switches, remove the trim panel from the affected door and check for voltage at the switch **(see illustration)** and at the motor while the switch is operated.

13   If voltage is reaching the motor, disconnect the glass from the regulator (see Chapter 11). Move the window up and down by hand while checking for binding and damage. Also check for binding and damage to the regulator. If the regulator is not damaged and the window moves up and down smoothly, replace the motor. If there's binding or damage, lubricate, repair or replace parts, as necessary.

14   If voltage isn't reaching the motor, check the wiring in the circuit for continuity between the switches and motors. You'll need to consult the wiring diagram for the vehicle. If the circuit is equipped with a relay, check that the relay is grounded properly and receiving voltage.

15   Test the windows after you are done to confirm proper repairs.

25   Power door lock and keyless entry system - description and check

Power door lock system

Refer to illustration 25.9

1   The power door lock system operates

12

the door lock actuators mounted in each door. The system consists of the switches, actuators, and associated wiring. Diagnosis can usually be limited to simple checks of the wiring connections and actuators for minor faults which can be easily repaired.

2   Power door lock systems are operated by bi-directional solenoids located in the doors. The lock switches have two operating positions: Lock and Unlock. On later models with keyless entry the switches activate a module which in turn connects voltage to the door lock solenoids. Depending on which way the switch is activated, it reverses polarity, allowing the two sides of the circuit to be used alternately as the feed (positive) and ground side. On earlier models with out keyless entry the switches directly activate the door lock motors.

3   If you are unable to locate the trouble using the following general steps, consult your a dealer service department.

4   Always check the circuit protection first. On these models the battery voltage passes through the 20 amp circuit breaker located in the passenger compartment fuse block.

5   Operate the door lock switches in both directions (Lock and Unlock) with the engine off. Listen for the faint click of the door lock solenoid (motor) or relay operating.

6   If there's no click, check for voltage at the switches. If no voltage is present, check the wiring between the fuse block and the switches for shorts and opens.

7   If voltage is present but no click is heard, test the switch for continuity. Replace it if there's no continuity in both switch positions.

8   If the switch has continuity but the solenoid doesn't click, check the wiring between the switch and solenoid for continuity. Repair the wiring if there's not continuity.

9   If all but one lock solenoids operate, remove the trim panel from the affected door (see Chapter 11) and check for voltage at the solenoid while the lock switch is operated **(see illustration)**. One of the wires should have voltage in the Lock position; the other should have voltage in the unlock position.

10   If the inoperative solenoid is receiving voltage, replace the solenoid. **Note:** *It's common for wires to break in the portion of the harness between the body and door (opening and closing the door fatigues and eventually breaks the wires).*

Keyless entry system

Refer to illustrations 25.13 and 25.14

11   The keyless entry system consists of a remote control transmitter that sends a coded infrared signal to a receiver which then operates the door lock system.

12   Replace the transmitter batteries when the red LED light on the side of the case doesn't light when the button is pushed.

13   Use a small screwdriver to carefully separate the case halves **(see illustration)**.

14   Replace the two 3-volt 2016 lithium batteries **(see illustration)**.

15   Snap the case halves together.

---

26   Electric side view mirrors - description and check

1   Most electric rear view mirrors use two motors to move the glass; one for up and down adjustments and one for left-right adjustments.

2   The control switch has a selector portion which sends voltage to the left or right side mirror. With the ignition ON but the engine OFF, roll down the windows and operate the mirror control switch through all functions (left-right and up-down) for both the left and right side mirrors.

3   Listen carefully for the sound of the electric motors running in the mirrors.

4   If the motors can be heard but the mirror glass doesn't move, there's probably a problem with the drive mechanism inside the mirror. Remove and disassemble the mirror to locate the problem.

5   If the mirrors don't operate and no sound comes from the mirrors, check the fuse (see Section 3).

6   If the fuse is OK, remove the mirror control switch from its mounting without disconnecting the wires attached to it. Turn the ignition ON and check for voltage at the switch. There should be voltage at one terminal. If there's no voltage at the switch, check for an open or short in the wiring between the fuse panel and the switch.

7   If there's voltage at the switch, disconnect it. Check the switch for continuity in all its operating positions. If the switch does not have continuity, replace it.

8   Re-connect the switch. Locate the wire going from the switch to ground. Leaving the switch connected, connect a jumper wire between this wire and ground. If the mirror works normally with this wire in place, repair the faulty ground connection.

9   If the mirror still doesn't work, remove the mirror and check the wires at the mirror for voltage. Check with ignition ON and the mirror selector switch on the appropriate side. Operate the mirror switch in all its positions. There should be voltage at one of the switch-to-mirror wires in each switch position (except the neutral "off" position).

10   If there's not voltage in each switch position, check the wiring between the mirror and control switch for opens and shorts.

11   If there's voltage, remove the mirror and test it off the vehicle with jumper wires. Replace the mirror if it fails this test.

---

27   Airbag - general information

Refer to illustrations 27.1a, 27.1b and 27.1c

All models are equipped with a Supplemental Restraint System (SRS), more commonly known as an airbag. This system is designed to protect the driver, and the front seat passenger, from serious injury in the event of a head-on or frontal collision. It consists of an airbag module in the center of the steering wheel and the right side of the instrument panel, two crash sensors mounted at the front of the vehicle and a diagnostic module which also contains a safing sensor is located inside the passenger compartment **(see illustrations)**.

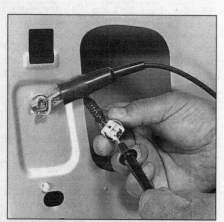

25.9 Check for voltage at the lock solenoid while the lock switch is operated

25.13 Use a small screwdriver to separate the transmitter halves

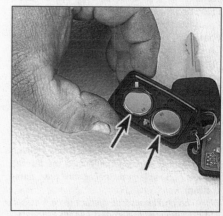

25.14 Replace the lithium batteries (arrows)

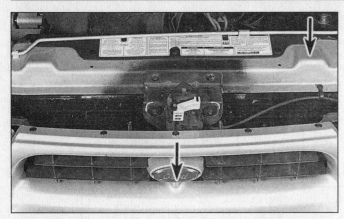

27.1a On 1995 through 1997 models, the left crash sensor (right arrow) is mounted just below the top of the radiator support and the center crash sensor (left arrow) is mounted behind the front bumper cover

27.1b On 1998 and later models, the front crash sensors (arrows) are mounted just below the top of the radiator support

Airbag module

Driver's airbag

The airbag inflator module contains a housing incorporating the cushion (airbag) and inflator unit, mounted in the center of the steering wheel The inflator assembly is mounted on the back of the housing over a hole through which gas is expelled, inflating the bag almost instantaneously when an electrical signal is sent from the system. A coil assembly on the steering column under the module carries this signal to the module.

This coil assembly can transmit an electrical signal regardless of steering wheel position. The igniter in the air bag converts the electrical signal to heat and ignites the sodium azide/copper oxide powder, producing nitrogen gas, which inflates the bag.

Passenger's airbag

The passenger's airbag is mounted above the glove compartment and designated by the letters SRS (Supplemental Restraint System). It consists of an inflator containing an igniter, an airbag assembly, a reaction housing and a trim cover.

The airbag is considerably larger that the steering wheel-mounted unit (8 cu. ft. vs. 2.3 cu. ft.) and is supported by the steel reaction housing. The trim cover is textured and painted to match the instrument panel and has a molded seam which splits when the bag inflates. As with the steering-wheel-mounted air bag, the igniter electrical signal converts to heat, converting sodium azide/iron oxide powder to nitrogen gas, inflating the bag.

Side airbags

On 1999 and later models, a side air bag is mounted within the front seat outboard bolster of both front seats. It consists of an inflator containing an igniter, an airbag assembly, a retention housing and the seat trim cover. The special seat cover is designed to allow air bag deployment. **Warning:** *Do not install an accessory seat cover over the existing seat cover as this material may prevent deployment of the air bag. As with the other air bags installed in the vehicle, the igniter electrical signal converts to heat, converting sodium azide/iron oxide powder to nitrogen gas, inflating the air bag.*

Sensors

The system has three sensors: two forward crash sensors at the front of the vehicle and a safing sensor mounted inside the airbag diagnostic module which is located behind the right kick panel.

The forward and passenger compartment sensors are basically pressure sensitive switches that complete an electrical circuit during an impact of sufficient G force. The electrical signal from these sensors is sent to the electronic diagnostic monitor which then completes the circuit and inflates the airbag(s).

Electronic diagnostic monitor

The electronic diagnostic monitor supplies the current to the airbag system in the event of the collision, even if battery power is cut off. It checks this system every time the vehicle is started, causing the "AIR BAG" light to go on then off, if the system is operating properly. If there is a fault in the system, the light will go on and stay on, flash, or the dash will make a beeping sound. If this happens, the vehicle should be taken to your dealer immediately for service.

Disabling the system

Whenever working in the vicinity of the steering wheel, steering column or near other components of the airbag system, the system should be disarmed. To do this, perform the following steps:

a) *Turn the ignition switch to Off.*
b) *Detach the cable from the negative battery terminal, then detach the positive cable. Wait two minutes for the electronic module backup power supply to be depleted.*

27.1c The airbag diagnostic module is mounted on top of the transmission tunnel behind the utility compartment

Enabling the system

a) *Turn the ignition switch to the Off position.*
b) *Connect the positive battery cable first, then connect the negative cable.*

---

28 Wiring diagrams - general information

Since it isn't possible to include all wiring diagrams for every year covered by this manual, the following diagrams are those that are typical and most commonly needed.

Prior to troubleshooting any circuits, check the fuse and circuit breakers (if equipped) to make sure they're in good condition. Make sure the battery is properly charged and check the cable connections (see Chapter 1).

When checking a circuit, make sure that all connectors are clean, with no broken or loose terminals. When unplugging a connector, do not pull on the wires. Pull only on the connector housings themselves.

12

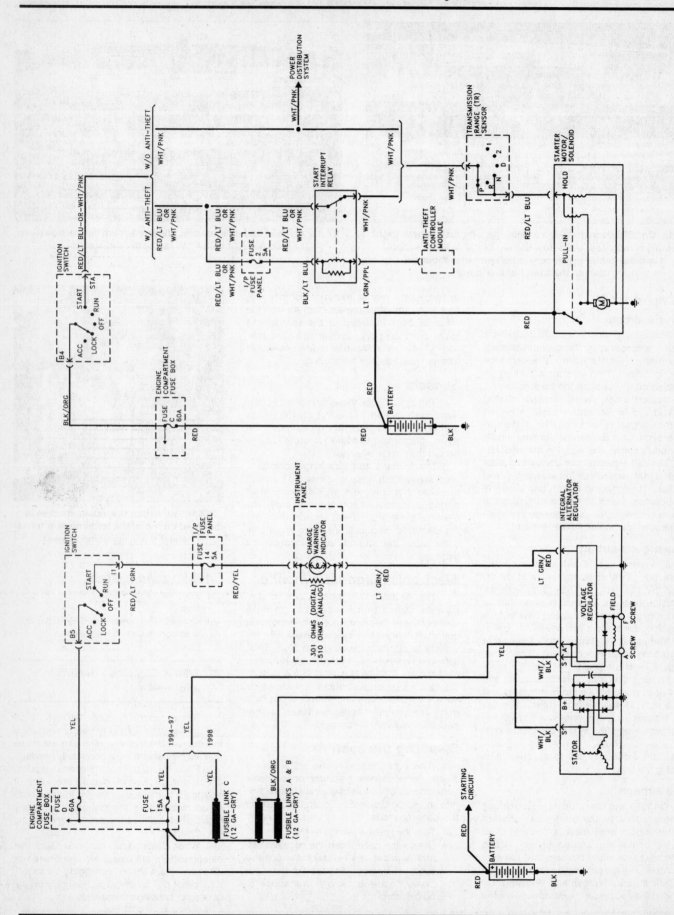

Starting and charging system

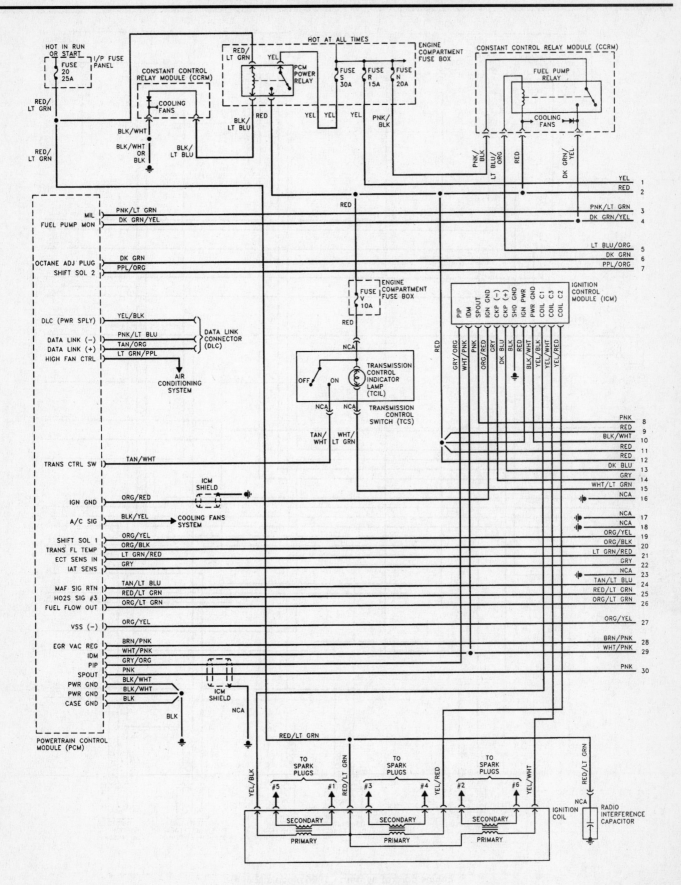

Engine control system - 1995 models (1 of 4)

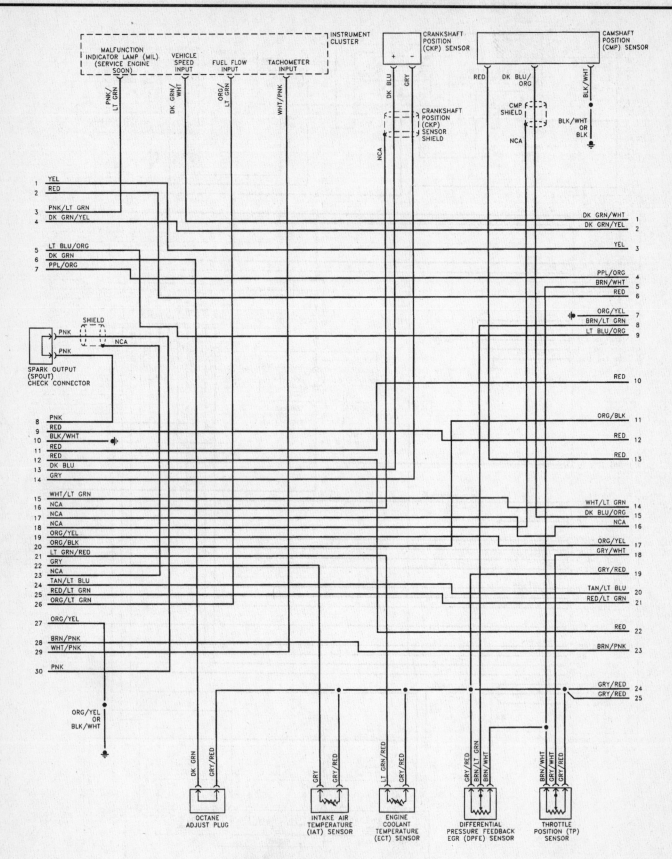

Engine control system - 1995 models (2 of 4)

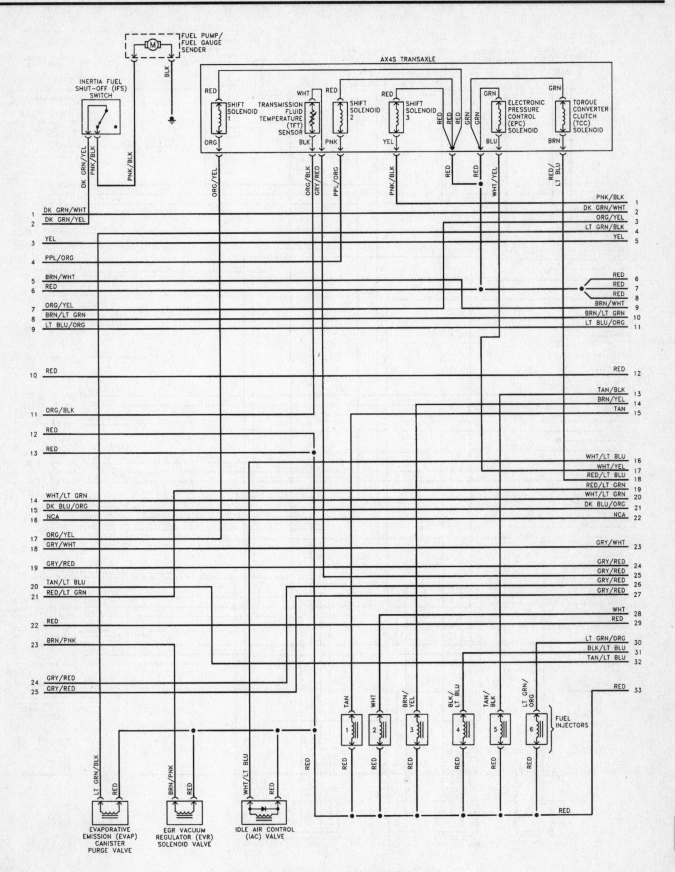

Engine control system - 1995 models (3 of 4)

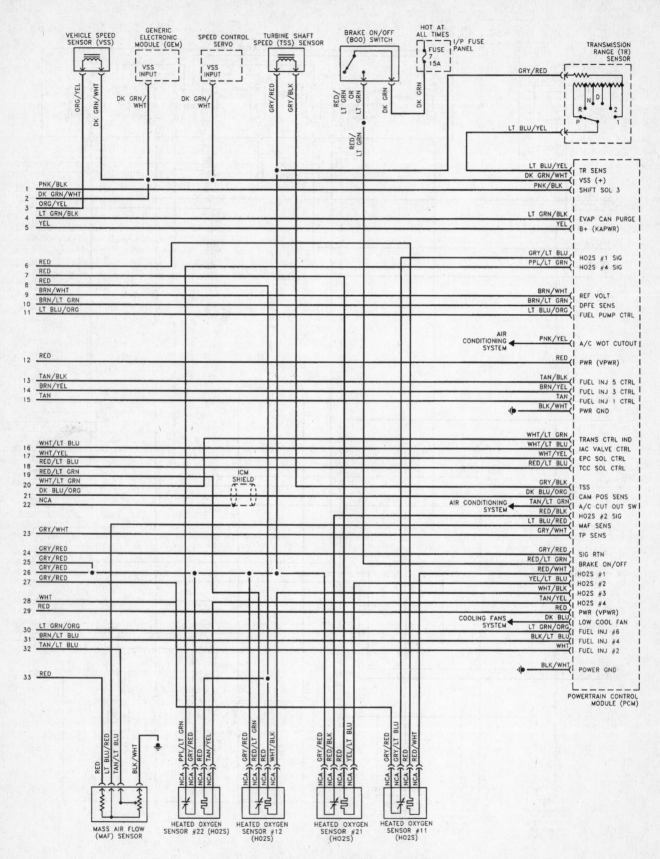

Engine control system - 1995 models (4 of 4)

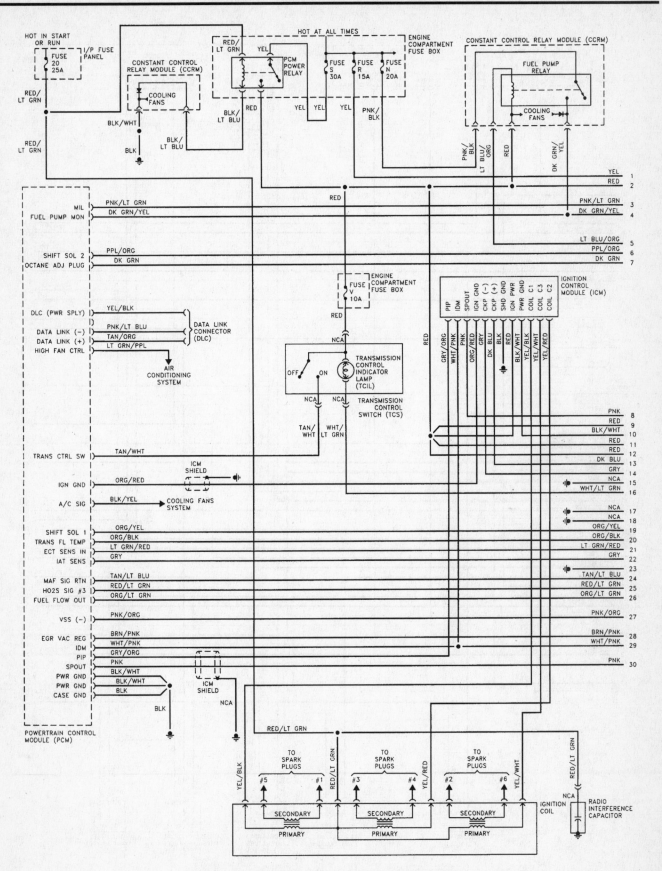

Engine control system - 1996 and 1997 models (1 of 4)

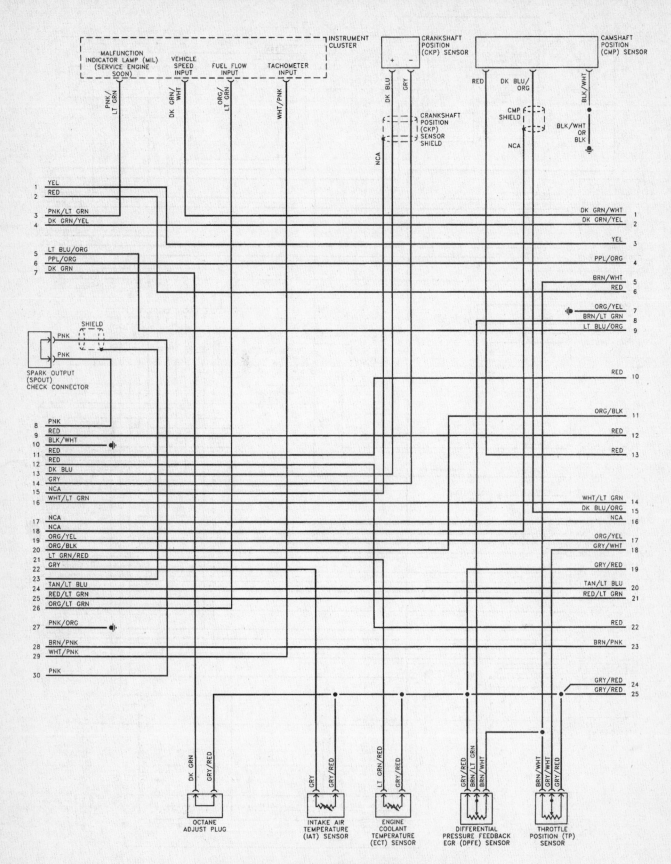

Engine control system - 1996 and 1997 models (2 of 4)

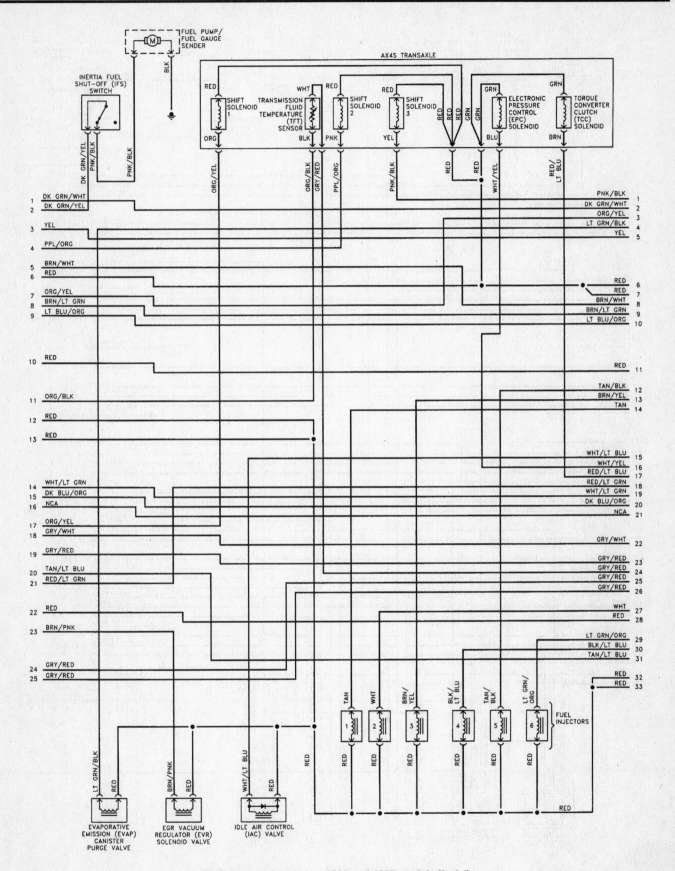

Engine control system - 1996 and 1997 models (3 of 4)

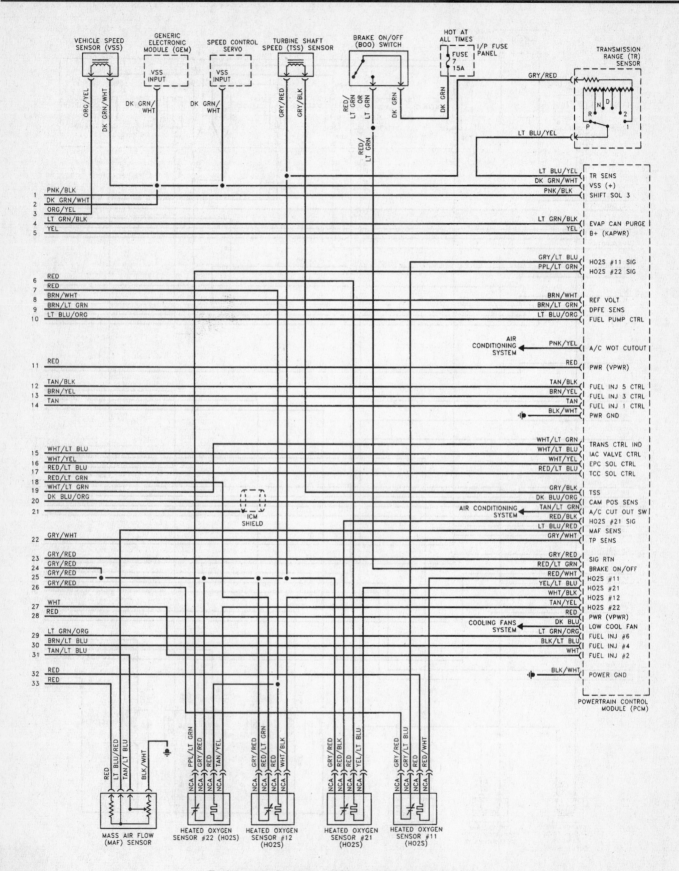

Engine control system - 1996 and 1997 models (4 of 4)

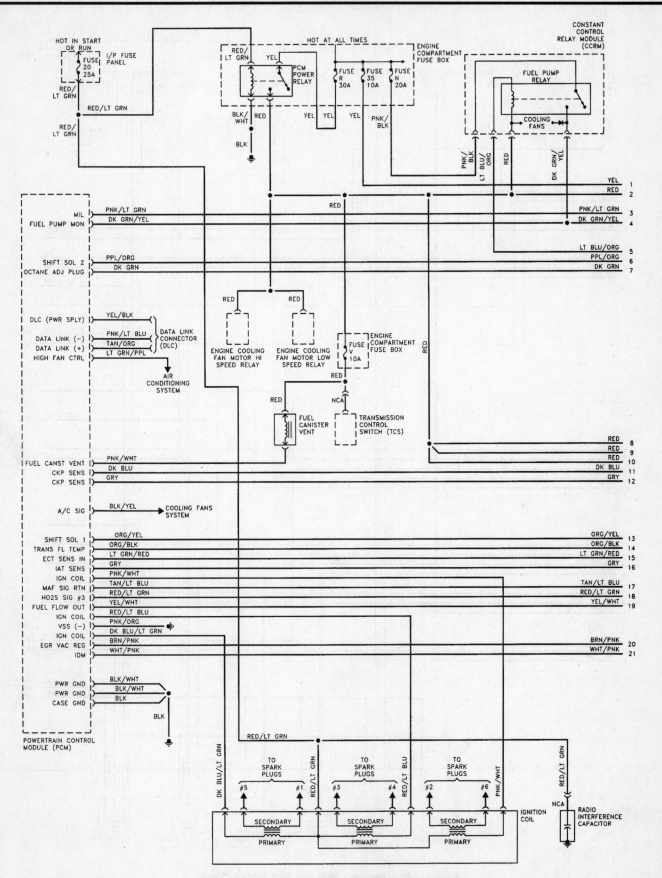

Engine control system - 1998 and later models (1 of 4)

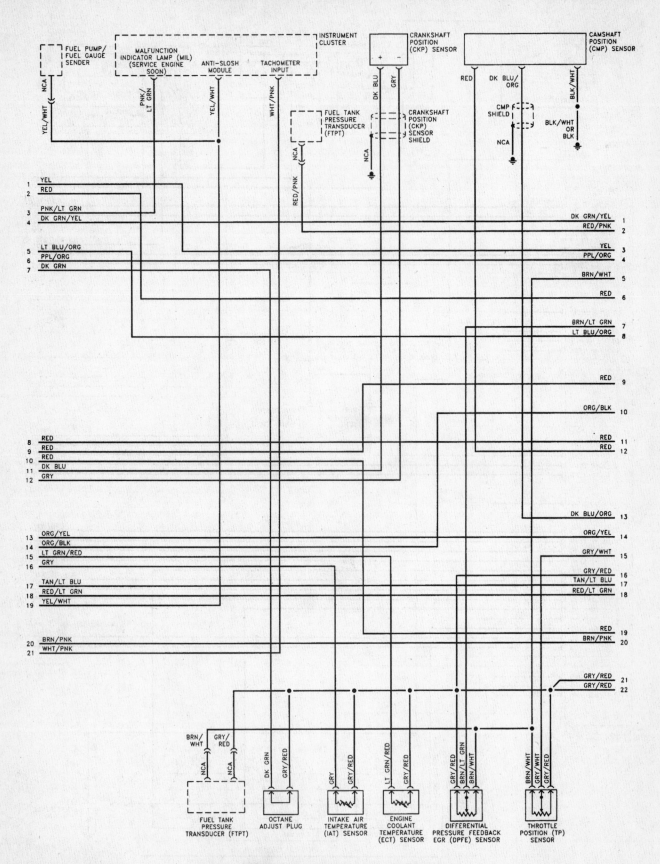

Engine control system - 1998 and later models (2 of 4)

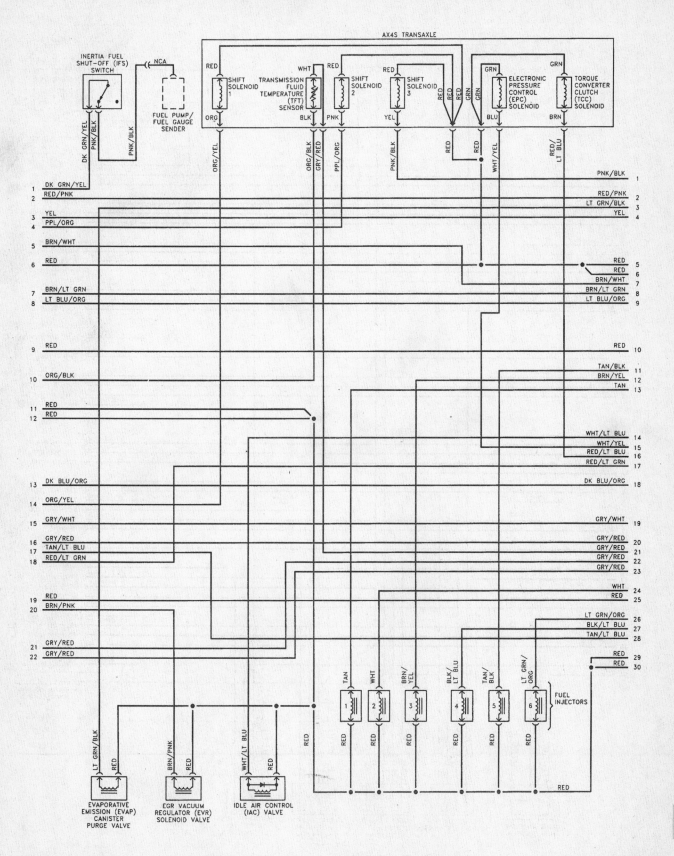

Engine control system - 1998 and later models (3 of 4)

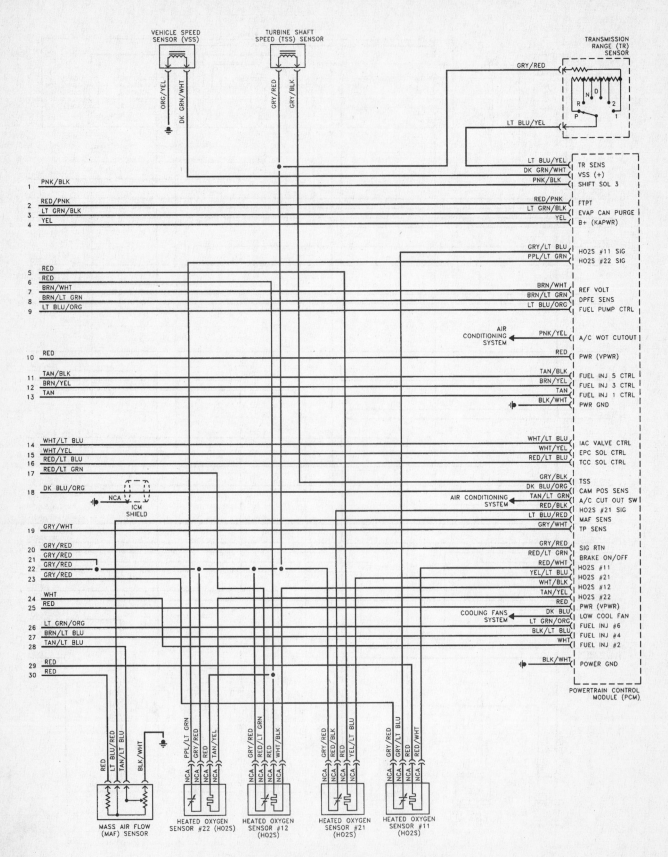

Engine control system - 1998 and later models (4 of 4)

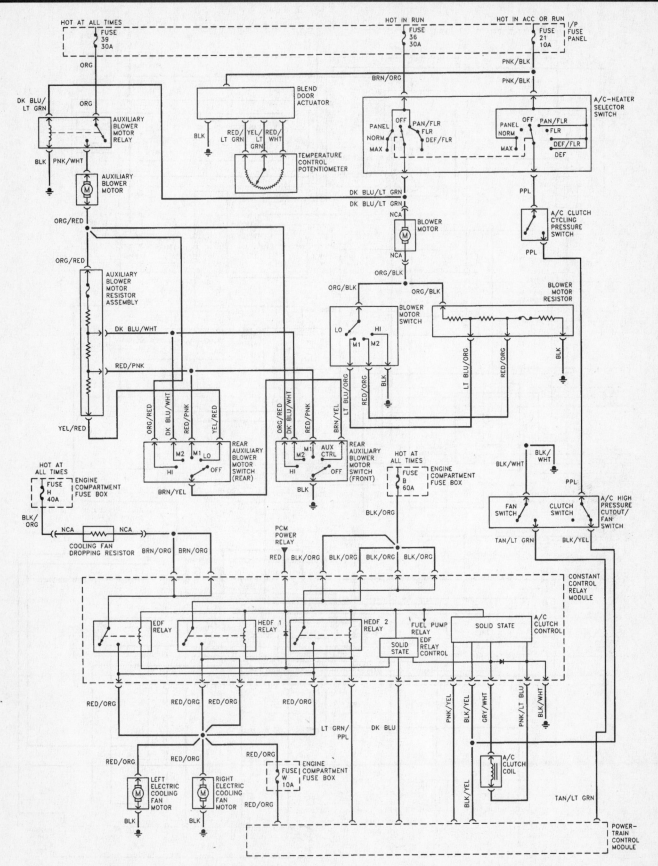

Heating , cooling and air conditioning system (includes engine cooling fan system) - 1995 models

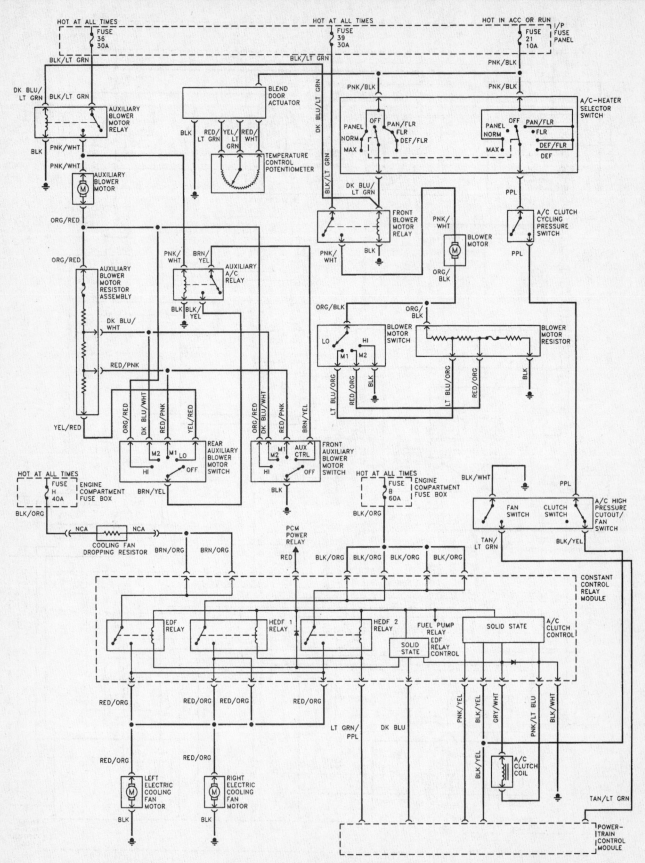

Heating , cooling and air conditioning system (includes engine cooling fan system) - 1996 and 1997 models

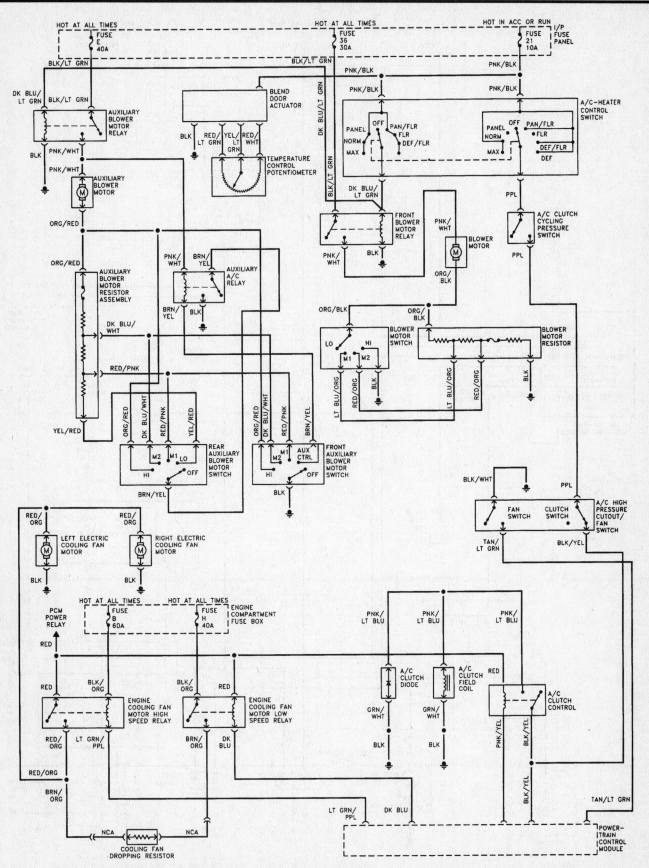

Heating , cooling and air conditioning system (includes engine cooling fan system) - 1998 and later models

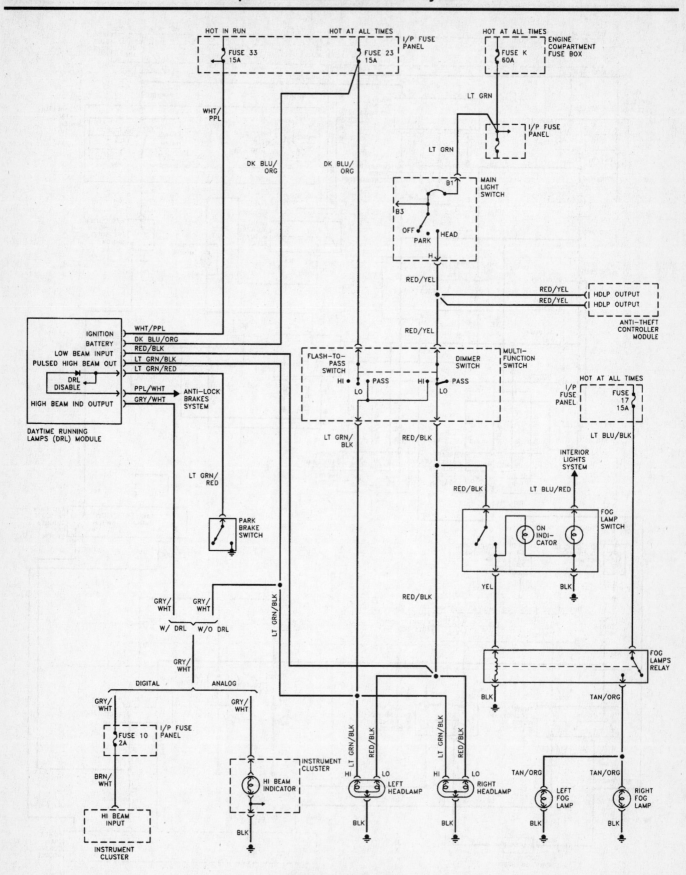

Headlight and foglight system

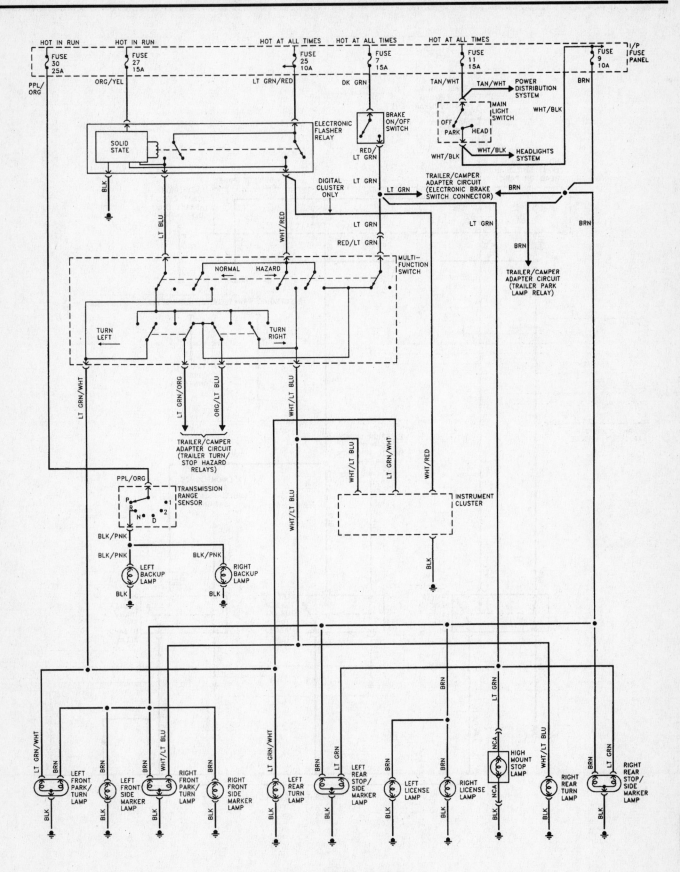

Exterior lighting system

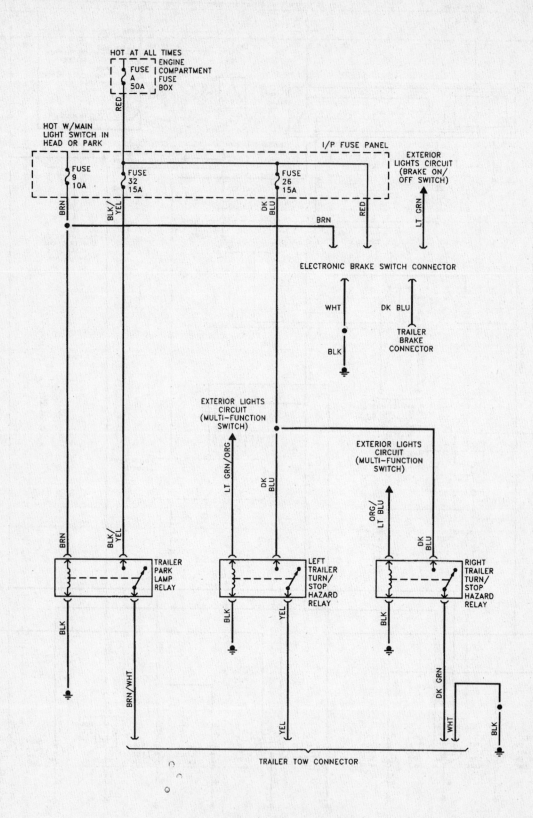

Optional trailering and towing system

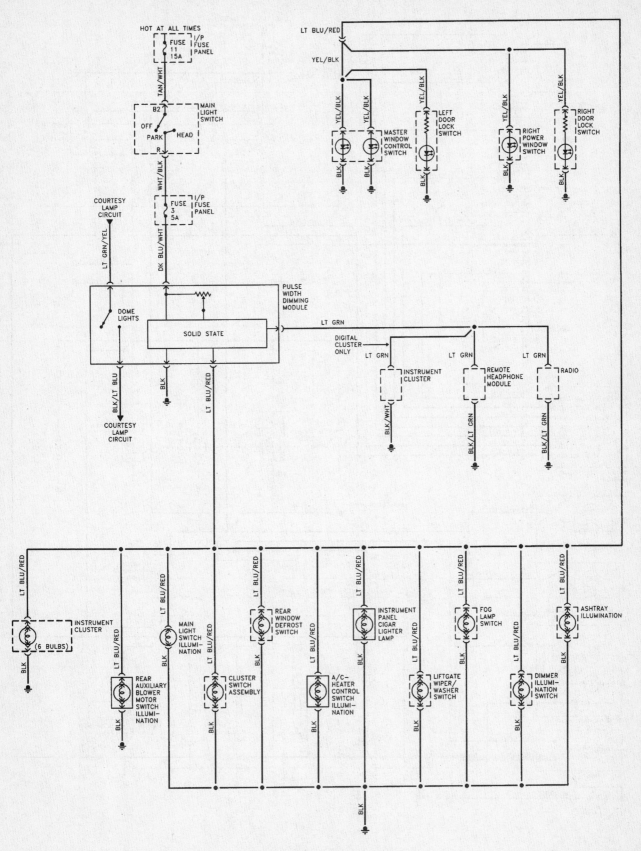

Interior lighting system (1 of 2)

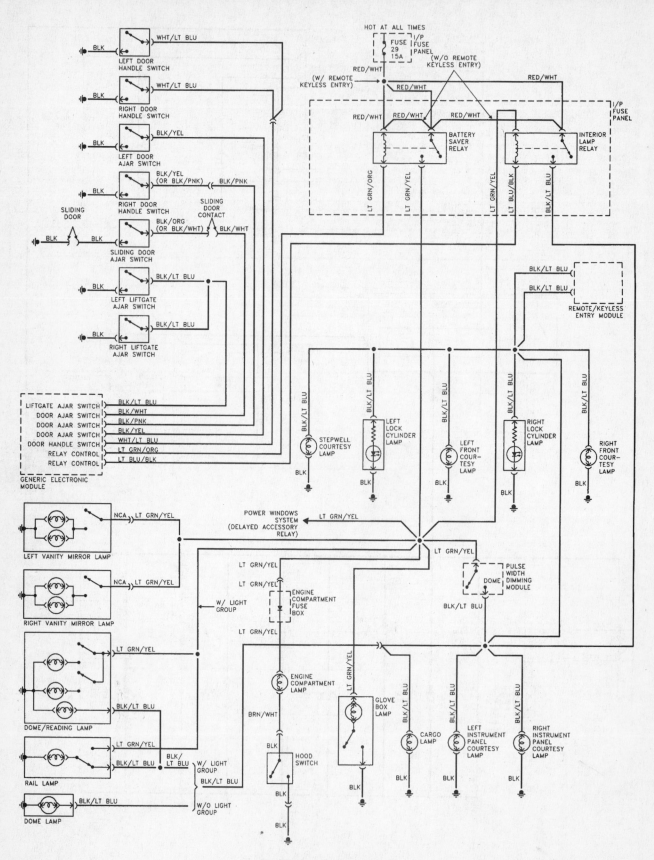

Interior lighting system (2 of 2)

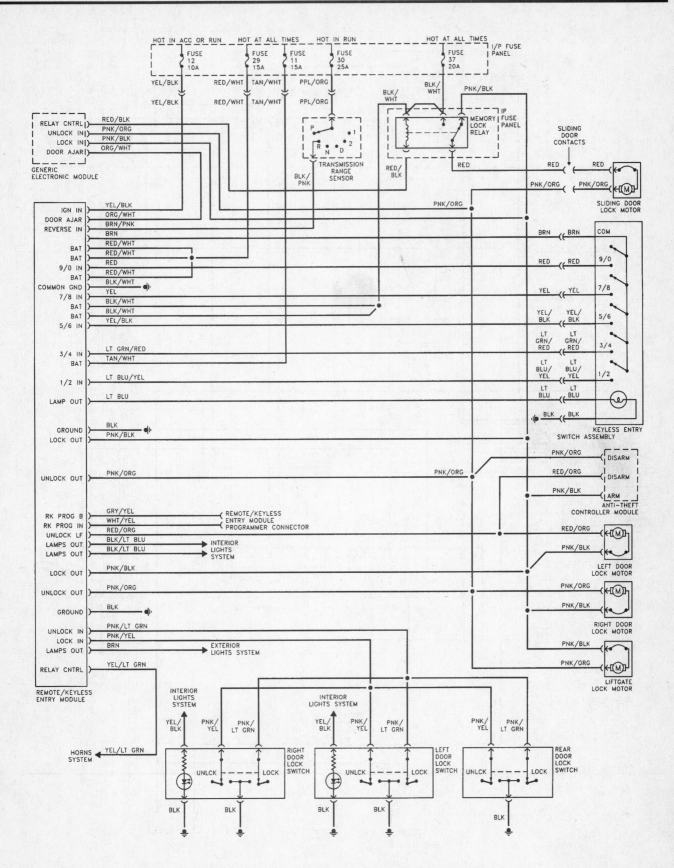

Power door lock system - 1995 through 1997 models (with keyless entry)

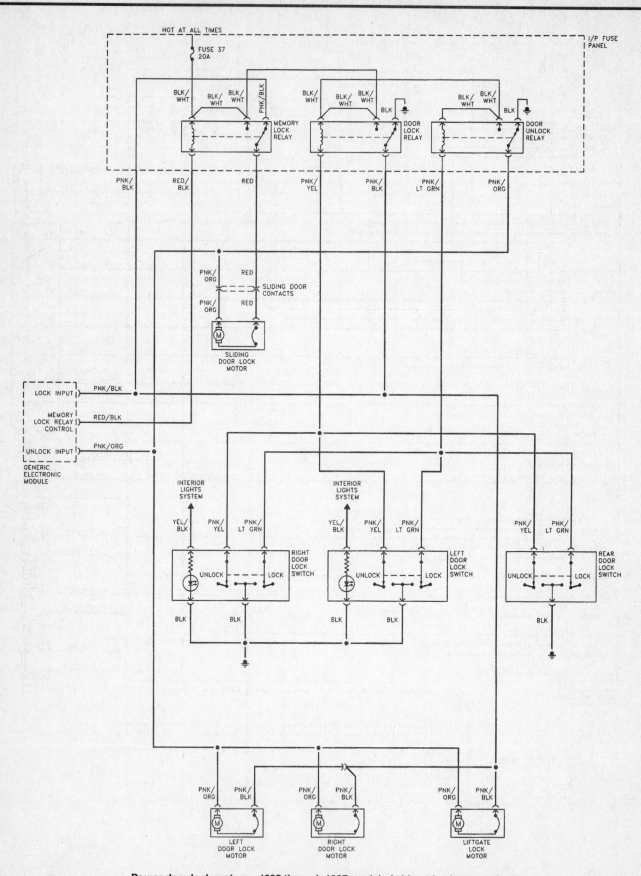

Power door lock system - 1995 through 1997 models (without keyless entry)

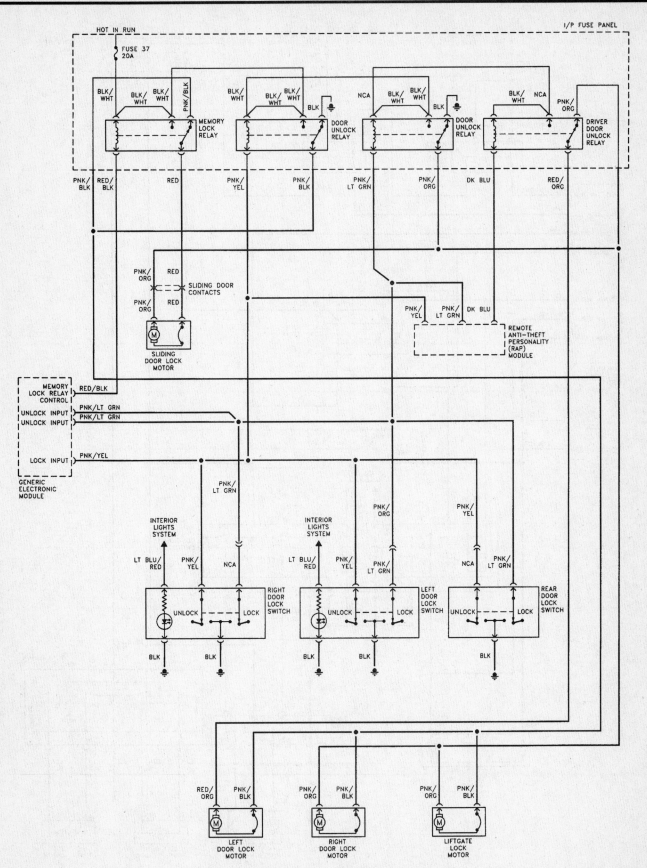

Power door lock system - 1998 and later models

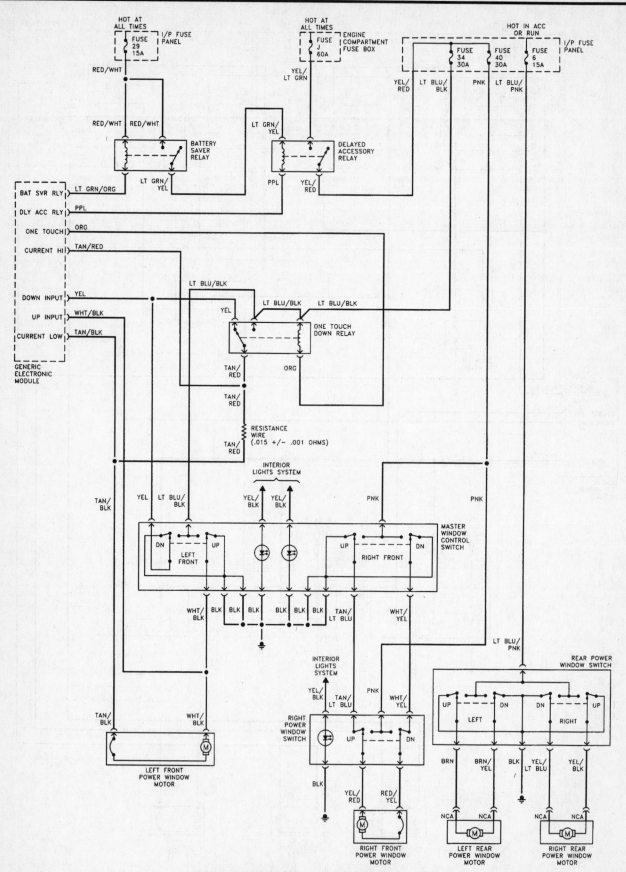

Power window system - 1995 through 1997 models

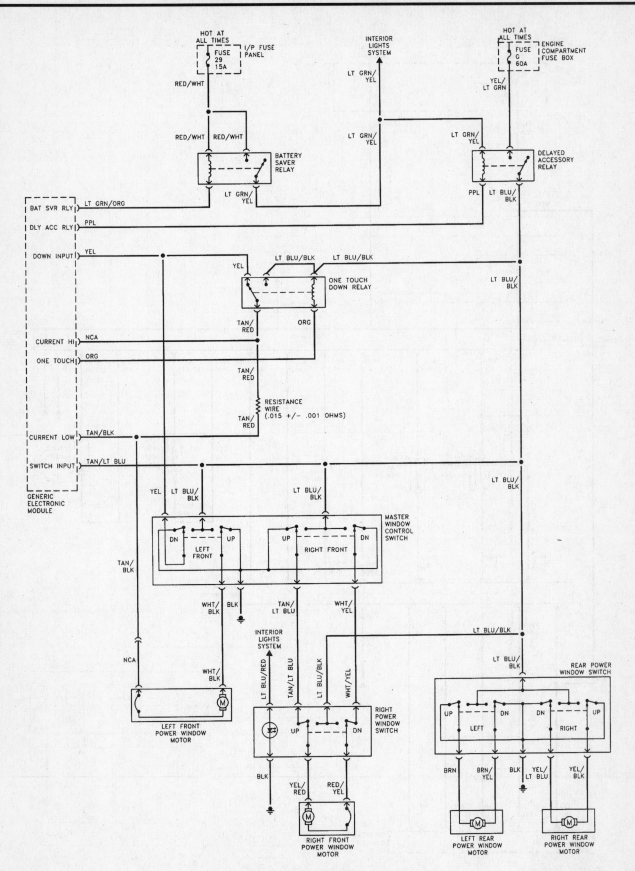

Power window system - 1998 and later models

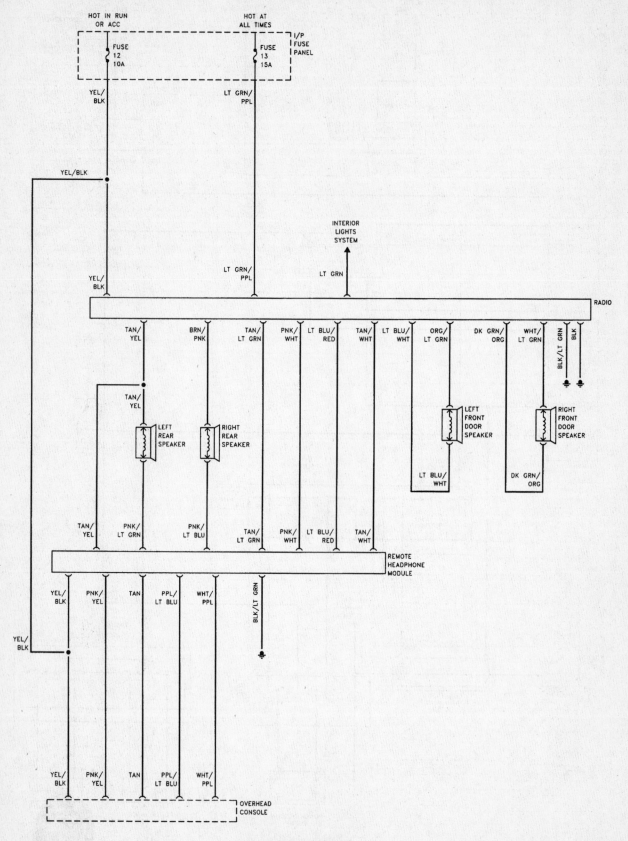

Typical audio system

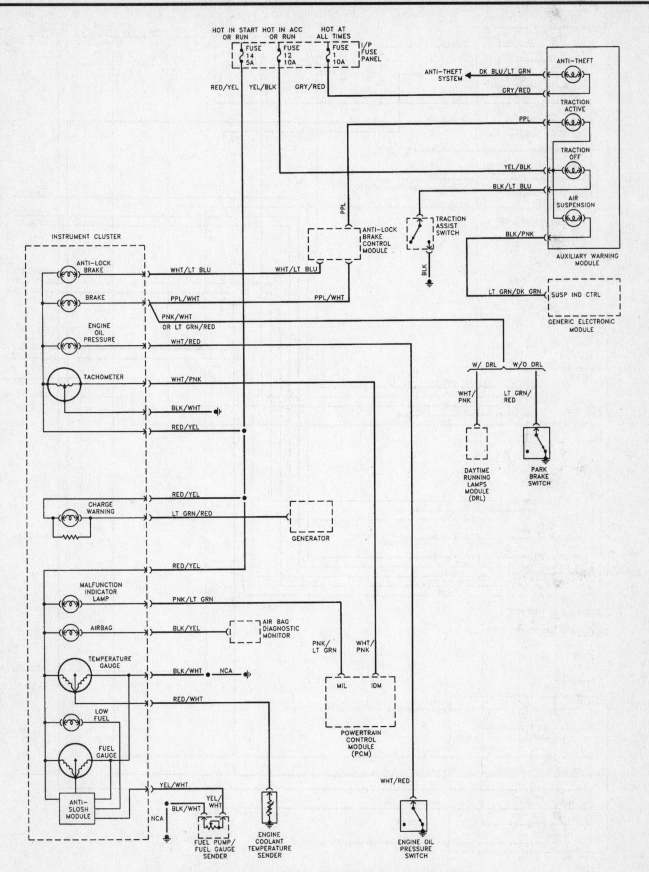

Instrument panel gauges and warning light system - analog gauges

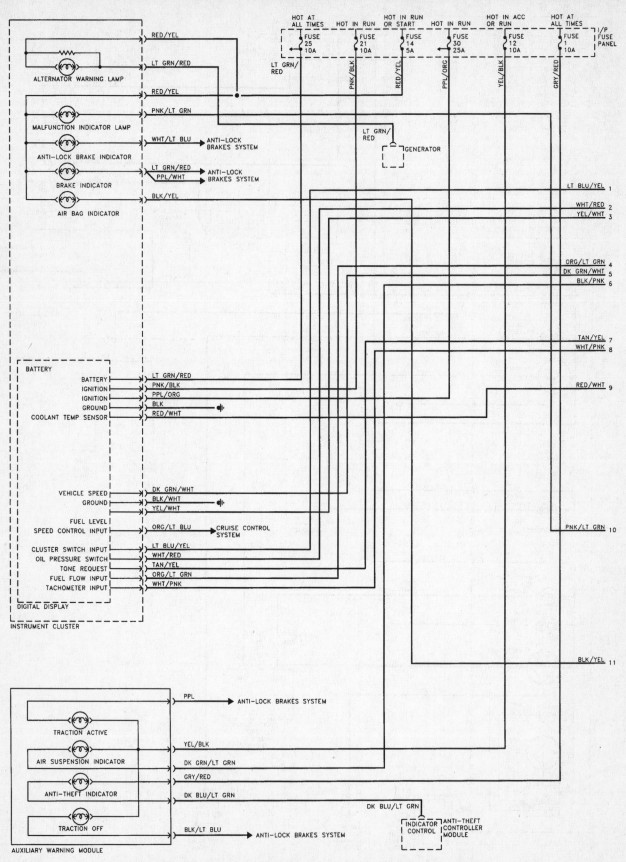

Instrument panel gauges and warning light system - digital gauges (1 of 2)

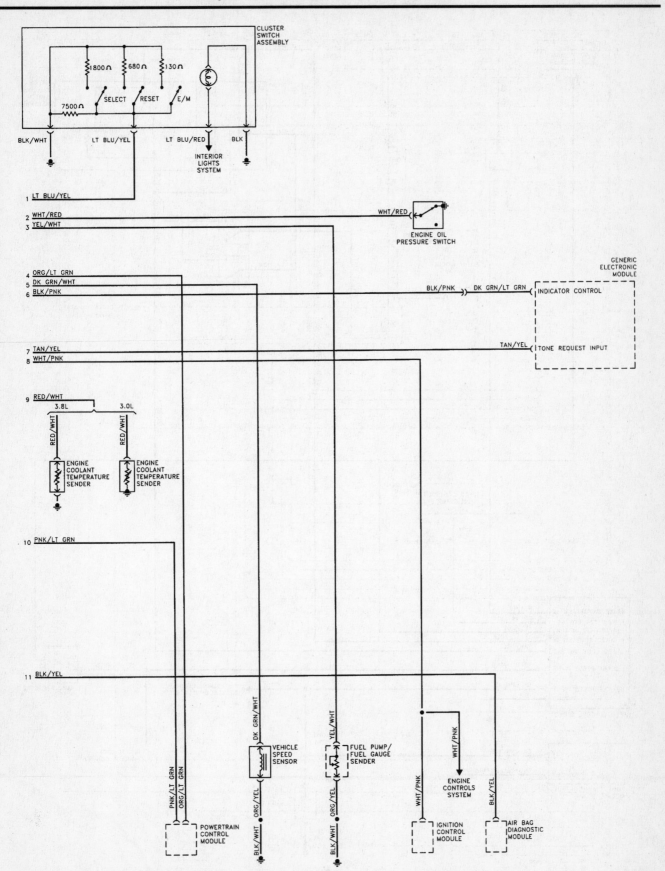

Instrument panel gauges and warning light system - digital gauges (2 of 2)

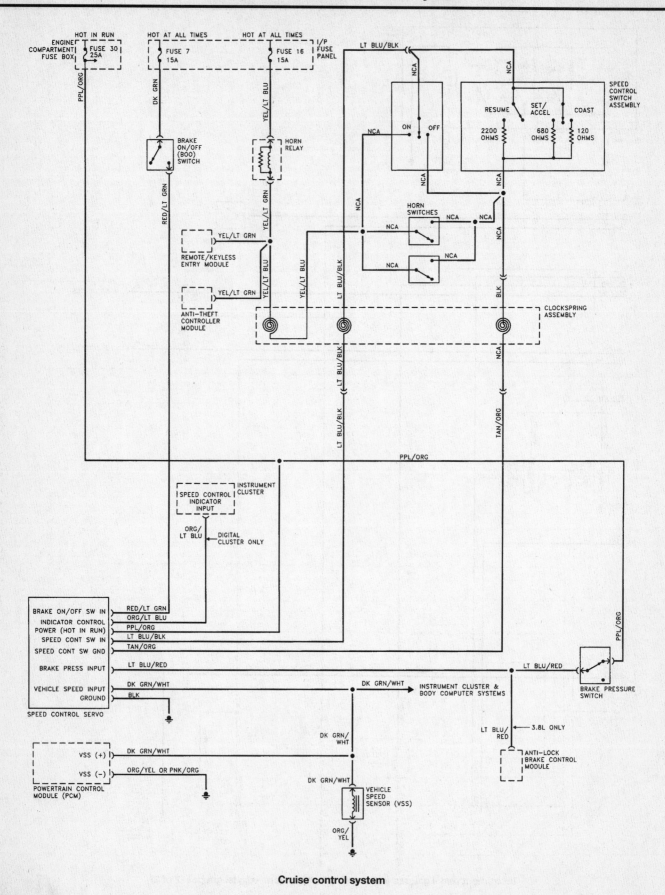

Cruise control system

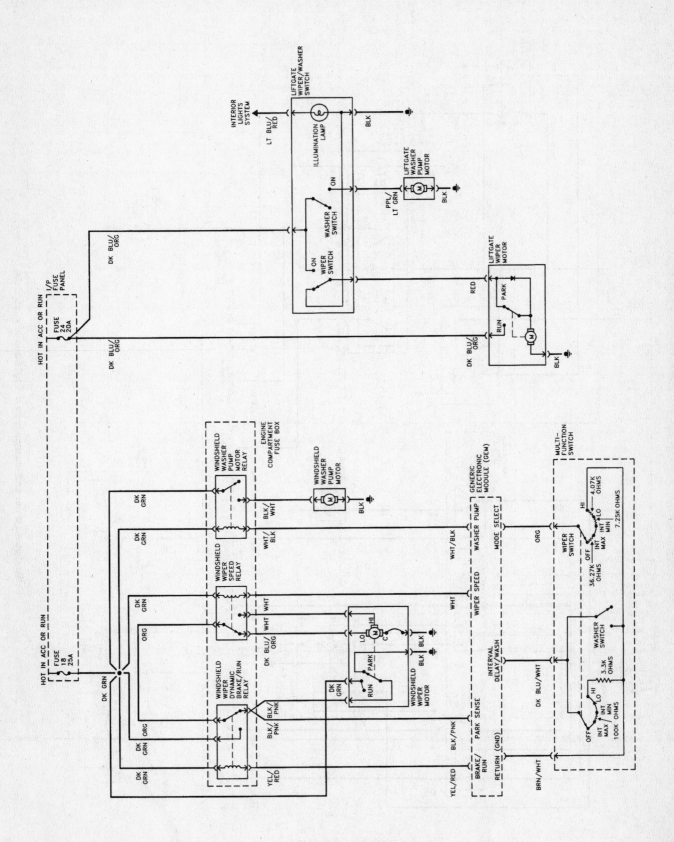

Windshield wiper and washer system - 1995 through 1997 models

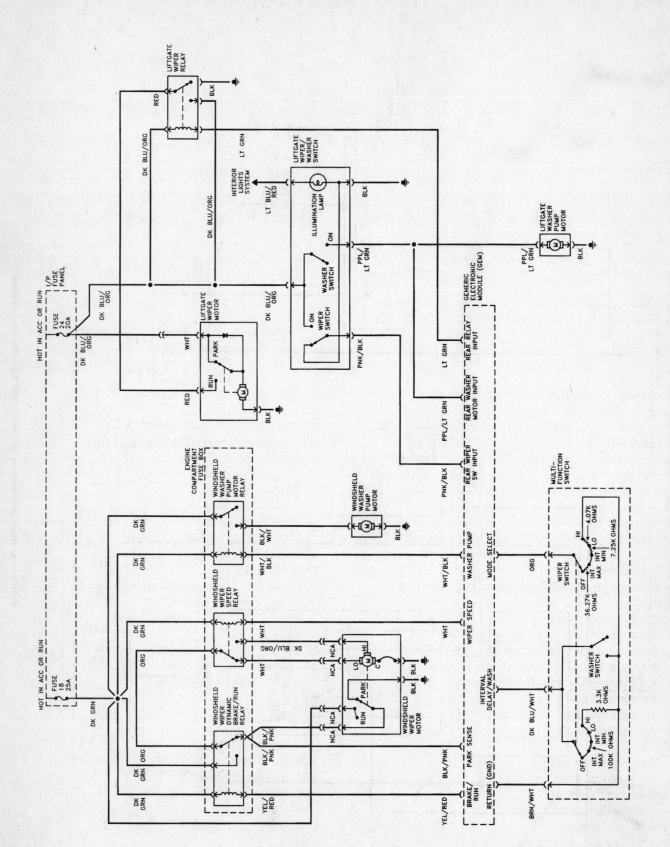

Windshield wiper and washer system - 1998 and later models

Index

Haynes Automotive Manuals

NOTE: New manuals are added to this list on a periodic basis. If you do not see a listing for your vehicle, consult your local Haynes dealer for the latest product information.

ACURA
*12020 Integra '86 thru '89 & Legend '86 thru '90

AMC
Jeep CJ - see JEEP (50020)
14020 Mid-size models, Concord, Hornet, Gremlin & Spirit '70 thru '83
14025 (Renault) Alliance & Encore '83 thru '87

AUDI
15020 4000 all models '80 thru '87
15025 5000 all models '77 thru '83
15026 5000 all models '84 thru '88

AUSTIN-HEALEY
Sprite - see MG Midget (66015)

BMW
*18020 3/5 Series not including diesel or all-wheel drive models '82 thru '92
*18021 3 Series except 325iX models '92 thru '97
18025 320i all 4 cyl models '75 thru '83
18035 528i & 530i all models '75 thru '80
18050 1500 thru 2002 except Turbo '59 thru '77

BUICK
Century (front wheel drive) - see GM (829)
*19020 Buick, Oldsmobile & Pontiac Full-size (Front wheel drive) all models '85 thru '98
Buick Electra, LeSabre and Park Avenue; Oldsmobile Delta 88 Royale, Ninety Eight and Regency; Pontiac Bonneville
19025 Buick Oldsmobile & Pontiac Full-size (Rear wheel drive)
Buick Estate '70 thru '90, Electra'70 thru '84, LeSabre '70 thru '85, Limited '74 thru '79
Oldsmobile Custom Cruiser '70 thru '90, Delta 88 '70 thru '85, Ninety-eight '70 thru '84
Pontiac Bonneville '70 thru '81, Catalina '70 thru '81, Grandville '70 thru '75, Parisienne '83 thru '86
19030 Mid-size Regal & Century all rear-drive models with V6, V8 and Turbo '74 thru '87
Regal - see GENERAL MOTORS (38010)
Riviera - see GENERAL MOTORS (38030)
Roadmaster - see CHEVROLET (24046)
Skyhawk - see GENERAL MOTORS (38015)
Skylark '80 thru '85 - see GM (38020)
Skylark '86 on - see GM (38025)
Somerset - see GENERAL MOTORS (38025)

CADILLAC
*21030 Cadillac Rear Wheel Drive all gasoline models '70 thru '93
Cimarron - see GENERAL MOTORS (38015)
Eldorado - see GENERAL MOTORS (38030)
Seville '80 thru '85 - see GM (38030)

CHEVROLET
*24010 Astro & GMC Safari Mini-vans '85 thru '93
24015 Camaro V8 all models '70 thru '81
24016 Camaro all models '82 thru '92
Cavalier - see GENERAL MOTORS (38015)
Celebrity - see GENERAL MOTORS (38005)
24017 Camaro & Firebird '93 thru '97
24020 Chevelle, Malibu & El Camino '69 thru '87
24024 Chevette & Pontiac T1000 '76 thru '87
Citation - see GENERAL MOTORS (38020)
*24032 Corsica/Beretta all models '87 thru '96
24040 Corvette all V8 models '68 thru '82
*24041 Corvette all models '84 thru '96
10305 Chevrolet Engine Overhaul Manual
24045 Full-size Sedans Caprice, Impala, Biscayne, Bel Air & Wagons '69 thru '90
24046 Impala SS & Caprice and Buick Roadmaster '91 thru '96
Lumina - see GENERAL MOTORS (38010)

24048 Lumina & Monte Carlo '95 thru '98
Lumina APV - see GM (38035)
24050 Luv Pick-up all 2WD & 4WD '72 thru '82
*24055 Monte Carlo all models '70 thru '88
Monte Carlo '95 thru '98 - see LUMINA (24048)
24059 Nova all V8 models '69 thru '79
*24060 Nova and Geo Prizm '85 thru '92
24064 Pick-ups '67 thru '87 - Chevrolet & GMC, all V8 & in-line 6 cyl, 2WD & 4WD '67 thru '87; Suburbans, Blazers & Jimmys '67 thru '91
*24065 Pick-ups '88 thru '98 - Chevrolet & GMC, all full-size pick-ups, '88 thru '98; Blazer & Jimmy '92 thru '94; Suburban '92 thru '98; Tahoe & Yukon '98
24070 S-10 & S-15 Pick-ups '82 thru '93, Blazer & Jimmy '83 thru '94,
*24071 S-10 & S-15 Pick-ups '94 thru '96 Blazer & Jimmy '95 thru '96
*24075 Sprint & Geo Metro '85 thru '94
*24080 Vans - Chevrolet & GMC, V8 & in-line 6 cylinder models '68 thru '96

CHRYSLER
25015 Chrysler Cirrus, Dodge Stratus, Plymouth Breeze '95 thru '98
25025 Chrysler Concorde, New Yorker & LHS, Dodge Intrepid, Eagle Vision, '93 thru '97
10310 Chrysler Engine Overhaul Manual
*25020 Full-size Front-Wheel Drive '88 thru '93
K-Cars - see DODGE Aries (30008)
Laser - see DODGE Daytona (30030)
*25030 Chrysler & Plymouth Mid-size front wheel drive '82 thru '95
Rear-wheel Drive - see Dodge (30050)

DATSUN
28005 200SX all models '80 thru '83
28007 B-210 all models '73 thru '78
28009 210 all models '79 thru '82
28012 240Z, 260Z & 280Z Coupe '70 thru '78
28014 280ZX Coupe & 2+2 '79 thru '83
300ZX - see NISSAN (72010)
28016 310 all models '78 thru '82
28018 510 & PL521 Pick-up '68 thru '73
28020 510 all models '78 thru '81
28022 620 Series Pick-up all models '73 thru '79
720 Series Pick-up - see NISSAN (72030)
28025 810/Maxima all gasoline models, '77 thru '84

DODGE
400 & 600 - see CHRYSLER (25030)
*30008 Aries & Plymouth Reliant '81 thru '89
30010 Caravan & Plymouth Voyager Mini-Vans all models '84 thru '95
*30011 Caravan & Plymouth Voyager Mini-Vans all models '96 thru '98
30012 Challenger/Plymouth Saporro '78 thru '83
30016 Colt & Plymouth Champ (front wheel drive) all models '78 thru '87
*30020 Dakota Pick-ups all models '87 thru '96
30025 Dart, Demon, Plymouth Barracuda, Duster & Valiant 6 cyl models '67 thru '76
*30030 Daytona & Chrysler Laser '84 thru '89
Intrepid - see CHRYSLER (25025)
*30034 Neon all models '95 thru '97
*30035 Omni & Plymouth Horizon '78 thru '90
*30040 Pick-ups all full-size models '74 thru '93
*30041 Pick-ups all full-size models '94 thru '96
*30045 Ram 50/D50 Pick-ups & Raider and Plymouth Arrow Pick-ups '79 thru '93
30050 Dodge/Plymouth/Chrysler rear wheel drive '71 thru '89
*30055 Shadow & Plymouth Sundance '87 thru '94
*30060 Spirit & Plymouth Acclaim '89 thru '95
*30065 Vans - Dodge & Plymouth '71 thru '96

EAGLE
Talon - see Mitsubishi Eclipse (68030)
Vision - see CHRYSLER (25025)

FIAT
34010 124 Sport Coupe & Spider '68 thru '78
34025 X1/9 all models '74 thru '80

FORD
10355 Ford Automatic Transmission Overhaul
*36004 Aerostar Mini-vans all models '86 thru '96
*36006 Contour & Mercury Mystique '95 thru '98
36008 Courier Pick-up all models '72 thru '82
36012 Crown Victoria & Mercury Grand Marquis '88 thru '96
10320 Ford Engine Overhaul Manual
36016 Escort/Mercury Lynx all models '81 thru '90
*36020 Escort/Mercury Tracer '91 thru '96
*36024 Explorer & Mazda Navajo '91 thru '95
36028 Fairmont & Mercury Zephyr '78 thru '83
36030 Festiva & Aspire '88 thru '97
36032 Fiesta all models '77 thru '80
36036 Ford & Mercury Full-size, Ford LTD & Mercury Marquis ('75 thru '82); Ford Custom 500,Country Squire, Crown Victoria & Mercury Colony Park ('75 thru '87); Ford LTD Crown Victoria & Mercury Gran Marquis ('83 thru '87)
36040 Granada & Mercury Monarch '75 thru '80
36044 Ford & Mercury Mid-size, Ford Thunderbird & Mercury Cougar ('75 thru '82); Ford LTD & Mercury Marquis ('83 thru '86); Ford Torino,Gran Torino, Elite, Ranchero pick-up, LTD II, Mercury Montego, Comet, XR-7 & Lincoln Versailles ('75 thru '86)
36048 Mustang V8 all models '64-1/2 '73
36049 Mustang II 4 cyl, V6 & V8 models '74 thru '78
36050 Mustang & Mercury Capri all models Mustang, '79 thru '93; Capri, '79 thru '86
*36051 Mustang all models '94 thru '97
36054 Pick-ups & Bronco '73 thru '79
36058 Pick-ups & Bronco '80 thru '96
36059 Pick-ups, Expedition & Mercury Navigator '97 thru '98
36062 Pinto & Mercury Bobcat '75 thru '80
36066 Probe all models '89 thru '92
36070 Ranger/Bronco II gasoline models '83 thru '92
*36071 Ranger '93 thru '97 & Mazda Pick-ups '94 thru '97
36074 Taurus & Mercury Sable '86 thru '95
*36075 Taurus & Mercury Sable '96 thru '98
*36078 Tempo & Mercury Topaz '84 thru '94
36082 Thunderbird/Mercury Cougar '83 thru '88
*36086 Thunderbird/Mercury Cougar '89 and '97
36090 Vans all V8 Econoline models '69 thru '91
*36094 Vans full size '92-'95
*36097 Windstar Mini-van '95-'98

GENERAL MOTORS
*10360 GM Automatic Transmission Overhaul
*38005 Buick Century, Chevrolet Celebrity, Oldsmobile Cutlass Ciera & Pontiac 6000 all models '82 thru '96
*38010 Buick Regal, Chevrolet Lumina, Oldsmobile Cutlass Supreme & Pontiac Grand Prix front-wheel drive models '88 thru '95
*38015 Buick Skyhawk, Cadillac Cimarron, Chevrolet Cavalier, Oldsmobile Firenza & Pontiac J-2000 & Sunbird '82 thru '94
*38016 Chevrolet Cavalier & Pontiac Sunfire '95 thru '98
38020 Buick Skylark, Chevrolet Citation, Olds Omega, Pontiac Phoenix '80 thru '85
38025 Buick Skylark & Somerset, Oldsmobile Achieva & Calais and Pontiac Grand Am all models '85 thru '95
38030 Cadillac Eldorado '71 thru '85, Seville '80 thru '85, Oldsmobile Toronado '71 thru '85 & Buick Riviera '79 thru '85
*38035 Chevrolet Lumina APV, Olds Silhouette & Pontiac Trans Sport all models '90 thru '95
General Motors Full-size Rear-wheel Drive - see BUICK (19025)

(Continued on other side)

---

* Listings shown with an asterisk (*) indicate model coverage as of this printing. These titles will be periodically updated to include later model years - consult your Haynes dealer for more information.

Haynes North America, Inc., 861 Lawrence Drive, Newbury Park, CA 91320-1514 • (805) 498-6703

Haynes Automotive Manuals (continued)

NOTE: New manuals are added to this list on a periodic basis. If you do not see a listing for your vehicle, consult your local Haynes dealer for the latest product information.

GEO
Metro - *see CHEVROLET Sprint (24075)*
Prizm - '85 thru '92 *see CHEVY (24060)*, '93 thru '96 *see TOYOTA Corolla (92036)*
*40030 Storm all models '90 thru '93
Tracker - *see SUZUKI Samurai (90010)*

GMC
Safari - *see CHEVROLET ASTRO (24010)*
Vans & Pick-ups - *see CHEVROLET*

HONDA
42010 Accord CVCC all models '76 thru '83
42011 Accord all models '84 thru '89
42012 Accord all models '90 thru '93
42013 Accord all models '94 thru '95
42020 Civic 1200 all models '73 thru '79
42021 Civic 1300 & 1500 CVCC '80 thru '83
42022 Civic 1500 CVCC all models '75 thru '79
42023 Civic all models '84 thru '91
*42024 Civic & del Sol '92 thru '95
*42040 Prelude CVCC all models '79 thru '89

HYUNDAI
*43015 Excel all models '86 thru '94

ISUZU
Hombre - *see CHEVROLET S-10 (24071)*
*47017 Rodeo '91 thru '97; Amigo '89 thru '94; Honda Passport '95 thru '97
*47020 Trooper & Pick-up, all gasoline models Pick-up, '81 thru '93; Trooper, '84 thru '91

JAGUAR
*49010 XJ6 all 6 cyl models '68 thru '86
*49011 XJ6 all models '88 thru '94
*49015 XJ12 & XJS all 12 cyl models '72 thru '85

JEEP
*50010 Cherokee, Comanche & Wagoneer Limited all models '84 thru '96
50020 CJ all models '49 thru '86
*50025 Grand Cherokee all models '93 thru '98
50029 Grand Wagoneer & Pick-up '72 thru '91 Grand Wagoneer '84 thru '91, Cherokee & Wagoneer '72 thru '83, Pick-up '72 thru '88
*50030 Wrangler all models '87 thru '95

LINCOLN
Navigator - *see FORD Pick-up (36059)*
59010 Rear Wheel Drive all models '70 thru '96

MAZDA
61010 GLC Hatchback (rear wheel drive) '77 thru '83
61011 GLC (front wheel drive) '81 thru '85
*61015 323 & Protegé '90 thru '97
*61016 MX-5 Miata '90 thru '97
*61020 MPV all models '89 thru '94
Navajo - *see Ford Explorer (36024)*
61030 Pick-ups '72 thru '93
Pick-ups '94 thru '96 - *see Ford Ranger (36071)*
61035 RX-7 all models '79 thru '85
*61036 RX-7 all models '86 thru '91
61040 626 (rear wheel drive) all models '79 thru '82
*61041 626/MX-6 (front wheel drive) '83 thru '91

MERCEDES-BENZ
63012 123 Series Diesel '76 thru '85
*63015 190 Series four-cyl gas models, '84 thru '88
63020 230/250/280 6 cyl sohc models '68 thru '72
63025 280 123 Series gasoline models '77 thru '81
63030 350 & 450 all models '71 thru '80

MERCURY
See *FORD Listing.*

MG
66010 MGB Roadster & GT Coupe '62 thru '80
66015 MG Midget, Austin Healey Sprite '58 thru '80

MITSUBISHI
*68020 Cordia, Tredia, Galant, Precis & Mirage '83 thru '93
*68030 Eclipse, Eagle Talon & Ply. Laser '90 thru '94
*68040 Pick-up '83 thru '96 & Montero '83 thru '93

NISSAN
72010 300ZX all models including Turbo '84 thru '89
*72015 Altima all models '93 thru '97
*72020 Maxima all models '85 thru '91
*72030 Pick-ups '80 thru '96 Pathfinder '87 thru '95
72040 Pulsar all models '83 thru '86
*72050 Sentra all models '82 thru '94
*72051 Sentra & 200SX all models '95 thru '98
*72060 Stanza all models '82 thru '90

OLDSMOBILE
*73015 Cutlass V6 & V8 gas models '74 thru '88
For other *OLDSMOBILE* titles, see *BUICK, CHEVROLET* or *GENERAL MOTORS* listing.

PLYMOUTH
For *PLYMOUTH* titles, see *DODGE* listing.

PONTIAC
79008 Fiero all models '84 thru '88
79018 Firebird V8 models except Turbo '70 thru '81
79019 Firebird all models '82 thru '92
For other *PONTIAC* titles, see *BUICK, CHEVROLET* or *GENERAL MOTORS* listing.

PORSCHE
*80020 911 except Turbo & Carrera 4 '65 thru '89
80025 914 all 4 cyl models '69 thru '76
80030 924 all models including Turbo '76 thru '82
*80035 944 all models including Turbo '83 thru '89

RENAULT
Alliance & Encore - *see AMC (14020)*

SAAB
*84010 900 all models including Turbo '79 thru '88

SATURN
87010 Saturn all models '91 thru '96

SUBARU
89002 1100, 1300, 1400 & 1600 '71 thru '79
*89003 1600 & 1800 2WD & 4WD '80 thru '94

SUZUKI
*90010 Samurai/Sidekick & Geo Tracker '86 thru '96

TOYOTA
92005 Camry all models '83 thru '91
92006 Camry all models '92 thru '96
92015 Celica Rear Wheel Drive '71 thru '85
*92020 Celica Front Wheel Drive '86 thru '93
92025 Celica Supra all models '79 thru '92
92030 Corolla all models '75 thru '79
92032 Corolla all rear wheel drive models '80 thru '87
92035 Corolla all front wheel drive models '84 thru '92
*92036 Corolla & Geo Prizm '93 thru '97
92040 Corolla Tercel all models '80 thru '82
92045 Corona all models '74 thru '82
92050 Cressida all models '78 thru '82
92055 Land Cruiser FJ40, 43, 45, 55 '68 thru '82
92056 Land Cruiser FJ60, 62, 80, FZJ80 '80 thru '96
*92065 MR2 all models '85 thru '87
92070 Pick-up all models '69 thru '78
*92075 Pick-up all models '79 thru '95
*92076 Tacoma '95 thru '98, 4Runner '96 thru '98, & T100 '93 thru '98
*92080 Previa all models '91 thru '95
92085 Tercel all models '87 thru '94

TRIUMPH
94007 Spitfire all models '62 thru '81
94010 TR7 all models '75 thru '81

VW
96008 Beetle & Karmann Ghia '54 thru '79
96012 Dasher all gasoline models '74 thru '81
*96016 Rabbit, Jetta, Scirocco, & Pick-up gas models '74 thru '91 & Convertible '80 thru '92
96017 Golf & Jetta all models '93 thru '97
96020 Rabbit, Jetta & Pick-up diesel '77 thru '84
96030 Transporter 1600 all models '68 thru '79
96035 Transporter 1700, 1800 & 2000 '72 thru '79
96040 Type 3 1500 & 1600 all models '63 thru '73
96045 Vanagon all air-cooled models '80 thru '83

VOLVO
97010 120, 130 Series & 1800 Sports '61 thru '73
97015 140 Series all models '66 thru '74
*97020 240 Series all models '76 thru '93
97025 260 Series all models '75 thru '82
*97040 740 & 760 Series all models '82 thru '88

TECHBOOK MANUALS
10205 Automotive Computer Codes
10210 Automotive Emissions Control Manual
10215 Fuel Injection Manual, 1978 thru 1985
10220 Fuel Injection Manual, 1986 thru 1996
10225 Holley Carburetor Manual
10230 Rochester Carburetor Manual
10240 Weber/Zenith/Stromberg/SU Carburetors
10305 Chevrolet Engine Overhaul Manual
10310

SPANISH MANUALS
98903 Reparación de Carrocería & Pintura
98905 Códigos Automotrices de la Computadora
98910 Frenos Automotriz
98915 Inyección de Combustible 1986 al 1994
99040 Chevrolet & GMC Camionetas '67 al '87 Incluye Suburban, Blazer & Jimmy '67 al '91
99041 Chevrolet & GMC Camionetas '88 al '95 Incluye Suburban '92 al '95, Blazer & Jimmy '92 al '94, Tahoe y Yukon '95
99042 Chevrolet & GMC Camionetas Cerradas '68 al '95
99055 Dodge Caravan & Plymouth Voyager '84 al '95
99075 Ford Camionetas y Bronco '80 al '94
99077 Ford Camionetas Cerradas '69 al '91
99083 Ford Modelos de Tamaño Grande '75 al '87
99088 Ford Modelos de Tamaño Mediano '75 al '86
99091 Ford Taurus & Mercury Sable '86 al '95
99095 GM Modelos de Tamaño Grande '70 al '90
99100 GM Modelos de Tamaño Mediano '70 al '88
99110 Nissan Camionetas '80 al '96, Pathfinder '87 al '95
99118 Nissan Sentra '82 al '94
99125 Toyota Camionetas y 4Runner '79 al '95

---

* Listings shown with an asterisk (*) indicate model coverage as of this printing. These titles will be periodically updated to include later model years - consult your Haynes dealer for more information.

Over 100 Haynes motorcycle manuals also available

5-98

Haynes North America, Inc., 861 Lawrence Drive, Newbury Park, CA 91320-1514 • (805) 498-6703

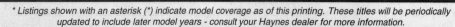